Embracing the Fog of War

ASSESSMENT AND METRICS IN COUNTERINSURGENCY

BEN CONNABLE

Prepared for the U.S. Department of Defense

Approved for public release; distribution unlimited

NATIONAL DEFENSE RESEARCH INSTITUTE

The research described in this report was prepared for the U.S. Department of Defense. The research was conducted within the RAND National Defense Research Institute, a federally funded research and development center sponsored by OSD, the Joint Staff, the Unified Combatant Commands, the Navy, the Marine Corps, the defense agencies, and the defense Intelligence Community under Contract W74V8H-06-C-0002.

Library of Congress Cataloging-in-Publication Data

Connable, Ben.
  Embracing the fog of war : assessment and metrics in counterinsurgency /
Ben Connable.
    p. cm.
  Includes bibliographical references.
  ISBN 978-0-8330-5815-7 (pbk. : alk. paper)
  1. Counterinsurgency—United States—Evaluation. 2. Counterinsurgency—
Afghanistan—Evaluation. 3. Counterinsurgency—Vietnam—Evaluation. I. Title.

  U241C63 2012
  355.02'180973—dc23

                                                                    2012001598

The RAND Corporation is a nonprofit institution that helps improve policy and decisionmaking through research and analysis. RAND's publications do not necessarily reflect the opinions of its research clients and sponsors.

RAND® is a registered trademark.

Cover design by Carol Earnest

Published 2012 by the RAND Corporation
1776 Main Street, P.O. Box 2138, Santa Monica, CA 90407-2138
1200 South Hayes Street, Arlington, VA 22202-5050
4570 Fifth Avenue, Suite 600, Pittsburgh, PA 15213-2665
RAND URL: http://www.rand.org/
To order RAND documents or to obtain additional information, contact
Distribution Services: Telephone: (310) 451-7002;
Fax: (310) 451-6915; Email: order@rand.org

Preface

In late 2009, RAND undertook an examination of military and nonmilitary assessment processes to help improve military methods for assessing counterinsurgency (COIN) operations. The original purpose of this research was to assist a military operational staff in shaping campaign design, in shifting priorities of effort during ongoing operations, and, ultimately, in meeting operational and strategic objectives in southern Afghanistan. The original intended audience for this resulting monograph was a single staff in Afghanistan. However, the study was expanded to include an examination of COIN assessment across the U.S. military. The findings of this broadened research effort will be of interest to military commanders and staffs who produce and use assessments, policymakers who rely on assessments to make decisions, and the U.S. public, the ultimate decisionmakers in any long-term U.S.-led COIN campaign.

An assessment is a report or series of reports intended to inform operational and strategic decisionmaking. Military staffs build campaign assessments, and military commanders use them to make operational decisions in the field. These commanders add their comments to the assessments and then pass them along to policymakers. Assessments can be presented in a variety of forms, including narrative papers, quantitative graphs, maps, and briefing slides. Senior military leaders and policy staffs use these materials to assess the progress of military campaigns, determine how to allocate (or reallocate) resources, identify trends that may indicate success or failure, or ascertain whether and when it may be necessary to alter a given strategy.

In support of this process, this monograph captures the complexity of the COIN operational environment, examines case studies of COIN assessment in context, explores critical weaknesses in the current assessment process, offers recommendations for improvement, and presents an alternative assessment process.

The manuscript for this publication was completed in mid-2011. Since that time, various elements in the U.S. Department of Defense have published new doctrine on assessment, some of which addresses criticisms raised here. Perhaps more importantly, the International Security Assistance Force in Afghanistan has completely revamped its assessment process, adopting many of the recommendations suggested in this monograph and in other subject-matter expert reports on COIN assessment.

This research was sponsored by the U.S. Department of Defense and conducted within the Intelligence Policy Center of the RAND National Defense Research Institute, a federally funded research and development center sponsored by the Office of the Secretary of Defense, the Joint Staff, the Unified Combatant Commands, the military services, the defense agencies, and the defense Intelligence Community.

For more information on the RAND Intelligence Policy Center, see http://www.rand.org/nsrd/ndri/centers/intel.html or contact the director (contact information is provided on the web page).

Contents

Figures

Tables and Text Boxes

Tables

Text Boxes

Summary

This monograph examines the U.S. military assessment process for counterinsurgency (COIN) campaigns. It focuses on the methods employed to develop and present to policymakers these theater-level assessments of ongoing campaigns. In support of this process, it captures the complexity of the COIN operational environment, examines case studies of COIN assessment in context, explores critical weaknesses in the current assessment process, and offers recommendations for improvement.

Sound strategic decisionmaking in military campaigns relies, in part, on the quality of campaign assessments. The quality of an assessment, in turn, reflects the methodology used to produce the assessment report or other materials: An assessment derived from a poorly conceived methodology might mislead decisionmakers, while a well-conceived assessment process might help shape a winning strategy. U.S. military campaign assessments are developed by a theater commander (for example, GEN William C. Westmoreland during the Vietnam War or Gen. John R. Allen during the current war in Afghanistan) for senior military and civilian decisionmakers. To produce a campaign assessment, military staffs currently collect and analyze information about a campaign's progress and then develop narratives, briefings, and other materials that can be provided to decisionmakers. The U.S. Department of Defense (DoD) and policymakers in Congress and the executive branch rely on these field reports to help them allocate (or reallocate) resources, identify trends that may indicate success or failure, and ascertain whether and when it may be necessary to alter a given strategy.

Figure S.1 provides an overview of the U.S. military's current approach to COIN campaign assessment. The assessment process begins with strategic direction from policymakers to the theater commander and staff.[1] The theater commander then interprets the request and produces operational orders for the units under his command.

---

[1] A more detailed schematic would show several layers of processing between the policymakers and the theater commander (e.g., the National Security Council, the combatant commander). This figure is purposefully simplified to convey the general concept and process of assessment throughout the U.S. military. This process, the different layers of command, the relationships among actors, and the characteristics of the COIN environment are discussed in greater detail later in this monograph.

Figure S.1
Overview of the COIN Campaign Assessment Process

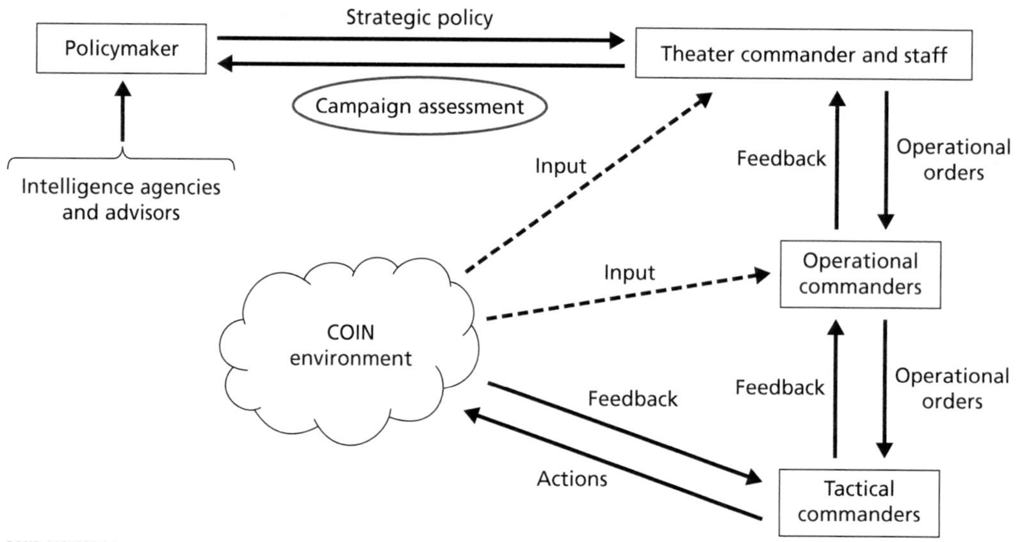

RAND *MG1086-S.1*

These orders are reinterpreted and executed at various organizational levels in theater, from the operational to the tactical. These actions in the COIN operating environment produce feedback. This feedback is then passed back up through the chain of command to the theater commander and staff, who use the information, along with other inputs, to produce a campaign assessment. The commander delivers this assessment to the policymakers, who combine it with other resources (e.g., intelligence agency reports, advisory briefs) and reshape policy, beginning the cycle again.[2]

The military is most capable of producing clear and useful campaign assessments when provided with clearly articulated and understandable strategic objectives. In the absence of such guidance, military staffs are left scrambling to design operational objectives, and then assessments, that seem reasonable—and assessments derived from an unclear strategy are unlikely to satisfy anyone. With a clear, well-defined strategic policy in place, the military can develop an assessment process to answer questions that relate specifically to policymaker requirements. In a best-case scenario, assessments should be tied directly to national policy objectives, and policymakers should be able to view such assessments with relative confidence.

The U.S. military has taken two broad approaches to assessing the progress of COIN campaigns. The first approach—*effects-based assessment* (EBA)—attempts to

---

[2]   The figure also shows how information, or "inputs," from other sources in the COIN environment (e.g., civilian agencies, host-nation organizations) can be incorporated into the assessment process. Feedback from policy action implemented in theater is only one type of information used for assessment; a great deal of information is derived from activities or observations that are not directly associated with friendly actions.

pinpoint and quantify events on the ground to produce centralized and highly accurate reports. The second approach—*pattern and trend analysis*—uses centralized quantitative analysis to produce a more-or-less impressionistic or, in some cases, deterministic understanding of campaign momentum. Both these approaches are centralized and rely to a great degree on quantitative measurement. No other comprehensive approach has been described or advocated in the literature on military COIN assessment.[3] In practice, the military has relied on an ad hoc approach to COIN assessment that lies somewhere between EBA and pattern and trend analysis.

Neither of these two centralized assessment methods is practical for COIN because, according to U.S. military doctrine, COIN is best practiced as a decentralized type of warfare predicated on "mission command." Indeed, doctrine envisions the COIN battlespace as a "mosaic" of localized situations, challenges, and possible solutions.[4] Decentralization and mission command necessitate a loosely structured, localized approach to prosecuting war. Such an approach allows commanders at the tactical level to assess local conditions and implement initiatives tailored to address local challenges. Due to the decentralized nature of COIN campaigns, few activities are generalizable or consistently implemented across a COIN theater of operations. This situation is exacerbated by the complexity and uncertainty inherent in the environment. It would be difficult (if not impossible) to develop a practical, centralized model for COIN assessment because complex COIN environments cannot be clearly interpreted through a centralized process that removes data from their salient local context. The incongruity between decentralized and complex COIN operations and centralized, decontextualized assessment has led military staffs to rely on ad hoc assessment methods that leave policymakers and the public dissatisfied with U.S. COIN campaign assessments.

This monograph examines and critiques EBA and pattern and trend analysis in an effort to explain why policymakers and the public tend to be dissatisfied with U.S. military assessments of COIN campaigns. That phase of the study involved 16 months of research, a detailed examination of two case studies and a review of a third case study, interviews and discussions with hundreds of operations and assessment officials from the battalion to the DoD level, and exchanges with policymakers and their staffs. Among those who offered their perspectives, there were significant concerns about the effectiveness and methodology of the current assessment process as it is being applied in support of COIN campaign assessment in Afghanistan. To provide historical con-

---

[3] The few alternatives to EBA that have been offered are centralized and quantitative and do not represent a significant departure from EBA. Most of the assessment literature relies on centralized, quantitative systems analysis methods to frame the COIN operating environment and to assess COIN campaigns. Although a review of COIN assessment methods did not uncover a viable alternative, this does not preclude the possibility that one exists but was simply not captured by the literature review.

[4] According to Joint Publication (JP) 3-24, *Counterinsurgency Operations* (U.S. Joint Chiefs of Staff, 2009c, p. III-19), "The mosaic nature of COIN is ideally suited to decentralized execution."

text for an examination of the current assessment process, the study also considered the case of COIN campaign assessment and outcomes during the Vietnam War era. The literature on the U.S. military's Vietnam-era assessment process was highly negative; even the most strident defenders of the assessment reports produced during that period expressed dissatisfaction with the process.

Although, as discussed earlier, the assessment literature contains no clearly articulated alternative to the two accepted forms of COIN assessment (EBA and pattern and trend analysis), advocates of a *traditionalist* approach to warfare have critiqued the theories behind EBA and have voiced doubt that any quantitative approach could produce assessment materials with greater utility at the policymaking level.[5] Traditionalists, including General James N. Mattis, adhere to strict interpretations of capstone service doctrine that describe warfare as inherently complex and chaotic. Some advocates of EBA and most advocates of pattern and trend analysis also acknowledge the complex and chaotic nature of war, but they argue that centralized assessment can sufficiently overcome these challenges to inform policy. Traditionalists argue that complexity and chaos cannot be adequately overcome through centralized and quantitative analysis. Despite this debate, no viable alternative to centralized assessment has emerged from these traditionalist critiques or from other quarters.

The study also produced a framework of standards for COIN assessment. These standards were derived from a detailed literature review, interviews with experts and practitioners, exchanges with subject-matter experts, and direct observation of field assessment activities. The first two standards—transparency and credibility—are deemed requisite for successful COIN assessment in a democracy. All assessment reports should be relevant to policymakers; this is clear in the campaign assessment process as presented in Figure S.1. The other four standards are practical requirements that should be met to establish an effective methodology. The seven standards are as follows:

1. *Transparent:* Transparent assessment reports are widely releasable to official consumers and, ideally, are unclassified and without distribution restrictions; the final iteration of the report is suitable for distribution on the Internet. Such reports reveal both the methods and data used at each level of assessment, from tactical to strategic, and allow for detailed and in-depth analysis without requiring additional research or requests for information. Any subjectivity or justification for disagreements between layers of command is explained clearly and comprehensively.

---

[5] "Traditionalist" is a nondoctrinal, unofficial term used to help frame the discussion presented here. It should not be confused with "traditional warfare" as described in JP 3-24, *Counterinsurgency Operations* (U.S. Joint Chiefs of Staff, 2009c, p. I-8). Here, traditional warfare is equated with conventional warfare. Traditionalists adhere to the theories of capstone doctrine and maneuver warfare, as discussed in greater detail in Chapter Two.

2. *Credible:* Credible assessment reports are also transparent, because opacity devalues any report to the level of opinion: If the data and methods used to produce the report cannot be openly examined and debated, consumers have no reason to trust the reporter. In credible reports, all biases are explained, and flaws in data collection, valuation, and analysis are clearly articulated. Data are explained in context. Such reports clearly explain the process used by commanders and staffs to select methods and are both accurate and precise to the greatest degree possible.

3. *Relevant:* Relevant reports effectively and efficiently inform policy consideration of COIN campaign progress. They are sufficient to help senior military leaders and policymakers determine resourcing and strategy, and they satisfy public demand for knowledge up to the point that they do not reveal sensitive or classified information.

4. *Balanced:* Balanced reports reflect information from all relevant sources available to military staffs and analysts, including both quantitative and qualitative data. Such reports reflect input from military commanders at all levels of command and are broad enough in scope to incorporate nonmilitary information and open-source data. Balanced reports include countervailing opinions and analysis, as well as data that both agree with and contradict the overall findings.

5. *Analyzed:* Ideally, finished reports do not simply reflect data input and the analysis of single data sets. They also contain thorough analyses of all available data used to produce a unified, holistic assessment. This analysis should be objective and predicated on at least a general methodological framework that can be modified to fit changing conditions and, if necessary, challenged by consumers. The requirement for holistic analysis is commonly voiced in the assessment literature.

6. *Congruent:* COIN assessment theory should be congruent with current U.S. joint and service understanding of warfare, the COIN environment, and the way in which COIN campaigns should be prosecuted. The standard for congruence was drawn from a combination of joint doctrine and service capstone doctrine, as discussed in greater detail in Chapter Two.

7. *Parsimonious:* Assessment cannot show all aspects of a COIN campaign, nor should it seek or claim to deliver omniscience. Collection and reporting requirements for assessment should be carefully considered relative to the demands and risk they may leverage on subordinate units. Assessment should rely to the greatest extent possible on information that is generated through intelligence and operational activities, without requiring additional collection and reporting.

This framework shaped the study's examination and critique of the current U.S. military COIN assessment process. It also provides a starting point for identifying potential alternatives or improvements to the current process. While the examination of EBA and pattern and trend analysis is critical, the ultimate purpose of this monograph is to help improve operational and strategic doctrine and practice for COIN assessment. The recommendations presented here are designed to support military commanders in their efforts to produce credible COIN assessments for policymakers and the public.

The most important recommendation to improve COIN assessment is to ensure that the overall process includes contextual assessment. The concept of contextual assessment as discussed here is framed by the seven criteria identified as useful standards for assessment. This monograph develops and uses this concept as a springboard for further exploration of alternative options to the current approach to COIN assessment. It is not intended to be the only option available to enhance the assessment process, nor is it necessarily the best option available. It is, however, intended to broaden the conceptual understanding of assessment methodology and pave the way for improvement.

Findings

The findings presented here call into question the use of centralized quantitative approaches to COIN assessment. An examination of theater-level assessments conducted in Vietnam, Iraq, and Afghanistan shows that U.S. COIN assessment has long relied on centralized and decontextualized quantitative methods at the expense of context, relevant qualitative data, and comprehensive analytic methods. This conclusion does not suggest, however, that quantitative data are not useful or that they are less valuable for COIN assessment than qualitative data. *Rather, quantitative and qualitative data should be assessed equally, in context, and according to appropriate criteria.* Further, because the findings presented here are narrowly focused on campaign COIN assessment, they have no proven bearing on the relative merit of quantitative methodologies used to assess other aspects of COIN operations (e.g., logistics), conventional warfare, or in any other military endeavor.

Key Finding

Context is critical: It is not possible to assess COIN campaign progress through centralized measurement because this approach cannot reflect the contextual nuance of a COIN campaign. Assessment tells policymakers what has happened so they can try to gauge progress or failure. Analysis, as a function of either assessment or intelligence, tells policymakers why events occurred and how they might unfold in the future. Efforts to measure (rather than assess or analyze) COIN campaigns using aggregated and centralized quantitative methods provide neither of these services effectively. Centralized quantitative reports (e.g., graphs, color-coded maps, aggregated tables) tend to contain data that are inaccurate to a significant degree; they also lack context and fail to account for qualitative inputs. Consequently, such reports often produce inaccurate or misleading findings. When they do identify recognizable patterns, these patterns may be misleading or insufficient to support policy decisions. Trends that are tied to mathematical thresholds can convey unrealistic accuracy and quantitative thresholds and milestones that are highly subjective. Patterns and trends can support assessment, but they cannot stand alone. Effective assessment depends on capturing and then relating contextual understanding in a way that is digestible to senior decisionmakers.

The Realities of Policy

At least for the duration of the Afghanistan campaign, policymakers will continue to require independent sets of aggregated quantitative metrics to support decisionmaking. Assessment staffs should make every effort to compensate for the inherent inadequacies and flaws in these metrics to support policymaker requirements. This monograph describes best practices in assessment as identified through empirical research, but it is important to note that best practices are not always attainable in the face of wartime friction and attendant bureaucratic realities. When required to measure the progress of a COIN campaign, civilian officials and military officers should strive to place metrics in context and to carefully tie visual representations of data to explanatory narratives. They should balance quantitative metrics with probability and accuracy ratings and also identify and explain gaps in the available information. All information should be clearly sourced and should be retrievable by policy staffs seeking an in-depth understanding of specific subjects. If possible, all quantitative reports should be presented as part of holistic, all-source analysis. Transparency and analytic quality might enhance the credibility of aggregated quantitative data.

General Findings on Assessment in Counterinsurgency

This research led to a number of findings and conclusions associated with the role of assessment in informing the future direction of COIN campaigns, as well as the benefits to those responsible for decisionmaking in support of such campaigns.

An effective theater-level assessment is transparent, credible, and relevant to policymakers. Western democracies depend on sustainable public support to prosecute what are often lengthy COIN campaigns. Building and sustaining trust in official government assessments requires a process that is both transparent and credible. Transparent reports are (preferably) unclassified and share with consumers the underlying data and the methods used at each level of assessment. Credibility is derived from transparency, accuracy, and the use of a sound methodology that captures the relevant context. Relevant reports are those that successfully and efficiently inform policy and satisfy the public need for information.

Centralized assessments built from aggregated data do not produce transparent or credible reports. Centralized, uniform, top-down assessment requirements, aggregation, and decontextualization can undermine transparency by obscuring data sources or the reasoning behind subordinate assessments. An assessment methodology that does not adhere to the tenet of transparency can undermine the credibility of a COIN campaign among consumers of the assessment report and the public.

There is general agreement that information and data sets in COIN assessment materials are inaccurate and incomplete to a significant yet unknown degree. U.S. military doctrine, the COIN literature, at least two U.S. Secretaries of Defense, and senior assessment analysts reviewing case studies from Vietnam, Iraq, and Afghanistan concur that much of the information that is collected and fed into the COIN assessment process is both inaccurate and incomplete to a significant yet unknown degree.

COIN information is best analyzed at the level at which it is collected. The COIN environment is complex and marked by sometimes-extreme variations in physical and human terrain, varying degrees of progress from area to area (and often from village to village), and varying levels of counterinsurgent presence and collection capability. Therefore, information can have very different meanings from place to place and over time.

Holistic military campaign assessment is not scientific research. Sometimes, campaign assessment is likened to scientific research. As this examination shows, assessment is not scientific research, and most assessment data and methods cannot be held to strict scientific standards. Because assessment is not scientific research, campaign assessment reports should not be held to scientific standards, nor should attendant data be presented as objective research findings. This conclusion relates to the study's key finding that COIN assessment must be considered within the appropriate context. What holds true for one COIN campaign may not for another campaign being conducted elsewhere.

It may be possible to identify some generalized patterns of behavior or trends in activity from aggregated quantitative metrics, but patterns and trends do not constitute assessments. Data can sometimes indicate very broad patterns of behavior or trends over time, but few patterns or trends can show clear cause and effect. Those that show correlation can be useful in some cases but usually only to help clarify more detailed assessments. Furthermore, inaccurate data can produce precise but inaccurate patterns and trends, which, in turn, can mislead both assessment staffs and policymakers. Many aggregated pattern and trend graphs can be interpreted as having diametrically opposed meanings by different consumers because these graphs do not reflect causal evidence.

Centralized, top-down assessment collection requirements are unparsimonious, they can unhelpfully shape behavior at the tactical level, and they undermine broadly accepted tenets of irregular warfare. The use of a centralized list of metrics (i.e., a "core metrics list") for assessment sets what is essentially a standard for performance at the tactical level of operation—a standard that is often used to gauge the success or failure of military commanders. This informal standard of performance may encourage a misdirection of COIN efforts and an inappropriate expenditure of resources because not all areas across a COIN theater of operations require the same application of techniques or resources at any one point in time or over time. For example, a core metric may require subordinate units to report the number of schools in each area of operation, but while schools may be critical to success in one area, they may be far less meaningful in another. Regardless of conditions, both units might feel compelled to focus on building or supporting schools due to the influence of the core metric.

Trying to find the "right" metrics is an exercise in futility. There are no inherently "good" or "bad" metrics in COIN. No single metric is necessarily good or bad. For example, it would not be correct to say, "A reduction in violence is always good," because such an outcome could be the result of coalition withdrawal leading to defeat, a change in insurgent tactics, or insurgent victory. In specific cases, the oft-maligned body count metric might be useful, but only if these data are supported by evidence of progress toward a specific objective (e.g., a unit identifies 50 foreign fighters and the objective is to eliminate that unit of foreign fighters). It may be helpful to describe the potential strengths and weaknesses of various metrics in various contexts, but these descriptions should not be taken as prescription.

Standing U.S. military doctrine on COIN assessment is, with few exceptions, incongruent with the literature and doctrine on COIN operations and environments. Joint and some service doctrine calls for the application of traditional COIN fundamentals in the form of bottom-up, distributed operations while simultaneously recommending a top-down, centralized assessment of those operations. Effects-based language that should have been removed by 2008 persists in joint doctrine and in practice in 2011. Effects-based theory has challenged the ability of the assessment process to usefully serve consumers at the tactical and policy levels.

COIN assessment is a poorly understood process. Few military officers, commanders, or policymakers have a strong understanding of the inner workings of COIN assessment. Neither military training nor military education adequately addresses COIN assessment, and there is little or no interagency collaboration on assessment in the U.S. government. Furthermore, assessment is rarely integral to U.S or NATO COIN campaign planning and execution.

Counterinsurgency Assessment in Afghanistan

Commanders and staffs must make due with the assessment tools they have on hand in the absence of a clearly defined, substantiated process. Because literature and experience with modern COIN assessment are limited and doctrine provides inadequate and sometimes contradictory guidance on assessment, commanders and staffs must make due with sometimes inventive but generally ad hoc solutions.

The drive to create large data repositories for core metrics to meet requirements in Afghanistan has proved to be ineffective and, in some ways, counterproductive. Inequities generated by "one-size-fits-all" collection requirements and weak communication infrastructure have prevented the development of a comprehensive, centralized report database. The centralized databases that exist are incomplete and contain an unknown number of inaccuracies. Even if these databases were sound, they would permit and encourage analysis of raw and aggregated data devoid of context. Assessment analysts who work with these databases spend an inordinate amount of time rectifying and categorizing data instead of analyzing data that have already been contextualized by subordinate units.

As of early 2011, assessment was not synonymous with analysis. At least until early 2011, campaign assessment reports tended to contain very little comprehensive analysis and often included summaries of subordinate analyses that did not follow a common methodology. Theater assessments were often little more than summations of data reporting combined with uncoordinated and inconsistent subordinate insights. The International Security Assistance Force (ISAF) in Afghanistan has recently attempted to incorporate analysis into assessment, but these efforts are limited and do not reflect a comprehensive shift in approach to COIN campaign assessment.

Assessments of the Afghanistan COIN campaign suffer from a lack of transparency and credibility. The various assessments of the Afghanistan campaign reflect a mix of EBA-like measurement and pattern and trend analysis. Reports tend to emphasize measurement. This emphasis, in turn, reflects U.S. effects-based doctrine that is heavily reliant on perception of effects, core metrics lists, and the presentation of aggregated or otherwise decontextualized data.

Reports tend not to explain inconsistencies inherent in COIN assessment. Methods of data collection, data analysis, and assessment reporting vary from level to level, laterally from unit to unit, and over time. While the realities of COIN may require variation in assessment, there has been little effort to capture or explain these variations for consumers.

Centralized military assessment in coalition and whole-of-government COIN warfare (e.g., ISAF in Afghanistan) does not and cannot reflect consistent or controlled information or data collection. Producing even a nonscientific centralized assessment based on pattern and trend analysis requires some degree of control over the collection of information and the inputs to the assessment process. But contemporary COIN demands a whole-of-government and, often, a coalition approach. With some exceptions, an external military force (e.g., the United States in Vietnam, Iraq, or Afghanistan) cannot explicitly task nonmilitary agencies—and can only indirectly task allies—to collect and report information or data. Similarly, host nation agencies may not respond to U.S. military assessment requirements or may respond to data requests selectively. Sets of data managed by U.S. or coalition forces often do not reflect input from all areas of the battlefield under coalition control or from all echelons of command.

Recommendations

The study's findings point to five primary recommendations for those involved with COIN campaign analysis at both the tactical and policy levels.

Conduct a thorough review of U.S. military assessment doctrine. Both U.S. joint and service staffs should conduct reviews with the purpose of rectifying contradictions and inconsistencies in doctrine.

Train and educate military staff officers to understand and incorporate assessment into COIN campaigns. The services should consider incorporating campaign assessment theory and practice into training and education programs that address COIN planning and operations.

Conduct a review of interagency COIN assessment. An interagency panel should address the lack of coordination between DoD and nonmilitary government agencies that are likely to participate in COIN campaigns (e.g., the U.S. Department of State, U.S. Department of Agriculture, U.S. Agency for International Development).

Incorporate all-source methodology as used by the intelligence community into the campaign assessment process. Analysis is generally absent from campaign assessment at least in part because there is no clearly defined assessment analysis process. All-source intelligence methodology provides a good framework for assessment and could be used to both improve and structure campaign assessment reports.

Implement a decentralized, objective-focused assessment process that incorporates sound analysis, or at least narrative context, at all levels. Any alternative to the current process must operate within the realities of the COIN environment and should aspire to transparency, credibility, and the other standards presented earlier. The data that it incorporates should be analyzed at the level at which they are collected, and this analysis should be balanced by in-depth analyses at higher levels. Assessments written to address campaign or theater-level objectives should be founded on layers of contextual analysis. Layered context will provide what might be described as "analysis in depth." In addition, all data sources should be clearly cited. The U.S. military should adopt or at least attempt to implement some variation of contextual COIN assessment.

A Note About Contextual Assessment

Contextual assessment is a framework for COIN assessment that better aligns it with capstone military doctrine and COIN operational doctrine. This comprehensive, bottom-up assessment process builds layers of contextual narrative and data from the battalion to the theater level. The assessment process should be decentralized, but the reporting process should be standardized across the theater to ensure consistency and continuity over time.[6] Every report—from the battalion to the theater level—would be wholly captured by the final theater report. *In the best-case scenario, analysis would be conducted at each level.* The creation of a long-form assessment would ensure that policymakers, the public, and senior leadership have access to detailed justification for the findings in executive summary reports. More importantly, a comprehensive long-form assessment would provide a sound basis for theater- and policy-level assessments. *Thus, consumers will not have to read several hundred pages of assessment, analysis, and data*, but the information will be available to provide depth, transparency, and credibility as needed.

Commanders' inputs are a critical element of holistic operational COIN assessment. A commander's *coup d'oeil*, or eye for the battlefield, is indispensable not only for sound decisionmaking in combat but also in helping to assess what has happened and what might happen in a complex combat environment. Military commanders are handpicked for command, and policymakers and the public necessarily rely on them to a significant degree in times of war. However, commanders' assessments are often incorrectly deemed wholly subjective and thus suspect. In the absence of layered contextual analysis, it is difficult for commanders to argue that their personal analysis is more substantive than subjective opinion. Contextual assessment is designed to reinforce commanders' assessments by giving weight to their analyses. All commanders, from the battalion to the theater level, would benefit considerably from a transparent, credible, and contextual approach to assessment.

Figure S.2 shows how battalion reports are folded into regiment- or brigade-level reports and then into regional reports in the proposed contextual analysis framework. The theater report contains each of these component reports and may be several hundred pages in length, but it does contain built-in summaries at each level. By comparison, the unclassified semiannual DoD assessment report on Afghanistan and Iraq is between 60 and 150 pages and includes only a single summary and almost no operational context or detailed citations. The theater-level report includes narrative justifications and superior analysis of subordinate reports at each level. These analyses, or critiques, help address the inherent bias of narrative assessment. To further counter bias in the narrative reports, the theater-level analytic team provides a comprehensive analysis of the entire theater report. Finally, the commander provides an overall assessment. The theater-level assessment, the theater-level analysis, and the commander's assessment contain summaries that can be used to build executive policy papers. The theater commander's assessment is the top layer of the report, as shown in the figure.

---

[6] A template for contextual assessment is presented in Appendix A. Appendix B provides a notional example of contextual assessment in practice.

Figure S.2
The Contextual Assessment Reporting and Analysis Process

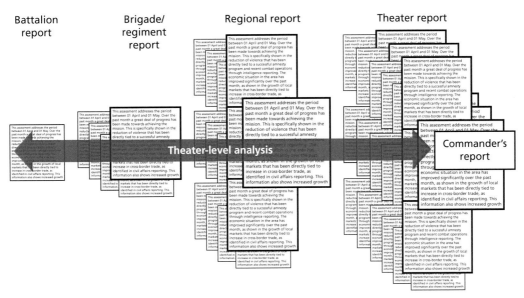

NOTE: Theater-level analysis considers all reporting from battalion up to the theater report. The commander's report addresses both the theater-level report and theater-level analysis.

Acknowledgments

I offer my sincere thanks to Stephen Downes-Martin of the U.S. Naval War College and Jonathan Schroden of the Center for Naval Analyses for providing valuable insight into my research and findings. Hriar Cabayan of the Office of the Secretary of Defense also provided support, access, and insight. The staff of the Information Dominance Center in ISAF Joint Command generously facilitated my research. LtCol Matthew Jones and LTC Vernon J. Bahm of ISAF Joint Command facilitated my research in Afghanistan, as did COL Robert Hume, who was director of the ISAF Afghan Assessment Group in early 2010. LtCol David LaRivee of the U.S. Air Force Academy and LTC Bret Van Poppel from the ISAF Afghan Assessment Group both facilitated my research and offered very helpful guidance on the current assessment process.

At RAND, Walter Perry provided mentoring on the subject of operations research and quantitative analysis and editorial input that greatly improved this monograph. Jerry M. Sollinger helped improve the organization and presentation of the study's findings. Jason Campbell offered helpful insight into the vagaries of COIN data sets.

Col (ret.) T. X. Hammes, U.S. Marine Corps, and Michael Hix and Martin Libicki at RAND provided critical review and mentorship. RAND colleague Arturo Muñoz also offered insight during the editing process. Christopher Paul at RAND provided a thorough critique that greatly improved the tone and approach of this monograph. Michael Baranick and Brian Efird at the National Defense University helped arrange forums for discussion and shared insight into modeling and simulations. James N. Bexfield at the Office of the Secretary of Defense helped facilitate discussions of my research and my participation in the NATO assessment review process.

I also owe thanks to Philip J. Eles of the Government of Canada, Kyle Pizzey of NATO, MAJ Tad Hunter, MAJ Sang M. Sok at the Center for Army Analysis, COL (ret.) Richard A. Starkey of U.S. Central Command, Joseph Soeters at the Netherlands Defence Academy, COL Thomas Cioppa and LTC David Hudak at the Army's Test and Evaluation Command Research and Analysis Center, MAJ Gary Kramlich of the 82d Airborne Division, and many others who helped provide insight into the U.S. military's current assessment practice. Gail Fisher at RAND helped brainstorm some of the crossover points between social science research and assessment, and Rebecca Thomasson at the Office of the Secretary of Defense served as a sounding board and contributed valuable insight regarding the national assessment process. I would particularly like to thank the librarians and archivists at Texas Tech University's Vietnam Center and Archive for the valuable service that they provide to researchers and historians of the Vietnam War.

Abbreviations

AAG	Afghanistan Assessment Group
ANSF	Afghan National Security Forces
ARVN	Army of the Republic of Vietnam
AWG	assessment working group
BLS	U.S. Bureau of Labor Statistics
CA	contextual assessment
CCIR	commander's critical information requirement
CIA	Central Intelligence Agency
CIDNE	Combined Information Data Network Exchange
CJIATF-435	Combined Joint Interagency Task Force–435
COIN	counterinsurgency
CORDS	Civil Operations and Revolutionary Development Support
CUAT	Commander's Unit Assessment Tool
DAM	District Assessment Model
DoD	U.S. Department of Defense
DoS	U.S. Department of State
EBA	effects-based assessment
EBO	effects-based operations
FM	field manual
GST	general system theory

GVN	Government of (South) Vietnam
HES	Hamlet Evaluation System
IBC	Iraq Body Count
IED	improvised explosive device
IJC	ISAF Joint Command
ISAF	International Security Assistance Force
JP	joint publication
MACV	Military Assistance Command, Vietnam
MCDP	Marine Corps doctrine publication
MOE	measure of effectiveness
MOP	measure of performance
MPICE	Measuring Progress in Conflict Environments
MtM	mission-type metrics
NATO	North Atlantic Treaty Organization
NC3A	NATO Consultation, Command, and Control Agency
NCO	noncommissioned officer
NCW	network-centric warfare
NGO	nongovernmental organization
NVA	North Vietnamese Army
ONA	operational net assessment
OSA	Office of Systems Analysis
OSD	Office of the Secretary of Defense
PAAS	Pacification Attitude Analysis System
PACES	Pacification Evaluation System
PRT	Provincial Reconstruction Team
RC	regional command
RMA	revolution in military affairs

SIGACT	significant activity
SoSA	system-of-systems analysis
TCAPF	Tactical Conflict Assessment and Planning Framework
USAID	U.S. Agency for International Development
VC	Viet Cong
VCI	Viet Cong infrastructure

CHAPTER ONE

Introduction

This monograph examines how combatant staffs and policymakers assess and describe the progress of counterinsurgency (COIN) campaigns and provides an in-depth analysis of the military assessment process for COIN. It also offers a critique of current doctrine and the performance of military campaign assessment across three case studies (Vietnam, Iraq, and Afghanistan). The purpose of this analysis and critique is twofold: (1) to identify the complexities and challenges of COIN assessment and (2) to frame standards and best practices to facilitate improvements to COIN assessment. The ultimate purpose is to help improve operational and strategic doctrine and practice for COIN assessment. Although the purpose of this research was derived from the sponsor's objectives and builds on recommendations developed to support units operating in Afghanistan in 2010, the recommendations are designed to support commanders across the U.S. military in their efforts to produce credible COIN assessments for policymakers and the public.

Assessment serves a dual purpose in that it tells the story of a campaign (the "up-and-out" assessment to policymakers and the public) while also helping to shape ongoing operations by informing operational and tactical commanders (the "down-and-in" assessment for operational units). These two goals are sometimes met through the production of separate assessment reports; in other cases, theater-level staffs attempt to address both audiences with a single type of report. This monograph focuses on the process used to produce theater-level "up-and-out" reports and recommends strategies to ensure that they provide credible and accurate input to national security strategy. Although it does not comprehensively address the "down-and-in" aspects of military assessment, the recommendations presented here should, if implemented, address both purposes.

A thorough review of the assessment literature available through early 2011 revealed no effort to comprehensively address COIN assessment. While no study of this nature can be fully comprehensive, this monograph attempts to break new ground by offering a holistic examination of COIN assessment. To this end, it describes and explains the theories that shape assessment, the environment in which assessments are developed, and the broad purposes of assessment while exploring the details of process and method. However, this monograph is not comprehensive in that it does not address all aspects of COIN assessment (see the section "Issues Not Addressed," later in this chapter). It does go into considerable detail on an array of issues pertaining to assessment. This detail is necessary to provide an in-depth understanding of a subject that has not received sufficient attention in current literature.

Although the findings presented here will be of interest to a broad audience, the research is specifically intended to inform the process used to assess ongoing COIN efforts in Afghanistan, and it uses Afghanistan as a central case study. Because COIN is prosecuted most often by the ground combat services (the Army and Marine Corps), the research, findings, and recommendations focus most directly on ground combat service or joint COIN assessment as it applies to the ground combat services.

Background

This section provides a brief description and explanation of assessment across the spectrum of warfare, outlines the difference between assessment and intelligence analysis, and examines policymaker and media dissatisfaction with the type of assessment reports issued during three relevant COIN campaigns: Vietnam, Iraq, and Afghanistan.

An *assessment* is a report or series of reports written to inform operational and strategic decisionmaking.[1] Military staffs build the assessments and, at least in theory, military commanders use them to make operational decisions in the field. These commanders add their comments to the assessment and then pass it along to policymakers. Assessments are typically delivered in the form of narratives, quantitative graphs, maps, or briefing slides.[2] Senior military leaders and policy staffs use these reports to track the progress of military campaigns and to determine how to allocate resources, whether specific operations or initiatives have met success or failure, and whether and when it may be necessary to alter strategy.

Differences Between Campaign Assessment and Intelligence Analysis
There are no clear standards for what a U.S. military campaign assessment should deliver or the form in which it should be presented. It is also not clear whether assessment should provide absolute clarity, reasonable understanding, an impression of events, or some combination thereof. A review of the assessment literature and interviews with staff and commanders indicated that assessments should give senior military leaders and policymakers a relevant understanding of what has happened on the

---

[1]   This monograph uses the terms *strategy* and *policy* interchangeably. Doctrinal or academic differences between the two concepts are not always consistent, and strategy and policy are often conflated in practice.

[2]   A number of military assessments (of all types) from Vietnam, Iraq, and Afghanistan are cited throughout this monograph. The following are examples of military assessments: General Westmoreland's Activities Reports, the Pentagon Papers, the ISAF Joint Command District Assessment map, the 1230-series reports on Iraq and Afghanistan, and General Stanley A. McChrystal's *COMISAF's Initial Assessment (Unclassified)*.

ground and why. In this way, a campaign assessment is very similar to intelligence analysis: It explains what has happened—and sometimes what is happening now—and why this information matters. However, it is different from intelligence analysis in that it is holistic and also addresses the performance of friendly organizations. The assessment explains how the entire campaign is progressing toward meeting the policymaker's strategic objectives. Table 1.1 lists some of the differences between campaign assessment and military intelligence analysis. These are broad generalizations drawn from multiple sources and thus intended for general comparison only.

The lines between assessment and intelligence analysis are sometimes blurred because military campaign assessment is poorly defined in doctrine and can involve the use of a wide array of methods, including many types of analysis. During the Vietnam War, at least one assessment staff clearly crossed the line into the realm of

Table 1.1
Campaign Assessment Versus Intelligence Analysis

Characteristic	Campaign Assessment	Intelligence Analysis
Primary purpose	Assess progress against operational and strategic objectives	Explain behavior and events and predict future behavior and events
Process	Describes and explains progress, recommends shifts in resources, strategy, informs operations	Describes and predicts behavior and actions in the environment, informs courses of action for operations and policy
Method	Any relevant and useful method	All-source analysis using structured, doctrinal methods within prescribed intelligence oversight limits
Sources	Any available sources, including friendly operations reports and completed intelligence reports	Limited to examination of enemy, foreign civilian, and environmental information[a]
Creators	Representatives of all military staff sections and military commanders[b]	Trained intelligence analysts
Time frame	Shows progress over long periods	Timely, degrades in value over time
Classification	Can be mostly or entirely unclassified	Almost always classified or restricted[c]

[a] Note that U.S. intelligence oversight regulations restrict the collection and analysis of information on U.S. citizens by designated intelligence organizations. While these policies do not fully preclude this kind of activity, they do regulate the activity to the point that participation is limited to a very few specific programs (e.g., those associated with international terrorism). There appears to be no clear regulation preventing U.S. military intelligence agencies from describing U.S. military activities in broad terms (i.e., without identifying individual U.S. service members), but in practice, military intelligence is confined to describing the ways in which U.S. military activities affect or shape the enemy, the environment, or foreign civilians.

[b] Trained assessment analysts should ideally play a role in the process at some point. As of early 2011, there was no "assessment analyst" position in the U.S. military. Operations researchers and red-team analysts sometimes fill this role.

[c] The exception to this general rule is open-source intelligence, but even many open-source products are restricted once analyzed.

intelligence analysis (a case described in greater detail in Chapter Six). As a result, this staff produced some reports that did not meet the standards or requirements for either assessment or intelligence analysis.[3] While intelligence often feeds assessment, military campaign assessment should not attempt to replicate or replace intelligence analysis. Military commanders and policymakers are best served by receiving both types of reports.

There is a final yet important distinction between campaign assessment and intelligence analysis. While assessment is intended primarily to support policy decision-making, democracies also depend on assessment to inform the general public about the progress of a war. At some point, campaign assessment must be published in a transparent and unclassified document. This unclassified report must be clear and simple enough to be understood by a wide range of consumers. Intelligence analyses are rarely written with the express intent of public release, and in most cases they are not officially declassified until long after the war has ended.

The Role and Purpose of Military Campaign Assessment

With these distinctions drawn, this chapter turns to a question that serves as the starting point for considering the assessment process: Why is military campaign assessment important and how is it used to shape strategy? The literature review identified three sources that offer a general framework of military assessment theory as it is applied to both conventional and irregular warfare. First, Scott Sigmund Gartner describes the essential nature of conventional warfare assessments in *Strategic Assessment in War*:

> Leaders assess and, if necessary, alter their strategies based on information they gather from the battlefield. . . . These decisions can have enormous impact. Decisions on strategy play a significant role in determining a war's nature, as well as its duration, intensity, and ultimately who wins and who loses.[4]

James Clancy and Chuck Crossett describe two broad approaches to conventional warfare assessment:

> Since World War II, the analysis of warfare has primarily been based upon two major concepts of effectiveness. In the grand movement of military forces, the gaining and control of territory is considered success. Those who control the land

---

[3] This refers to a report on the situation along Highway 4 in South Vietnam by Office of Systems Analysis Southeast Asia Intelligence Analysis and Force Effectiveness Division. As discussed in Chapter Six, the report was too detailed to serve as a stand-alone campaign assessment for policymakers and simultaneously insufficiently sourced, cited, or assigned appropriate caveats to meet contemporaneous or current standards for intelligence analysis.

[4] Gartner, 1997, p. 2. Gartner takes the position that strategic assessment is achieved through the use of quantitative indicators. For additional discussion of the distinction between conventional and COIN assessment, see Clancy and Crossett, 2007.

control the resources, population, and legal structures within it. . . . Physical space is the battlefield. The other traditional metric of success is the order of battle (OOB). . . .Such metrics assume large force-on-force battles in a Clausewitzian-style engagement.[5]

Clancy and Crossett assert that this is not a useful paradigm in irregular warfare or in COIN, specifically. For COIN, they recommend a model that focuses on "sustainability, legitimacy, and environmental stability."[6] U.S. Army COL Bobby Claflin and panel co-chairs offered the following definition of military assessment at an April 2010 conference hosted by the Military Operations Research Society. He refers to COIN assessment, but his remarks could be applicable across the spectrum of conflict:

> Assessments are critical for an organization to understand how well the organization performs its mission; both what it produces as an output and what it achieves as an outcome. Assessments provide a rigorous analytic bridge between the functioning activities of the organization and those decisions necessary to make improvement; often articulated in policies and plans.[7]

U.S. military publications contain a range of definitions for assessment. *The United States Army Commander's Appreciation and Campaign Design* highlights five "opportunities to learn" from an assessment that define the purposes of the assessment process. Campaign assessments tell commanders, staffs, and policymakers

1. How to execute the planned course of action for a specific operation;

2. Whether another course of action needs to be adopted;

3. Whether the operational design based on the problem frame is producing results;

4. Whether the problem framing needs adjusting; and

5. Whether the learning mechanisms of the organization are tuned to the particular operational problem.[8]

The pamphlet states that the Army has focused on the first two functions but has put insufficient effort into developing functions 3 through 5. The U.S. Army field manual (FM) 5-0, *The Operations Process*, reinforces the idea that assessment is a tool for fine-tuning current and prospective operations. It also states that assessment is a continuous process. The manual defines assessment fundamentals for the Army as follows:

---

[5] Clancy and Crossett, 2007, p. 90.

[6] Clancy and Crossett, 2007, p. 96.

[7] Claflin, Sanders, and Boylan, 2010, p. 2.

[8] U.S. Army Training and Doctrine Command, 2008, p. 18.

> Assessment is the continuous monitoring and evaluation of the current situation, particularly the enemy, and progress of an operation (FM 3-0). Assessment is both a continuous activity of the operations process and an activity of battle command. Commanders, assisted by their staffs and subordinate commanders, continuously assess the operational environment and the progress of the operation. Based on their assessment, commanders direct adjustments thus ensuring the operation remains focused on accomplishing the mission.[9]

This description seems to be focused on the down-and-in or operational purpose of assessment. Assessments are intended to feed operational analysis, but they are also tools for strategic decisionmaking and the primary means by which policymakers in the executive branch explain the progress of a campaign to lawmakers and the public. At least in theory, holistic theater-level assessments, such as the ones produced by the International Security Assistance Force (ISAF) in Afghanistan, are used by the United States and the North Atlantic Treaty Organization (NATO) to gauge campaign progress, determine COIN policy, dedicate resources, shape training and force transformation, and support strategic communication messaging.

Dissatisfaction with Counterinsurgency Campaign Assessment

Producing a truly holistic and effective COIN campaign assessment has proven to be an elusive goal. The Army points out in *The Commander's Appreciation and Campaign Design* that the military has paid insufficient attention to certain aspects of campaign assessment. This conclusion is supported by the findings presented here: The research shows gaps in the process that affect not only strategic assessments but also operational, "down-and-in" assessments. After describing the intent of assessment, Claflin et al. describe some of the weaknesses in the assessment process as has been applied in Afghanistan:

> Despite the critical role that assessments play, organizations frequently treat assessments as an afterthought. Assessment capabilities are often recognized as lacking well after deployment and are subsequently generated out of the institutional force as a temporary loan. A lack of "operating force" assessment doctrine and analytic structure at echelons above corps may contribute to this assessment lag.[10]

Shortcomings in campaign assessments delivered to the public have not gone unnoticed in policymaking circles or in the public domain. Obvious gaps and inconsistencies in various reports in all three cases examined in this report (Vietnam, Iraq, and Afghanistan) have made consumers wary of military assessments. A few well-publicized manipulations of the Vietnam-era COIN reports by military and civilian

---

[9] HQDA, 2010, p. 6-1.

[10] Claflin, Sanders, and Boylan, 2010, p. 2.

leaders sowed an underlying layer of distrust among some policymakers and members of the press and public, and this distrust has carried over to Iraq and Afghanistan assessments. H. R. McMaster (*Dereliction of Duty*), James William Gibson (*The Perfect War*), and Guenter Lewy (*America in Vietnam*) have documented the extraordinarily contentious and often disingenuous Vietnam-era assessment process. In his book, McMaster describes the foundation of these misgivings, particularly then–Secretary of Defense Robert S. McNamara's predilection for shaping and (at least in McMaster's view) whitewashing Vietnam assessments, beginning as early as 1963:

> Although he acknowledged that "statistics received over the past year or more from the GVN [Government of (South) Vietnam] officials and reported by the U.S. mission . . . were grossly in error," he firmly believed that tracking quantitative indices would give him a clear picture of how the war against the Viet Cong was going. Despite MACV's [Military Assistance Command, Vietnam] protest that it was "impossible to measure progress in any meaningful way on a weekly basis," McNamara insisted on "Weekly Headway Reports" that included "measurable criteria" to help chart the progress of the war. The very title of the report revealed his eagerness to demonstrate the South's improvement under his program.[11]

In 1969, Congressman John V. Tunney of California delivered a scathing report on the Hamlet Evaluation System (HES)—one of the pillars of the Vietnam-era campaign assessment process—to the Committee on Foreign Affairs in which he used a quote from Lewis Carroll's *Through the Looking-Glass* to describe hamlet assessments. In his conclusion, Tunney stated, "It is difficult, after studying the HES, to understand how our officials could have put so much uncritical faith in [that system] in the face of opposing facts."[12] Distrust of Vietnam-era military briefings became so endemic that members of the press corps referred to the daily military press briefings as the "five o'clock follies."[13] Assessments performed during Operation Iraqi Freedom drew critical review from Congress and the press, at least until violence dropped precipitously in 2008, at which point interest in the Iraq War and Iraq policy plummet-

---

[11] McMaster, 1997, p. 58. In 2010, U.S. Secretary of Defense Robert M. Gates told the Senate Appropriations Committee, "This is not something where we do ourselves any favors by tearing ourselves up by the roots every week to see if we're growing," in reference to the assessment of U.S. operations in Afghanistan (for written testimony, see Gates, 2010). McNamara explained his actions in the books *In Retrospect: The Tragedy and Lessons of Vietnam* (McNamara, 1995) and *Argument Without End: In Search of Answers to the Vietnam Tragedy* (McNamara, Blight, and Brigham, 1999).

[12] Tunney, p. 9. HES is examined in some detail in Chapter Six of this monograph. A Brookings Institution study comparing public statements regarding Vietnam assessments and the assessments themselves showed remarkable congruence between the two. The report stated that concerns tended to revolve around the assessments' optimistic emphasis and denial of specific actions rather than the perception that there was a deliberate attempt to reshape information coming from the field (see Gelb and Betts, 1979, p. 320).

[13] Hammond, 1988, p. 239.

ed.[14] More recently, some congressional staff members and members of Congress have expressed a deep and sustained dissatisfaction with the military assessments of the Afghanistan campaign.[15]

The findings presented in this monograph show a distinct linear association between Vietnam-era assessment processes and methods and the assessment processes used to gauge COIN campaign progress in Iraq and Afghanistan. For example, media criticism of contemporary assessment is very similar to that articulated during the Vietnam War. A widely promulgated 2009 commentary by Tom Engelhardt stated,

> The problem was that none of the official metrics managed to measure what mattered most in Vietnam. History may not simply repeat itself, but there's good reason to look askance at whatever set of metrics the Obama Administration manages to devise [for Afghanistan]. . . . By the time they reach Washington, they are likely to have the best possible patina on them.[16]

The lack of confidence in military and political assessments of Vietnam, Iraq, and Afghanistan, at times, severely strained or (in the case of Vietnam) helped shatter national consensus and undermined a sustained focus on objectives and an honest debate on strategy. Such a loss of credibility can negate what FM 3-24, *Counterinsurgency*, refers to as a critical requirement in COIN: gaining and maintaining public support for a prolonged deployment.[17] While assessments should not be written with the express intent of shaping public opinion, the failure to design and implement a transparent and credible assessment process can directly undermine national strategy.

Literature Review

This section discusses the sources used in the literature review portion of the study and also briefly examines the state of literature on COIN assessment. The bibliography reflects most (if not all) of the literature on COIN assessment published through early 2011. In addition, the literature review included a range of published, unpublished, official, and unofficial work across a number of fields. To build a comprehensive picture of assessment, it is necessary to capture lessons from a diverse range of fields, such as systems analysis, military operational art, time-series analysis, COIN doctrine, policy decisionmaking, intelligence analysis and policy, and military leadership. The initial

---

[14] An Associated Press survey of press reporting showed a dramatic decline in press coverage of Iraq after 2007 (see Ricchiardi, 2008).

[15] According to discussions with congressional staff members between November 2009 and February 2011, as well as discussions with analysts in Afghanistan and the United States who worked on Afghanistan assessment.

[16] Engelhardt, 2009.

[17] HQDA, 2006c, p. 1-24.

phase of the review considered a wide array of assessment and measurement publications, including business literature, professional journals on assessment and measurement, systems analysis literature, and publications on operations research.[18] The second phase of the literature review had a narrower focus, on assessment theory and doctrine for military and stabilization operations, and included the few published books on military assessment and the somewhat more extensive corpus of journal articles on irregular warfare assessment.[19] Relevant doctrinal publications included most joint, Army, and Marine Corps publications on operations and intelligence. A review of the literature on stabilization and development assessment tapped into this narrow field and included lessons-learned publications from the U.S. Agency for International Development (USAID) and Canadian government sources.[20]

The examination of effects-based operations and assessment (EBO and EBA) relied extensively on official publications and doctrine, as well as journal articles and research reports.[21] Analysis of the "wicked problem" was derived primarily from Rittel and Webber's work and supplemented by the literature on complexity, chaos theory, and systemic operational design.[22] The review of the policymaking and leadership literature focused on works associated with systems analysis, warfare, and COIN.[23] The review of the intelligence analysis literature relied on official sources and (to a lesser extent) journal articles.[24]

---

[18] For example, *Measurement in the Social Sciences: Theories and Strategies* (Blalock, 1974), *Handbook of Research Design and Social Measurement* (Miller and Salkind, 2002), *Measure Theory* (Halmos, 1950), and *Selected Methods and Models in Military Operations Research* (Zehna, 2005). The review also included selections from the *Journal of Economic and Social Measurement*, *Measurement: Interdisciplinary Research and Perspectives*, the *Journal of Business Cycle Measurement and Analysis*, *Measurement Techniques*, *Applied Psychological Measurement*, the *Journal of Quantitative Analysis in Sports*, *Advances in Data Analysis and Classification*, *Advances in Statistical Analysis*, the *Journal of the Royal Statistical Society*, *Statistical Science*, and the *Journal of the American Statistical Association*.

[19] The books included *Strategic Assessment in War* (Gartner, 1997) and *Analysis for Military Decisions* (Quade, 1964). Articles included "Measuring Effectiveness in Irregular Warfare" (Clancy and Crossett, 2007), "A Will to Measure" (Murray, 2001), and "How to Measure the War" (Campbell, O'Hanlon, and Shapiro, 2009b).

[20] For example, "Military Operational Measures of Effectiveness for Peacekeeping Operations" (Anderson, 2001), *Measuring Fragility: Indicators and Methods for Rating State Performance* (USAID, 2005), and *Measuring Effectiveness in Complex Operations: What Is Good Enough* (Meharg, 2009).

[21] Examples here include Army FM 5-0.1, *The Operations Process* (HQDA, 2006a); *Lifting the Fog of War* (Owens, 2000); *Operational Assessment—The Achilles Heel of Effects-Based Operations?* (Bowman, 2002); and *Effects-Based Operations (EBO): A Grand Challenge for the Analytic Community* (Davis, 2001).

[22] These texts included "Dilemmas in a General Theory of Planning" (Rittel and Webber, 1973), "Chaos Theory and Its Implications for Social Science Research" (Gregersen and Sailer, 1993), and "Systemic Operational Design: Learning and Adapting in Complex Missions" (Wass de Czege, 2009).

[23] Specifically, *A Question of Command: Counterinsurgency from the Civil War to Iraq* (Moyar, 2009); *How Much is Enough? Shaping the Defense Program, 1961–1969* (Enthoven and Smith, 1971/2005); and FM 3-24, *Counterinsurgency* (HQDA, 2006c).

[24] Examples here include Joint Publication (JP) 2-0, *Joint Intelligence* (U.S. Joint Chiefs of Staff, 2007); FM 34-3, *Intelligence Analysis* (HQDA, 1990); *A Compendium of Analytic Tradecraft Notes* (CIA, 1997); and *A Tradecraft*

The approach to the case-study literature differed for the two primary cases considered in this monograph (Vietnam and Afghanistan). The literature review for the case of Vietnam necessarily focused on historical works and data, beginning with books that included both secondary source material and original information.[25] It then narrowed to focus on work conducted by the Office of Systems Analysis and on historical accounts of the HES, body counts, and policymaking.[26] The subsequent detailed analysis of the Vietnam case relied heavily on historical documents from the Vietnam Center and Archive at the Texas Tech University, which provided access to a breadth of original source material, including obscure military documents, regulations, and reports.[27] Unfortunately, Gregory A. Daddis's examination of U.S. Army assessments in Vietnam, *No Sure Victory: Measuring U.S. Army Effectiveness and Progress in the Vietnam War*, was published after this study had concluded.

The literature review for the Afghanistan case included official reports by NATO and the U.S. government, as well as published and unpublished journal articles and reports.[28] The number of available publications on Afghanistan assessment paled in comparison to the ready availability of material on Vietnam assessment, but the literature review on Afghanistan was intended primarily to frame the observation and interview process.

Primer: Structured Analytic Techniques for Improving Intelligence Analysis (U.S. Government, 2009).

[25] For example, *Vietnam: A History* (Karnow, 1984), *America in Vietnam* (Lewy, 1978), and *The 25-Year War: America's Military Role in Vietnam* (Palmer, 1984).

[26] Accounts of Office of Systems Analysis research include *War Without Fronts: The American Experience in Vietnam* (Thayer, 1985) and *A Systems Analysis View of the Vietnam War 1965–1972*, Vols. 1, 9, and 10 (Thayer, 1975a, 1975b, 1975c). Details about HES were found in *The American Experience with Pacification in Vietnam*, Vols. 1 and 2 (Cooper et al., 1972a, 1972b); *Analysis of Vietnamization: Hamlet Evaluation System Revisions* (Prince and Adkins, 1973); and *Measuring Hamlet Security in Vietnam: Report of a Special Study Mission* (Tunney, 1968). Sources for body counts include *Report on the War in Vietnam* (Sharp and Westmoreland, 1968) and selected debriefings. Accounts on policymaking include *Dereliction of Duty: Lyndon Johnson, Robert McNamara, the Joint Chiefs of Staff, and the Lies That Led to Vietnam* (McMaster, 1997); *In Retrospect: The Tragedy and Lessons of Vietnam* (McNamara, 1995); and *Lessons in Disaster: McGeorge Bundy and the Path to War in Vietnam* (Goldstein, 2008).

[27] For example, *MACCORDS-OAD Fact Sheet: RD Cadre Evaluation System* (MACV, 1968b), *Commander's Summary of the MACV Objectives Plan* (MACV, 1969a), and "General Westmoreland's Activities Report for September" (Westmoreland, 1967a).

[28] Sources include selections from the series *Report on Progress Toward Security and Stability in Afghanistan* (DoD, 2009a, 2010a, 2010b), "Unclassified Metrics" (ISAF Headquarters Strategic Advisory Group, 2009), "Transfer of Lead Security Responsibility Effect Scoring Model" (ISAF AAG, undated), "Measuring Progress in Afghanistan" (Kilcullen, 2009b), *COMISAF's Initial Assessment* (ISAF Headquarters, 2009), and *Afghanistan in 2009: A Survey of the Afghan People* (Rennie, Sharma, and Sen, 2009).

The review of the Iraq assessment literature was limited by classification restrictions, but it included journal articles, official documents and reports, and subject-matter expert reports and data indexes.[29]

Compared with the detailed and exhaustive study of measurement or assessment that is characteristic of the hard and soft sciences, little effort has been made to develop a comprehensive assessment model for military operations.[30] Much of the existing scholarship on the assessment of warfare is dedicated to historical review, strategic analysis of conventional war, or specific technical assessment methods. The work on assessment methodology specifically for COIN also has been limited in scope and, in many cases, merely explores basic theory or narrow mathematical processes. Thomas C. Thayer led the most comprehensive analyses of COIN metrics to date, published in the 12-part series *A Systems Analysis View of the Vietnam War* and in *War Without Fronts*, a retrospective examination of his work during the Vietnam War for the Office of Systems Analysis, Southeast Asia Intelligence and Force Effectiveness Division. There are few thorough recommendations for practical application of assessment methods outside the scope of the U.S. experience in Vietnam, however, and only a handful of very recent academic efforts on Afghanistan. Standing U.S. military doctrine has not yet adequately addressed this gap, although various U.S. joint and service manuals attempt to explain assessment. But these manuals generally provide only a brief overview and, on occasion, offer contradictory perspectives. This monograph attempts to address these gaps and inconsistencies in U.S. military assessment doctrine.

Two themes emerged from the literature review. First, assessment is often conflated with measurement or some form of centralized quantitative analysis. Of the sources cited here that specifically address military campaign assessment, nearly all assume or accept that assessment is a process of centralized measurement or pattern analysis. None of the works reviewed for this study, including the most aggressive critiques of centralized effects-based assessment, offer a comprehensive, decentralized alternative or a way to comprehensively incorporate nonquantitative data. Second, because the literature accepts the premise of centralized assessment, it tends to focus on detailed disagreement over centralized assessment issues, such as the selection of individual core metrics or the best method for aggregating data. While some sources do address the broader issues of assessment, few have done so comprehensively or in a way that might widen the aperture for the inclusion of other assessment theories. Professional debate over COIN campaign assessment is bounded in a way that seems to preclude the consideration of viable alternatives to doctrine or current practice.

---

[29] See "Measures for Security in a Counterinsurgency" (Schroden, 2009), "Assessing Iraq's Sunni Arab Insurgency" (Eisenstadt and White, 2006), selected *Measuring Stability and Security in Iraq* quarterly reports (DoD, 2009b, 2010c), the memorandum "Long-Term Plan for IRMO Metrics" (Sullivan, 2004), and *Iraq Index: Tracking Reconstruction and Security in Post-Saddam Iraq* (Brookings Institution, 2010).

[30] This conclusion is derived from the research conducted for this study but also echoes Darilek et al., 2001, p. 98, and comments by senior U.S. military officers made as recently as early 2010.

While this monograph considers only two existing approaches to COIN campaign assessment, there are three very general schools of thought on how to address the complexities and challenges of assessing a war (see Chapters Three and Four for a discussion of these approaches). The first school of thought prioritizes EBO/EBA, while the second prioritizes pattern and trend analysis. The third school of thought is represented by traditionalists like General James N. Mattis and Milan Vego.[31] Traditionalists, or advocates of capstone *maneuver warfare* theory, accept complexity and chaos as inevitable realities in warfare. They believe that some of the structured methods designed and used to penetrate complexity are ill conceived and not applicable to COIN. However, a generally agreed-upon and tested alternative to EBA doctrine or pattern and trend analysis has yet to emerge from either the traditionalists or other advocates or experts. In the literature, debate among these three loosely defined groups is inconsistent, insufficiently documented, and as yet unresolved.

It is distinctly possible that the analysis in this monograph has failed to capture other groups or schools of thought on the subject of COIN assessment, but it does attempt to add some clarity to the discussion. The first step to understanding all three arguments is to clarify the nature of the debate: how best to address the fog of war to assess COIN campaigns.

Research Methodology

The methodology for this study consisted of a literature review, direct observation, interviews, and case-study research. The research was conducted in two phases, with the first phase addressing COIN assessment in Afghanistan and the second addressing the broader issue of U.S. military COIN campaign assessment. The resulting monograph attempts to answer the following questions:

- What challenges does the COIN environment present to assessment?
- How does recent doctrine address these challenges, and what are the points of disagreement regarding doctrinal approaches to assessment?
- What do the lessons of Vietnam hold for contemporary assessment?
- What can recent COIN assessment contribute to an analysis of other assessment approaches?
- What are the strengths and weaknesses of the current approach?
- How could doctrine and practice be improved?

---

[31] In 2008, Gen. James N. Mattis, U.S. Marine Corps, issued a memo to U.S. Joint Forces Command stating, "It is my view that EBO has been misapplied and overextended to the point that it actually hinders rather than helps joint operations" (Mattis, 2008). See also Vego, 2006.

The first phase of this research aimed to develop alternative assessment methods for a combat element that was preparing to deploy to Afghanistan. This phase included a literature review, interviews, and observations derived from my participation in conferences and workshops on assessment. Preliminary findings provided a limited-scope analysis of assessment processes in Afghanistan and offered limited recommendations. These recommendations focused on the development of a process that would add context to assessment reporting and incorporate qualitative data. As the scope of the research expanded in the second phase of the study to include all U.S. COIN assessment, a framework for assessment criteria began to evolve. Although it would have been ideal to develop this framework at the outset of the project, it was necessary to dissect the existing theories of assessment first.

A number of conflicting and partial lists of standards have been proposed by various experts and in assessment manuals, as discussed in Chapter Five.[32] With a few exceptions, nearly all of the criteria described in the literature applied only to centralized quantitative assessment. However, two generalized criteria for successful COIN assessment emerged from the initial phase of the research: transparency and credibility. These two standards were introduced by ISAF in 2009 and are not necessarily consistent with assessment literature. Nonetheless, this study and prior RAND research showed that the two standards are well aligned with the theory, doctrine, and case-study literature on COIN as it is currently practiced by Western democracies. These standards may or may not be universally applicable (e.g., to COIN operations supported by dictatorships), but they are relevant for democracies that must sustain willing popular support for ongoing campaigns.[33]

Further examination of policymaker requirements, all-source analysis methodologies, COIN theory and doctrine, and scientific method revealed five additional criteria, or standards, that could be generalized for the production of holistic campaign assessment. Each was selected based on the examination of the literature, interviews with subject-matter experts, direct observation of assessment in Iraq and Afghanistan, and a close examination of the historical record on Vietnam. The first two standards—transparency and credibility—are deemed requisite for successful COIN assessment in a democracy. All assessment reports should be relevant to policymakers. The other

---

[32] These lists tend to focus only on certain aspects of assessment, typically the selection of core metrics for centralized assessment. There are very few recommendations in the literature that might help the military establish standards for an overarching assessment methodology.

[33] One could argue that transparency and credibility are more universally relevant since the advent of the Internet and a near-pervasive media environment. It has become more difficult for dictatorships, anocracies, and oligarchies to sustain unpopular external military operations when atrocities, costs, and opinions of military actions are widely promulgated on the Internet. For additional RAND research supporting the inclusion of transparency and credibility as assessment criteria, see the examination of 89 case studies in *How Insurgencies End* (Connable and Libicki, 2010).

four standards are practical requirements that should be met to establish an effective methodology. The seven standards are as follows:

1. *Transparent:* Transparent assessment reports are widely releasable to official consumers and, ideally, are unclassified and without distribution restrictions; the final iteration of the report is suitable for distribution on the Internet. Such reports reveal both the methods and data used at each level of assessment, from tactical to strategic, and allow for detailed and in-depth analysis without requiring additional research or requests for information. Any subjectivity or justification for disagreements between layers of command is explained clearly and comprehensively.

2. *Credible:* Credible assessment reports are also transparent, because opacity devalues any report to the level of opinion: If the data and methods used to produce the report cannot be openly examined and debated, consumers have no reason to trust the reporter. In credible reports, all biases are explained, and flaws in data collection, valuation, and analysis are clearly articulated. Data are explained in context. Such reports clearly explain the process used by commanders and staffs to select methods and are both accurate and precise to the greatest degree possible.

3. *Relevant:* Relevant reports effectively and efficiently inform policy consideration of COIN campaign progress. They are sufficient to help senior military leaders and policymakers determine resourcing and strategy, and they satisfy public demand for knowledge up to the point that they do not reveal sensitive or classified information.

4. *Balanced:* Balanced reports reflect information from all relevant sources available to military staffs and analysts, including both quantitative and qualitative data. Such reports reflect input from military commanders at all levels of command and are broad enough in scope to incorporate nonmilitary information and open-source data. Balanced reports include countervailing opinions and analysis, as well as data that both agree with and contradict the overall findings.

5. *Analyzed:* Ideally, finished reports do not simply reflect data input and the analysis of single data sets. They also contain thorough analyses of all available data used to produce a unified, holistic assessment. This analysis should be objective and predicated on at least a general methodological framework that can be modified to fit changing conditions and, if necessary, challenged by consumers. The requirement for holistic analysis is commonly voiced in the assessment literature.

6. *Congruent:* COIN assessment theory should be congruent with current U.S. joint and service understanding of warfare, the COIN environment, and the way in which COIN campaigns should be prosecuted. The standard for con-

gruence was drawn from a combination of joint doctrine and service capstone doctrine, as discussed in greater detail in Chapter Two.

7. *Parsimonious:* Assessment cannot show all aspects of a COIN campaign, nor should it seek or claim to deliver omniscience. Collection and reporting requirements for assessment should be carefully considered relative to the demands and risk they may leverage on subordinate units. Assessment should rely to the greatest extent possible on information that is generated through intelligence and operational activities, without requiring additional collection and reporting. Parsimony is a common theme in COIN assessment literature and is often identified as a requirement by assessment experts.[34]

Notably absent from this list are the standards of scientific method described in Chapter Three. That chapter explains how some assessment analysts have attempted to apply scientific rigor to assessment and suggests ways in which scientific standards might be applied to the various phases of COIN assessment. However, this monograph maintains that assessment is not scientific research. Because assessment is not research, none of these standards were deemed to be *generally* applicable to holistic campaign assessment. This does not negate their value in helping to shape assessment. Certainly, if a staff chooses to approach assessment as scientific research, it should pursue methods that are reliable and produce valid findings.

This study drew on several sources of information in addition to the literature review, including more than 20 interviews, both in Afghanistan and elsewhere. Questions for the interviews were crafted specifically for the position and experience of the interviewee; some of these interviews were planned in advance, while others were conducted in the field as opportunities presented themselves. Interview subjects included military personnel at all echelons of command and civilians involved in the assessment process or who had related responsibilities. In addition, I participated in the assessment process during three tours in Iraq between 2003 and 2006 and used this knowledge to frame questions and identify research sources. I also participated in and helped lead several conferences and workshops on COIN assessment in Afghanistan between late 2009 and early 2011, such as the official NATO Systems Analysis and Studies conference series designed to help build and validate Afghanistan metrics. Finally, I was briefly embedded with the ISAF Joint Command assessment team in May 2010. The in-depth case-study analysis of the Vietnam-era assessment processes was conducted over the course of a calendar year, and I obtained additional insight into the current assessment process while providing informal and official support to the ISAF Afghanistan Assessment Group (AAG) from mid-2010 through early 2011.

---

[34] However, this call for parsimony tends to be focused only on the selection of core metrics for centralized assessment. The standard for parsimony in the framework developed for this study is intended to address the assessment process more broadly.

Issues Not Addressed

This research focused specifically on the military assessment process for COIN campaigns. Both in doctrine and in recognized best practices as detailed in the literature, COIN assessment should (in theory) incorporate input from civilian agencies, nongovernmental organizations (NGOs), and host-nation officials and should therefore reflect a holistic civilian-military approach. Joint doctrine on COIN recommends that civilian agencies take the lead in U.S. and coalition COIN operations, and it stresses unity of effort between military and nonmilitary activities.[35] But doctrine has not translated neatly into assessment practice. For example, the U.S. Department of Defense (DoD) and U.S. Department of State (DoS) build separate theater-level assessments, and as of early 2011, they do not effectively communicate on assessment reporting or methodology. DoS and USAID use (or are in the process of testing) methods like MPICE (Measuring Progress in Conflict Environments) and TCAPF (Tactical Conflict Assessment and Planning Framework).[36] While the military has tested TCAPF, it has not been fully incorporated into the theater assessment process. Although this monograph recognizes the gap between civilian and military assessment, a detailed examination was beyond the scope of the research effort.[37] It does, however, address the incorporation of Provincial Reconstruction Team (PRT) information in the assessment process.

Research on the Vietnam-era assessment process was extensive, but the analysis and recommendations presented here focus on contemporary COIN doctrine and practice because their purpose is to help inform current policy debate. It also bears mentioning that while a great deal of information on the Afghanistan assessment process is unclassified, most of the official documentation of the Iraq war remains classified. Although this monograph cites research on Iraq, it was as much by necessity as by design that it focuses on Vietnam and Afghanistan as its principal case studies. Of course, the United States has been involved in many other COIN operations over the past 60 years, but most of these cases have less relevance to current campaign assessment than Vietnam, Iraq, and Afghanistan because they were primarily short-term, advisory, or covert-action missions and not what GEN David H. Petraeus calls "industrial-strength insurgencies."[38] One might use very different processes to assess the advisory mission to El Salvador and the COIN campaign in Iraq. Furthermore, while some historical cases prior to Vietnam informed this research (e.g., U.S. activities

---

[35] JP 3-24, *Counterinsurgency Operations*, states, "It is always preferable for civilians to lead the overall COIN effort" (p. IV-11); see also pp. IV-1–IV-22 (U.S. Joint Chiefs of Staff, 2009c).

[36] As of October 2010, TCAPF was being remodeled and expected to be replaced by the District Stability Framework tool. Statement by a U.S. USAID representative, October 14, 2010.

[37] Armstrong and Chura-Beaver, 2010, address some of these other reports and processes.

[38] Rubin, 2010b.

in the Philippines, Nicaragua, Greece), these cases were not sufficiently germane to the current process to warrant in-depth examination and comparison.

To ensure that the findings presented here were sufficiently streamlined for use by policymakers and those involved in the assessment process, it was necessary to contain the initial scope of the research and focus on methodology and U.S. COIN operations since, and including, Vietnam. The disadvantage of this approach was that it excluded potentially valuable information on British, Canadian, Soviet/Russian, and other international efforts.[39] Finally, NATO assessment doctrine is published at an unclassified but restricted level and cannot be cited in public documents. Therefore, this monograph does not refer to specific NATO documents or doctrine.

This monograph does not specifically address each and every aspect of campaign assessment. Most noticeable to assessment staffs and experts will be the absence of security force assessment and a detailed treatment of opinion polling. Both of these subsets of assessment are relevant to holistic campaign assessment. Security force assessment is shaped by many of the same concerns that shape holistic assessment, but it is usually more technical in nature than campaign assessment (in that it focuses on manpower, training, and logistics). Opinion polling is integral to campaign assessment, but it is a complex subset of holistic assessment that would require distinct and detailed treatment, beyond the time and resources available for this project.

Organization of This Monograph

This monograph provides a thorough examination of not only the practice of COIN assessment but also the theoretical roots underlying the various practical approaches to the assessment of contemporary COIN warfare. Its structure is intended to take the issue down to its roots and then build back up to a series of recommendations, with a thorough appreciation of the complex and contested issues in play. The narrative is designed to work from theory to practical application by delivering the broad themes underlying assessment, background on assessment theory and doctrine, a detailed historical case study (Vietnam), an examination of current practice and findings from that review, and recommendations that can be incorporated in the short and long terms. This progression is intended to put current practice in clear context to inform contemporary policy.

Because COIN assessment is controversial and inadequately addressed in the literature, this monograph introduces and examines the seven framework concepts for assessment presented earlier in this chapter, in the section "Research Methodology."

---

[39] Ample information on international efforts is available online. The bibliography at the end of this monograph also includes many references examined for this study that ultimately did not figure prominently in the findings presented here.

The document is organized as follows. Chapter Two examines policymaker requirements for assessment, the need for transparency and credibility for policymaking, and the impact of the COIN environment on assessment. Chapter Three introduces the concepts behind centralized assessment, including systems analysis, scientific rigor, and time-series analysis, and describes how pattern and trend analysis figures into the current assessment process. Chapter Four provides an overview of effects-based theories and the doctrine of EBA, and Chapter Five explains the considerations for selecting core metrics, with examples from Afghanistan. It also describes some of the challenges of centralized assessment and the necessity for context in assessment analysis. Chapter Six presents the historical case study of COIN assessment during the Vietnam War, examining how assessment failed in Vietnam and why and placing pattern and trend analysis in a case context. Chapter Seven provides an overview of the Afghanistan assessment process as of early 2011 and identifies some concerns about the assessments being produced in that contingency. Chapter Eight explains how and why current assessment approaches have failed to deliver adequate support to policymakers, and Chapter Nine offers several recommendations and options to improve the current process. Finally, Chapter Ten proposes an alternative to centralized assessment processes.

This document also contains five appendixes that provide resources and background information intended to supplement the discussions presented in the body of the document. Appendix A offers a step-by-step framework for contextual assessment, including templates for assessment at the battalion, brigade/regiment, regional command, and theater levels. Appendix B provides a detailed hypothetical example of contextual assessment in practice, using notional data. Appendixes C and D present, as background to the historical case study outlined in Chapter Six, excerpts from a declassified province-level narrative assessment report and a declassified theater-level narrative assessment report from Vietnam, respectively. Appendix E concludes the monograph with a brief outline of the debate over effects-based operations and discusses its relevance to effects-based assessment.

CHAPTER TWO

Concepts That Shape Counterinsurgency Assessment

The purpose of this chapter is to lay the groundwork for a focused examination of centralized assessment processes (EBA and pattern and trend analysis), as well as the Vietnam and Afghanistan case studies. To determine best practices for assessment, it is necessary to understand policymaker requirements. It is also important to know how the COIN environment will affect the ability to collect and analyze information. Ideally, this understanding should provide a common foundation for the development of a more effective assessment process that could be adapted to meet the unique challenges of any current or prospective campaign.

This chapter describes the complex balance between policymaker requirements and the ability of military staffs to provide relevant assessment input to the decision-making process. This relationship foundered during the Vietnam War, and the patterns established during that period have posed ongoing challenges in both the Afghanistan and Iraq campaigns. The complexities and chaos of the COIN environment exacerbate the challenges that military staffs face in attempting to develop a relevant campaign assessment; these environmental impediments also complicate the process of strategic decisionmaking.

Policy Requirements and Counterinsurgency Assessment

National Public Radio: Is there a number that you track that makes you feel confident that you have made progress against the insurgents? That you are winning militarily?

Secretary of Defense Donald H. Rumsfeld: No one number is determinative, and the answer is no. We probably look at 50, 60, 70 different types of metrics, and come away with [sic] them with an impression. It's impressionistic more than determinative.[1]

---

[1] Rumsfeld, 2005.

For practical purposes, the end user of a military campaign assessment is the *policy-maker*.[2] A policymaker could be the President, a cabinet member or advisor, an executive office staff member (e.g., on the National Security Council), a senior DoD official, or a legislator. What do policymakers require from COIN assessment?[3] How will policy and policymaker concerns and behavior shape the U.S. military's ability to provide useful assessments? How can the military shape assessment to best inform policy? Specifically, what should the relationship between COIN policy and assessment look like?

The military is best positioned to produce a transparent, credible, and relevant campaign assessment when provided with clear strategic objectives or "strategic guidance" by policymakers. According to U.S. joint doctrine, this guidance should contain a "national strategic end state" and "termination criteria." The former is defined as "the broadly expressed conditions that should exist at the end of a campaign or operation," while the latter is "the specified standards approved by the President or the SecDef [Secretary of Defense] that must be met before a joint operation can be concluded."[4] For the purposes of simplicity, this monograph refers to these criteria as national strategic "objectives."[5] Without clear or understandable objectives, military staffs are left scrambling to design operational objectives—and then assessments—that seem reasonable. This monograph addresses the confusion about assessment in Vietnam; some of that confusion can certainly be tied to the absence of clear objectives early in the war. A 1974 survey of U.S. general officers showed that only 29 percent felt that pre-1969 strategic objectives for the Vietnam War were clear and understandable, while 35 percent thought that they were "rather fuzzy." A full 91 percent of these officers listed "defining the objectives" as a key recommendation if they had to fight the war again.[6] The commanding general of MACV in 1969 described the U.S. objectives as

---

[2] Although policymakers are the "end users," the public is the ultimate recipient of assessments. Public opinion cannot directly alter strategy, but it does influence the course of events, the willingness of policymakers to commit forces and resources, and the decision to sustain or withdraw from a campaign.

[3] This section focuses specifically on COIN policy requirements, but a broader understanding of policy decisionmaking would also inform this subject. The literature on policy decisionmaking requirements is voluminous and its focus ranges from technical aspects of intelligence production (Sherman Kent, e.g., CIA, 1966) to defense budget analysis (Charles J. Hitch and E. S. Quade) and strategic assessment (Scott Sigmund Gartner).

[4] U.S. Joint Chiefs of Staff, 2006, p. III-5.

[5] Military terminology regarding strategic guidance, objectives, and end states is complex and sometimes contradictory. JP 5-0, *Joint Operation Planning*, states that "strategic direction encompasses the processes and products by which the President, SecDef, and CJCS [Chairman of the Joint Chiefs of Staff] provide strategic guidance" to military forces (U.S. Joint Chiefs of Staff, 2006, p. II-1). In Army doctrine, strategic guidance is translated into military commanders' intent, mission statements, end states, and intermediate objectives, which are, in turn, used to shape operations. FM 5-0, *The Operations Process* (HQDA, 2010), provides a more in-depth analysis of the operational-level terminology. This is somewhat contradictory to joint doctrine, which describes strategic and operational objectives, and these terms seem equivalent to Army mission statements and end states.

[6] Kinnard, 1977, pp. 169, 176.

ill defined and misunderstood.[7] A Brookings Institution report on the Vietnam War stated, "Administration leaders persistently failed to clarify U.S. objectives in concrete and specific terms. Uncertainty and ambiguity in reports were therefore bound to emerge, for no one could be certain what he was measuring progress against or how victory could be defined."[8] Similar concerns dogged President George W. Bush's strategy for both Iraq and Afghanistan, and criticism of both strategies has carried over into the current administration.[9]

With a clear and understandable policy in place, the military can develop an assessment process to answer questions that relate specifically to policymaker requirements. This envisions a best-case scenario: Assessments should be tied directly to clear and understandable national policy objectives. Policymakers can turn to such assessments with relative confidence. However, it seems to be more likely that policy on COIN campaigns will not be clear. U.S. Army doctrine describes COIN as an "ill-structured" problem that is likely to create disagreements over the formulation of a clear end state.[10] Of the three cases examined in this monograph—Vietnam, Iraq, and Afghanistan—it could be argued that none had a clear or generally agreed-upon end state, either throughout the campaign or at any one point in time. In cases without a clear end state, or "termination criteria," the military assessment process is likely to be less clear, but it must still support policy decisionmaking.

Whether policy provides clear or less clear strategic objectives, assessment will have to be flexible. By design, no assessment process is static or immutable, because the nature of war is fluid. Just as policymakers will have to adjust policy to meet changing objectives, they will also have to anticipate and accept some changes in assessment methods and outputs. If the assessment process starts from the ideal point—a clear policy and a closely aligned assessment design that are agreed to prior to engaging in conflict—then these shifts will have only a marginally negative impact on the ability of policymakers to ascertain progress. If policy direction fails to stay abreast of the truth on the ground, then assessments will slide out of alignment as the military reports on events that seem to have little bearing on the original (and possibly outdated) policy. For example, in 2004, the multinational forces in Iraq were still focused on assessing post-invasion transition objectives as the country slid deeper into insurgency and civil violence.

Fred C. Iklé describes another layer of complexity in the relationship between policy and military assessment. He states that the military is so wrapped up in day-to-

---

[7] MACV, 1969a, p. 4.

[8] Gelb and Betts, 1979, p. 307.

[9] See, for example, Mark Schrecker's 2010 article in *Joint Force Quarterly*, "U.S. Strategy in Afghanistan: Flawed Assumptions Will Lead to Ultimate Failure."

[10] HQDA, 2010, p. 2-4. In defining an ill-structured problem, the manual states, "At the root of this lack of consensus is the difficulty in agreeing on what is the problem."

day operations that it has trouble seeing the big picture and often fails to produce adequate strategic assessment. Conversely, policymakers often do not have the background or context to understand the minutiae of military campaigns.[11] There may be some undue generalization in this analysis, but it is practical: If the military fails to provide an adequate strategic assessment, it encourages micromanagement by policymakers, and policymakers waste valuable time by immersing themselves in arcane military detail with little to show for their efforts but frustration. It behooves the military to produce a report that is tuned to strategic decisionmaking, and it is incumbent upon policymakers to facilitate assessment by providing clear objectives and by remaining somewhat flexible as the military adjusts to meet changing ground conditions.

This monograph describes many examples of military assessments that failed to adequately support policy decisionmaking. Figure 2.1 provides a visual account of extreme policymaker micromanagement. In the photo, President Lyndon B. Johnson (second from left) examines a detailed model of the Khe Sanh combat base in Vietnam. The battle for Khe Sanh preoccupied Johnson for much of early 1968; he had the model built in the White House Situation Room so he could track the location

Figure 2.1
President Johnson and the Khe Sanh Model in the White House Situation Room

SOURCE: February 15, 1968, photo via the National Archives, Archival Research Catalog, courtesy of the Lyndon Baines Johnson Library.
RAND MG1086-2.1

---

[11] Iklé, 2005, pp. 18–19.

and status of individual military units.[12] Johnson was both reacting to a gap in useful assessment from the military and indulging his own proclivity to personally manage military operations. Chapter Six shows the effect that this kind of aggressive micromanagement had on the accuracy and integrity of military assessments during the Vietnam War.

A relevant, or effective, assessment should preclude the need for this kind of micromanagement. But what does "effective" entail? What should assessment show and why? Previous RAND research on insurgency endings has showed that lasting victory in COIN comes not by military action alone but ultimately by addressing the root causes of the conflict to achieve naturally occurring stability.[13] Most experts cited in *How Insurgencies End* concurred that addressing root causes is most often the key to victory, and FM 3-24 states, "Long-term success in COIN depends on the people taking charge of their own affairs and consenting to the government's rule."[14] JP 3-24, *Counterinsurgency Operations*, describes an end state for COIN:

> COIN is successful when three general conditions are met. First, the [host-nation] government effectively controls legitimate social, political, economic, and security institutions that meet the population's general expectations, including adequate mechanisms to address the grievances that may have fueled support of the insurgency. Second, the insurgency and its leaders are effectively co-opted, marginalized, or separated physically and psychologically from the population, with the voluntary assistance and consent of the population. Third, armed insurgent forces have been destroyed or demobilized and reintegrated into the political, economic, and social structures of the population.[15]

Assessing whether a COIN campaign is progressing toward long-lasting stability demands a clear assessment of human factors, or human terrain. However, this joint definition of success does not apply universally to all COIN operations. COIN policy varies, and national objectives may be more limited in scope. Policy may require an outcome ranging from the short-term disruption of an insurgent cadre to long-lasting stability. Because there is such a range of possible strategic objectives in COIN, it is difficult to identify a singular approach to assessment. Some generalization is necessary.

---

[12] Johnson requested and received direct cables on the situation in Khe Sanh on a near-daily basis from February 3 to March 30, 1968 (see, e.g., Wheeler, 1968).

[13] See, especially, *How Insurgencies End* (Connable and Libicki, 2010) and *Victory Has a Thousand Fathers: Sources of Success in Counterinsurgency* (Paul, Clarke, and Grill, 2010).

[14] HQDA, 2006c, p. 1-1.

[15] U.S. Joint Chiefs of Staff, 2009c, p. III-5.

COIN objectives *tend* to be long-term objectives.[16] Therefore, it would be reasonable to assume that COIN assessment should show progress over time toward a clear (or at least definable) end state.[17] Policymakers who are accustomed to executive summaries and briefings tend to expect concise and often quantitative reports that show a near-mathematical path toward an end state. Unfortunately, as already discussed, defining a clear end state for a full-scale COIN campaign is difficult. "Addressing root causes" is not necessarily a quantifiable or even visible process over time or even at any one point in time. National strategy and the military operations conducted to address that strategy are likely to be complex, dynamic, and subject to regular review and sometimes drastic revision. Therefore, while policymakers may desire a clear and easily digestible assessment report, a realistic assessment will necessarily be intricate, because it will reflect complex and dynamic circumstances. *Policymakers should be wary of simple and concise assessments of complex COIN operations.* Both military officers and policymakers will have to absorb some level of detail if they are to understand the arc of a specific COIN campaign.

Assessments of Vietnam, Iraq, and Afghanistan have produced reports that support what former Secretary of State Donald Rumsfeld referred to as either impressionistic or determinative decisionmaking. In one sense, these two terms describe a thought process: Rumsfeld asserts in the interview that he drew only impressions from assessment reports and did not take them as literal interpretations of events on the ground. In other words, he would absorb massive quantities of data and reporting from multiple sources and gain a broad impression of the war. Specifically, a policymaker using an impressionistic approach might see broad trends in violence going up, troop levels going up, and popular opinion generally rising and thus determine that a troop surge is working. This approach relies on one of two assumptions. On one hand, the policymaker may assume that it is unnecessary to know "ground truth" because the aggregated data provide adequate indication of broad patterns or trends; thus, impressions are not based on fact but are instead vague and indistinct.[18] Chapter Six examines whether such patterns and trends are accurate enough to feed effective impressionistic decisionmaking for COIN and how the impressionistic approach fared in a specific case during the Vietnam War. On the other hand, if the policymaker assumes or is assured that the assembled data are accurate, he or she must determine whether a determinative decision is possible. Typically, determinative decisions require highly accurate

---

[16] According to U.S. Joint Chiefs of Staff, 2009c, p. III-16, "Insurgencies are protracted by nature, and history demonstrates that they often last for years or even decades."

[17] There was some debate over the value of having an end state versus an exit strategy in Afghanistan as of early 2011. That debate may be useful to some readers, but because it is a specific policy debate it was not clearly germane to this research effort.

[18] In practice, however, trends are viewed as both precise and accurate, and policymakers tend to gravitate toward mathematical threshold–driven campaigns.

quantitative data, a sound theory of the problem at hand, and a clear understanding of the second- and third-order effects likely to emerge from the decision.

With few exceptions, most policy decisions are impressionistic rather than determinative, and wartime decisionmaking tends to be especially so because the issues at stake are complex and often nebulous. All policymakers rely on impressions to make wartime decisions, but they must consider whether these impressions are derived from layered, transparent, and credible reporting or merely single-layer subjective analysis of aggregated and often inaccurate and incomplete data. Figure 2.2 is a simplified depiction of the process of impressionistic decisionmaking using notional COIN data. It shows an array of data on aspects of the complex COIN environment, including economic performance, insurgent activity, and level of popular support. While these types of data might be presented in sequence for an executive assessment briefing, in practice, they are typically presented with little to no association or correlation. These data might be used to determine force deployments and resourcing, or they might be used to describe campaign progress to Congress or the public. This is the process that Secretary Rumsfeld described in his interview.

Policymakers taking a determinative approach might try to gather all this information, attempt to find cause and effect between each data source (or have a mili-

Figure 2.2
Impressionistic Decisionmaking Using a Broad Array of Counterinsurgency Data

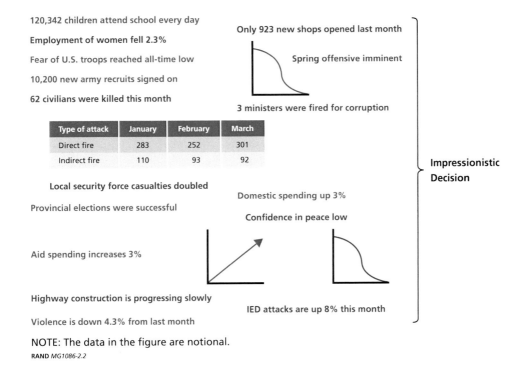

NOTE: The data in the figure are notional.
RAND *MG1086-2.2*

tary staff analyze the data), and produce finely tuned policy decisions based on what tends to be mostly quantitative analysis. Much of the Vietnam-era COIN reporting (e.g., HES) had the veneer of accuracy, and because it was presented in a heavily quantitative format, it fed determinative policymaking. But in the case of HES, significant data inaccuracies and gaps led to suboptimal determinative decisions. Policymakers can be pulled into making determinative decisions based on data that appear to be precise and accurate but that are in actuality imprecise and inaccurate, or precise but inaccurate. An overly aggressive drive to reduce uncertainty can be counterproductive when precise and accurate data are not available or cannot reasonably be collected.

Determinative assessment does not have to be primarily quantitative. Detailed and layered contextual assessment can also support determinative decisionmaking if it is properly written and presented. Assessments can be used to create specific, finely tuned policy, but military assessment in general is intended as only one input into policy decisionmaking. Policymakers should seek other inputs as well.

Although military campaign assessments might recommend shifts in military strategy or national policy, these recommendations tend to be limited in scope or focused on resource allocation. Only rarely do they recommend major shifts in U.S. national security policy.[19] Assessment can be useful in identifying the failure of the military, civilian agencies, or the host nation to execute a good policy, and also in restructuring a failing military strategy. But, ultimately, a poorly conceived COIN strategy will live or die on its merits. Inadequate or inaccurate assessment can help lose wars, but even the best assessment cannot by itself rescue a bad plan.

Finally, some policymakers require assessment to be predictive. They want to know what to expect in the months ahead as they determine how to allocate resources and shore up political support for the campaign. But because assessment is not intelligence analysis, COIN campaign assessments are not designed to provide well-analyzed predictive reporting. Some prediction is possible, of course, but prediction is an analytic quagmire and likely to produce inaccurate results. Compounding the inherent challenges of predictive analysis, most assessment staffs are not trained for the task. Policymakers must be cautious when requesting predictive analysis from a military assessment, at least until the gap in military analytic capability for assessment is remedied. Chapter Six examines these issues within the context of the Vietnam case.

This section concludes with a summary of some of the aforementioned considerations, as well as other considerations drawn from the policy decision literature and primary-source research. No list of policy requirements can be comprehensive, but these are informative and provide a foundation for the discussion that follows.

---

[19] See *Commander's Summary of the MACV Objectives Plan* for an example of a top-level report that suggested changes in strategic approach (MACV, 1969a).

- What the policymaker requires from COIN assessment:[20]
  - a close approximation of ground truth upon which decisions can be based
  - an appreciation for resource requirements (current and predictive)
  - transparent and credible reporting that can be used to inform Congress and the public
  - relevance to decisionmaking
  - predictive analysis (sometimes).
- What the military requires from policymakers to facilitate sound assessment:
  - clear policy objectives
  - a description of what the policymaker wishes to learn from the assessment
  - an agreed-upon reporting format that is useful but also flexible and realistic.
- Basic considerations for developing assessments for policymakers:
  - Ensure that the assessment reports on progress toward policy objectives.
  - Explain the assessment methodology in writing and in detail.
  - Communicate clearly with policymakers when assessment methods need to be changed.
  - The objectives should anchor the assessment, no matter what else changes; the link between current objectives and assessment should be immediate and unbreakable.
  - Because COIN campaigns are lengthy, assessment should help policymakers understand current and prospective timelines.
  - If policy objectives describe an end state, assessment should describe both how the campaign is progressing toward that end state and how events on the ground have informed changes to the described end state.
  - To establish credibility and, ultimately, deliver sound policy, it is important to understand the limitations that the COIN environment places on the ability to deliver precise and accurate assessments.
  - Policymakers must understand that, at best, assessment should be used to *inform* decisions. An assessment should not be relied upon as an independent means to provide clear answers to complex strategic challenges.

Wicked Problems and Counterinsurgency

A brief explanation of the complexities of the COIN environment is a necessary backdrop to the more focused examination of the case studies later in this monograph. The

---

[20] This does not include responses to the question, "What do policymakers *want* from COIN assessment?" *Want* and *need* are not necessarily synonymous. For example, a policymaker might want explicit detail about a specific military operation (e.g., President Johnson and the Khe Sanh model). More recently, policymakers tend to want quantifiable information. However, as discussed in greater detail later in this monograph, when such information is provided without context, it cannot help policymakers comprehend ground truth to any degree of accuracy.

issue of complexity is closely tied to policymaker requirements: If policymakers do not understand or appreciate the complexity of COIN they cannot extend realistic guidance on assessment to the military. If military staffs fail to account for complexity in an assessment or in the caveats of their reports, they do policymakers and the public a disservice. The term *wicked problem* is commonly used in both policy and military circles to describe COIN, so it is appropriate to use the concept to introduce the idea of complexity in COIN assessment. An understanding of wicked problems in assessment also highlights the daunting challenges facing counterinsurgents and COIN assessment staffs on the ground.

In their landmark 1973 article in *Policy Sciences*, Horst Rittel and Melvin Webber use the term "wicked problem" to describe planning processes for governance, teaching, housing development, policing, and other real-world social challenges.[21] One could read their article as a polemic or simply as an effort to draw a line between the hard and soft sciences. Perhaps the best interpretation is that it identifies a spectrum of complexity in research environments and shows that highly complex environments tend to complicate research and reduce the accuracy of findings. Therefore, these complex, real-world problems are "wicked" because they do not offer simple, finite, or necessarily replicable solutions. The concept is best captured in this excerpt:

> The problems that scientists and engineers have usually focused upon are mostly "tame" or "benign" ones. As an example, consider a problem of mathematics, such as solving an equation; or the task of an organic chemist in analyzing the structure of some unknown compound; or that of the chessplayer attempting to accomplish checkmate in five moves. For each the mission is clear. It is clear, in turn, whether or not the problems have been solved.

> Wicked problems, in contrast, have neither of these clarifying traits; and they include nearly all public policy issues—whether the question concerns the location of a freeway, the adjustment of a tax rate, the modification of school curricula, or the confrontation of crime.[22]

A scientist working on a cure for cancer might not agree. Indeed, there are some overstatements in Rittel and Webber's argument. However, in their effort to describe a chaotic, hectic, unrewarding, and (in some cases) nearly opaque civil planning environment, they have also described COIN.[23] Rittel (a systems analyst) and Webber (a city planner) lay the groundwork for several themes that run through this mono-

---

[21] Rittel and Webber, 1973. As discussed earlier, Army doctrine refers to this as an "ill-structured" problem (HQDA, 2010, p. 2-4; U.S. Army Training and Doctrine Command, 2008, p. 9).

[22] Rittel and Webber, 1973, p. 160.

[23] For a different and more in-depth exploration of the wicked problem construct as applied to military planning and operations, see Greenwood and Hammes, 2009, and U.S. Army Training and Doctrine Command, 2008.

graph. First, COIN is complex and unpredictable; second, COIN is both art and science, but mostly art; and third, context is critical to understanding the COIN environment. While "wicked problem" is now thrown about so freely that it has lost some meaning, COIN and therefore COIN assessment can still be described within the framework of Rittel and Webber's concept:

- War assessment does not have a definitive formulation—there are many methods.
- COIN has no "stopping rule"—there is rarely a clear or concise ending.[24]
- There are no black/white, true/false assessments, only better or worse.
- One cannot easily test COIN assessment for validity.
- There are no second chances; every assessment is consequential to strategy.
- No exhaustible or knowable set of COIN endings or outcomes exists.
- Every COIN campaign—and, consequently, assessment process—is a unique problem.
- Tracing cause and effect and "weighting" various factors is a subjective process that makes it nearly impossible to scientifically determine solutions.
- Anyone assessing COIN operations is subject to harsh and sometimes personal critique because assessors' work is often politicized and has immediate, real-world impact in a high-stakes environment.

Kenneth Menkhaus argues that state failure exacerbates the symptoms of wicked problems, and he presents varying degrees of failure in a series of typologies on a spectrum. He places Afghanistan within the typology of state failure, describing it as a case of "shorter term state collapse."[25] Iraq would also fall within this typology. One could infer from this analysis that these two active cases of U.S. COIN are not simply wicked problems but *exacerbated* wicked problems that make assessment particularly challenging.

Wicked problems tend to emerge from or describe *complex adaptive systems*, a term used in mathematics, in the social sciences, and, more recently, by military practitioners to describe battlefield environments. Wicked problem theory incorporates, or perhaps depends on, the idea that societies (or, in this case, populations in COIN), like groups on a battlefield, form complex adaptive systems. In the social sciences, a complex adaptive system is defined by individual actors in a society—people or groups—changing behavior in response to an intricate and interdependent web of actions, reactions, and happenstance.[26] The next section explains the complex adaptive system in the context of military assessment.

---

[24] See Connable and Libicki, 2010.

[25] Menkhaus, 2010, p. 89.

[26] For a lay introduction to complex adaptive systems and mathematical approaches to understanding these systems, see Miller and Page, 2007.

Complex Adaptive Systems, Nonlinearity, and Chaos

Wicked problem is a general, nonscientific term that is useful to frame the notion of complexity. To fully appreciate the degree to which complexity affects assessment, it is necessary to delve into a more structured discussion of the COIN environment. There is something about a battlefield—or an operational environment—that confounds accurate assessment. A conventional battle between two opposing armies is certainly complex, but when one side is hidden among the population, the degree of complexity faced by the opposing side (the counterinsurgent) is magnified. The counterinsurgent must locate and separate the insurgents without damaging the population, all the while attempting to address a complex web of root causes. Efforts to assess progress against these confounding requirements often fall short, particularly since complexity is anathema to accuracy. This section introduces three broad concepts—complex adaptive systems, nonlinearity, and chaos—each of which should help build a more comprehensive understanding of the challenges posed to assessment staffs and policymakers.

Assessment of COIN is commonly likened to the assessment of a system. Most literature refers to the COIN environment, the society within that environment, or all the groups and actors in the operational environment (e.g., friendly, insurgent, civilian) as parts of a larger and more complex type of system. Noted operations researcher and systems analyst Russell L. Ackoff describes a system as follows:

> A whole consisting of two or more parts (1) each of which can affect the performance or properties of the whole, (2) none of which can have an independent effect on the whole, and (3) no subgroup of which can have an independent effect on the whole. In brief, then, a system is a whole that cannot be divided into independent parts or subgroups of parts.[27]

A *complex system* is a system that is typically self-contained but consists of a number of different moving and interacting sections or parts. A car is a complex system in that it is made up of thousands of intricate moving parts that interact with each other. It is possible to predict what will happen when someone turns the key in the ignition of a functioning car: It will start.[28] It is also possible to know what each specific part of the car will and will not do and what will happen when that part ceases to function. For instance, if the starter motor breaks, the car will not start. It is possible to isolate, deconstruct, and understand a complex system in great detail using a wide array of scientific and nonscientific methods. Thorough and persistent technical examination over many decades has led to steady and sometimes revolutionary improvements in the car, and millions of people have a good working understanding of the car's complex system.

---

[27] Ackoff, 1994, p. 175.

[28] *Commander's Appreciation and Campaign Design* (U.S. Army Training and Doctrine Command, 2008, p. 6) also uses the analogy of the car to explain the difference between a complex system and a complex adaptive system.

All cars have many predictable elements in common, and cars within one model are nearly identical in construct.

People, on the other hand, are adaptive and not binary. No two people are alike; they can and often do react unpredictably. The reasons for these reactions are often unknown or cannot be easily discerned. People also adapt and change the way they think and behave based on an infinite number of possible inputs, which, in the absence of a controlled scientific study, are unknowable to anyone but the individual. When people interact with each other, they form a broader complex adaptive system. It is possible to envision a society or population in a COIN environment as a complex adaptive system consisting of interrelated but individual complex adaptive systems. Indeed, this is how doctrine describes human terrain and how Ackoff describes social systems.[29] Because COIN depends on popular support and the behavior of individuals and small groups, some argue that success in COIN can be gauged only through an understanding of the complex adaptive social system that affects it.[30]

But complex adaptive systems are also self-adapting in that they change and react without the need for external stimuli. If a car were a complex *adaptive* system, it might be able to start and drive itself. Complex adaptive systems are "marked by self-organization and something called 'emergence'—the capability to generate system changes without external input."[31] This adds a layer of complexity to the assessment problem: Measuring inputs and outputs may be insufficient to understand emergent behavior that is, for all intents and purposes, unpredictable.

Another way of looking at the relative complexity of an environment is to assess its linearity. In general, simple, closed systems are linear, and open, complex systems, like COIN environments, are nonlinear. This implies that a complex adaptive social system is not a truly integrated system. In many cases, the parts, or people, are not interconnected in any clear or definitive way. Small rural villages in Vietnam, Iraq, or Afghanistan might be completely unaffected by violence or other activity hundreds of miles away. There are no physical boundaries to a non-linear complex adaptive social system because people can move in and out, often at

---

[29] See Ackoff, 1994, p. 176.

[30] There is some ongoing debate on this point and disagreement with the premise that COIN is a population-centric endeavor. This monograph is primarily intended to inform U.S. government consumers and was written to address standing U.S. policy and doctrine on COIN. Both JP 3-24 and FM 3-24 contend that COIN is population-centric. This section cites several sources that link COIN and the concept of the complex adaptive system.

[31] Wass de Czege, 2009, p. 7. Bousquet (2009, pp. 175, 181) uses similar terms to describe the same dynamic. While Bousquet argues that there may be patterns to self-organization, or emergence, he does not claim that these patterns are easily identified in the kind of open complex adaptive system we see in COIN. The systems analysis literature on emergence blends concepts and approaches from a wide array of scientific theories and methods and uses a similar range of methods to attempt to reduce uncertainty in such systems.

will.[32] There are also no definitive psychological boundaries to such a system: People can engage or disengage from the system while remaining physically in place. Barring physical necessity (e.g., threat of violence or requirement for essential services), they can also choose whether or not to act or react to any specific input or output at any given point in time.

The less linear a system is, the more complex and *chaotic* it tends to be. Chaos theory is closely related to the wicked problem proposition. In scientific terms, *chaos* describes something akin to the complex adaptive system: A chaotic system is highly complex, interconnected, and dynamic. However, a chaotic system is also "wildly unpredictable" and susceptible to dramatic and (arguably) unpredictable changes that can alter the entire system.[33] This is the concept of emergence taken to an extreme. Chaos theorists Hal Gregersen and Lee Sailer believe that "systems exhibiting chaotic behaviors can only be understood, whereas non-chaotic systems [like a car] can be understood, predicted, and perhaps controlled."[34] Understanding, of course, assumes that accurate and relatively complete data are available, but this is rarely the case in COIN. Kelly and Kilcullen claim not only that chaos (and therefore nonlinearity) is linked to complex adaptive systems but that chaos, in fact, "makes war a complex adaptive system, rather than a closed or equilibrium-based system."[35]

This monograph argues that the COIN environment can be loosely analogized to a complex adaptive system but that the nonlinearity of the COIN environment, a lack of adequate data, and the chaos of war prevent analysts from accurately dissecting a COIN campaign as they would dissect a system.

A complex adaptive social system marked by openness, emergence, nonlinearity, and some degree of chaos would be difficult enough to assess objectively. However, assessment staffs are part of a military organization that works within the COIN operating environment and has effectively become part of that environment. Therefore, when analysts or assessment officers refer to the COIN environment as a complex adaptive system, they are (or should be) referring to everyone and everything in the environment, including their own organization. Every U.S. infantry unit, PRT, advisor, and aid worker adds complexity to the assessment challenge. Analysts also need to consider any exogenous factors that might affect the environment. Exogenous factors that are not easily linked to behavior through systems analysis might include foreign

---

[32] Ackoff (1994, p. 176) describes social systems as "open systems that have a purpose of their own; at least some of whose essential parts have purposes of their own; and are parts of larger (containing) systems that have purposes of their own."

[33] Gregersen and Sailer, 1993, p. 779. The science behind chaos theory focuses on finding underlying patterns in what appears to be completely random behavior.

[34] Gregersen and Sailer, 1993, p. 798.

[35] Kelly and Kilcullen, 2006, p. 66. They also believe that "generating an analysis sophisticated enough to derive coherent and rational whole-of-government inputs that are required by EBO is probably unattainable."

media broadcasts that shape public opinion, individual communiqués with external influencers (e.g., emails or phone calls to expatriate organizers), hard-to-trace financial transactions in both directions across a border (e.g., hawala transfers), the impact of insurgent or refugee groups operating from across a border (e.g., Afghan Taliban members and refugees in Pakistan), the impact of foreign sponsors of insurgents or counterinsurgents (e.g., China in Vietnam or NATO in Afghanistan), and so on.

Assessments are most accurate and precise when describing environments that are simple, closed, linear, and predictable. Assessments are least accurate and precise when describing environments (or systems) marked by complexity, adaptation, emergence, and nonlinearity. The violence, the threat of violence, economic displacement, and social upheaval associated with COIN all contribute to chaos, making the COIN environment both complex and chaotic. When environments slide into chaos, they become increasingly difficult to understand and assess and, arguably, next to impossible to comprehend through systems analysis.

Military Doctrine and the COIN Environment

To understand the complex COIN environment, the military relies on doctrine, and all military COIN assessment is shaped to some extent by military doctrine. While doctrinal manuals are not necessarily intended to be literal "how-to" guides, doctrine plays a dominant role in training, education, and operations. It shapes the way military officers think about assessment, and it can serve as an official arbiter to establish unified policy and process in a joint environment. Therefore, to understand past and current military assessment methods, some knowledge of doctrine is not only helpful but also necessary. And while the majority of the relevant literature on COIN was written in the 20th and early 21st centuries, the U.S. military has been studying complexity and the chaos generated by extreme violence for more than 200 years. Its amassed knowledge on warfare also benefits from more than two millennia of wisdom on the subject, from Julius Caesar and Thucydides to Clausewitz and contemporary experts. *Capstone* military service doctrine reflects the sum of this knowledge as well as cutting-edge philosophy.

In their capstone literature, both the Army and Marine Corps use a pair of very simple terms to concisely describe wicked problems, complex adaptive systems, and chaos theory: *friction* and *the fog of war*. Both terms (as applied to warfare) are attributed to Carl von Clausewitz and describe the chaotic, confusing, and often unpredictable nature of battle.[36] The seminal Marine Corps Doctrine Publication (MCDP) 1, *Warfighting*, defines friction as

> the force that resists all action and saps energy. It makes the simple difficult and the difficult seemingly impossible. The very essence of war as a clash between opposed wills creates friction. In this dynamic environment of interacting forces, friction abounds.[37]

The concept of the fog of war reflects the inherent uncertainty of battle as described by Clausewitz, B. H. Liddell Hart, and other noted experts. Friction and the fog of war are closely linked concepts in that uncertainty contributes to friction and allows it to flourish. MCDP 1 provides more detailed guidance:

> All actions in war take place in an atmosphere of uncertainty, or the "fog of war." Uncertainty pervades battle in the form of unknowns about the enemy, about the environment, and even about the friendly situation. While we try to reduce these unknowns by gathering information, we must realize that we cannot eliminate them—or even come close. *The very nature of war makes certainty impossible; all actions in war will be based on incomplete, inaccurate, or even contradictory information.* War is intrinsically unpredictable. . . . At best, we can hope to determine possibilities and probabilities.[38]

These final propositions—that certainty is impossible; that information is incomplete, inaccurate, and even contradictory; and that war is intrinsically unpredictable—are also the most contentious in terms of applying the approaches found in any kind of science (including social science) to the wicked problem or to COIN assessment. MCDP 1 goes on to describe the battlefield as nonlinear in that causes and effects are often disproportionate. The battlefield is also complex in that the millions of moving pieces, or people in the environment, decide whether to act, react, or not to react to each other. It is also fluid in that "each episode merges with those that precede and follow it—shaped by the former and shaping the conditions of the latter—creating a continuous, fluctuating flow of activity replete with fleeting opportunities and unfore-

---

[36] While *fog of war* is widely attributed to Clausewitz, Eugenia C. Kiesling (2001) argues that he never actually uses the phrase in a way that matches doctrinal interpretation. According to Kiesling, he refers to friction more than a dozen times but to fog only three. Nevertheless, she agrees that Clausewitz intended to describe the fog of war in the way that it is now used. Not all interpretations of *On War* agree with this analysis. For example, Alan Beyerchen (1992–1993, p. 77) states that, "his famous metaphor of the 'fog' of war is not so much about a dearth of information as how distortion and overload of information produce uncertainty as to the actual state of affairs."

[37] Headquarters, U.S. Marine Corps, 1997b, p. 5.

[38] Headquarters, U.S. Marine Corps, 1997b, p. 7 (emphasis added). MCDP 2, *Intelligence*, says, "We must continually remember that intelligence can reduce but never eliminate the uncertainty that is an inherent feature of war" (Headquarters, U.S. Marine Corps, 1997a, p. 19).

seen events."[39] Friction, fog of war, nonlinearity, complexity, and fluidity all result in *disorder* and *uncertainty*, conditions in warfare that, according to the Marine Corps, can never be eliminated. The authors of MCDP 1 conclude the discussion on the nature of war by placing warfare somewhere between art and science. This statement describes the principal challenge of the wicked problem:

> Human beings interact with each other in ways that are fundamentally different from the way a scientist works with chemicals or formulas or the way an artist works with paints or musical notes. . . . *We thus conclude that the conduct of war is fundamentally a dynamic process of human competition requiring both the knowledge of science and the creativity of art but driven ultimately by the power of human will.*[40]

Whether or not one agrees with this statement, it is the philosophical foundation of one of two major land combat services in the U.S. military.[41] All doctrine—including assessment doctrine—should stem from this understanding of war. U.S. Army doctrine is similarly direct and follows these same themes:

> The operational environment will become extremely fluid, with continually changing coalitions, alliances, partnerships, and actors. . . . Finally, complex cultural, demographic, and physical environmental factors will be present, adding to the fog of war. Such factors include humanitarian crises, ethnic and religious differences, and complex and urban terrain, which often become major centers of gravity and a haven for potential threats. The operational environment will be interconnected, dynamic, and extremely volatile.[42]

The complexity and chaos inherent in the COIN environment is also acknowledged in COIN doctrine. Two documents describe the military's approach to COIN: JP 3-24, *Counterinsurgency Operations*, and the Army/Marine Corps publication FM 3-24/MCWP 3-33.5, *Counterinsurgency*. The latter prescribes the application of distributed operations and *mission command*:

---

[39] Headquarters, U.S. Marine Corps, 1997b, p. 9.

[40] Headquarters, U.S. Marine Corps, 1997b, p. 19 (emphasis in original).

[41] In the foreword to MCDP-1, then–Commandant of the Marine Corps Charles C. Krulak describes warfighting as a philosophy:

> Very simply, this publication describes the philosophy which distinguishes the U.S. Marine Corps. The thoughts contained here are not merely guidance for action in combat but a way of thinking. This publication provides the authoritative basis for how we fight and how we prepare to fight. This book contains no specific techniques or procedures for conduct. Rather, it provides broad guidance in the form of concepts and values. It requires judgment in application. (Headquarters, U.S. Marine Corps, 1997b, Foreword)

[42] FM 3-0, 2008, p. 1-3. *Commander's Appreciation and Campaign Design*, an experimental pamphlet published by U.S. Army Training and Doctrine Command (2008), specifically describes war as a wicked problem.

> Mission *command* is the conduct of military operations through decentralized execution based upon mission orders for effective mission accomplishment. . . . Mission command is ideally suited to the mosaic nature of COIN operations. Local commanders have the best grasp of their situations. . . . Thus, effective COIN operations are decentralized, and higher commanders owe it to their subordinates to push as many capabilities as possible down to their level.[43]

Mission command and distributed operations are intended to shape campaigns in a way that accepts and takes advantage of complexity and chaos. Perhaps the most salient line in the quote above is the one that describes the "mosaic nature of COIN operations." This description not only acknowledges the existence of complexity and chaos but it also establishes the idea that local context is particularly relevant in COIN. According to FM 3-24, challenges and operational solutions are asynchronous from area to area across a COIN theater.

Chapter Summary

COIN is a wicked problem that presents unique challenges to assessment. Most literature on COIN supports the notion that the COIN environment can be loosely analogized to a complex adaptive system that is shaped in large part by unknown or unknowable events and actions. But complex adaptive system theory is not sufficient to define or explain the COIN environment for campaign assessment, because COIN is marked by chaos, friction, and the fog of war. Joint doctrine recognizes these challenges, as does the capstone doctrine of the U.S. Army and Marine Corps. Doctrine also recognizes that COIN is perhaps the most complex, opaque, and disorderly *type* of warfare. This monograph argues that the already daunting challenges faced by military staffs assessing conventional war are exacerbated many times over in COIN. In theory, military doctrine on COIN assessment should reflect and account for the Army and Marine Corps' positions on complexity and chaos. However, as discussed later, military doctrine instead proscribes a one-size-fits-all approach to assessment, an approach originally crafted to address conventional air campaigns.

---

[43] HQDA, 2006c, p. 1-26 (emphasis in original). The concept of mission command for COIN is deeply rooted in U.S. Army doctrine. FM 31-20, *Operations Against Guerrilla Forces*, published in 1951, was intentionally written to allow for maximum flexibility, initiative, and adaptation at the tactical level. See also Birtle, 2006, pp. 134–138.

CHAPTER THREE

Centralized Assessment Theory and Pattern and Trend Analysis

The previous two chapters described the process and requirements of assessment, as well as the challenges that the complexity and chaos of COIN pose to assessment analysts. This chapter introduces the concepts and theories that have shaped the two existing approaches to assessment examined in this study: EBA and pattern and trend analysis. Both of these approaches are centralized and almost entirely quantitative. Each relies on the idea that the people, groups, and physical places and things in the COIN environment constitute a system that can be framed and assessed through systems analysis. EBA is explicitly linked to systems analysis theory, but it also incorporates some aspects of pattern and trend analysis. The salient difference between the two approaches is in the degree to which they assume the availability of complete and accurate data. As it is described in U.S. military doctrine, EBA requires the availability of plentiful and accurate data that can be used to develop a near-comprehensive picture of the battlespace. Pattern and trend analysts tend to accept that data on COIN campaigns will be incomplete and inaccurate to a rather significant degree but that these data are still sufficient to produce relevant centralized, quantitative analyses.

The first part of this chapter examines the general propositions that shape both EBA and pattern and trend analysis. As mentioned, both approaches rely to varying degrees on systems analysis to help commanders and policymakers understand the COIN environment. Thus, the first part of this chapter introduces the concept of systems analysis. Subsequent sections address and dispel the notion that COIN assessment can be equated with scientific research. This distinction—that military assessment is not scientific research or analysis—has repercussions for all aspects of centralized assessment. This discussion provides a backdrop for an examination of the most commonly used centralized analytic method in COIN assessment: time-series analysis. Because time-series charts are so prevalent in assessment reporting, it behooves practitioners and consumers (policymakers) to understand the strengths and weaknesses of time-series analysis in the context of COIN assessment.

The final two sections of the chapter present the concepts of pattern analysis and trend analysis for COIN assessment. The more complex theories and concepts of EBO and EBA are addressed in Chapter Four.

Centralized Quantitative Assessment

As discussed, both EBA and pattern and trend analysis depend to varying degrees on the idea that the COIN environment is a system, and both approaches apply varying interpretations of *systems analysis* to the problem: EBA explicitly refers to the COIN environment as a "system of systems," while pattern and trend analysts, such as Thayer, have attempted to apply variations of systems analysis to campaign assessment.[1] Therefore, a basic appreciation for the theory and practice of systems analysis (in its various incarnations) is necessary to understand both EBA and other centralized assessment methods.

There is no universally agreed-upon definition of systems analysis; it is more of a general approach to problem-solving than a specified set of tools or methods.[2] Systems analysis is, in some ways, self-defining—it is an analysis of systems—but it encompasses a wide array of theories, methods, and applications. Business analysts use systems analysis to improve the bottom line of a company. The 1960s saw the first dedicated application of systems analysis to military decisionmaking.[3] Then–Secretary of Defense Robert S. McNamara created the Office of Systems Analysis (OSA) to conduct cost-benefit studies of big-ticket weapon procurement projects and manpower issues. These studies led to the development of the Planning-Programming-Budgeting System, which is still taught to students of national security and used to shape DoD policy.[4] In the mid-1960s, OSA expanded its application of systems analysis to help assess the Vietnam War. This expansion was predicated on the notion that systems analysis was a useful approach not only for procurement but also for policy analysis. McNamara and the OSA team believed that although systems analysis could not necessarily provide a definitive quantitative answer to the challenges of strategy, it could be used to present a reasonably accurate summary to support decisionmaking.

Alain C. Enthoven and K. Wayne Smith made this argument in their defense of military systems analysis, *How Much Is Enough? Shaping the Defense Program 1961–*

---

[1] Some EBO advocates argue that EBO and EBA are not reliant on system-of-systems analysis. See Appendix E for an overview of these arguments.

[2] This assertion is based on a review of the systems analysis literature and discussions with systems analysts at RAND. There are several existing definitions of systems analysis, but they all differ from each other to some extent.

[3] Operations research was first applied systematically to U.S. defense issues during World War II, and some aspects of systems analysis may have been applied prior to the 1960s. But it seems that the creation of the Office of Systems Analysis was the first *institutionalized* application of systems analysis to defense decisionmaking in the U.S. defense community.

[4] See Hitch, 1965, for a general description of efforts by the Office of the Secretary of Defense to apply systems analysis to budgeting. Hitch was the DoD comptroller under McNamara.

1969.[5] Both Enthoven and Smith were leading members of OSA in the 1960s. They argued,

> Even the best studies leave much to be desired. And no study can account for all the variables or quantify all the factors involved. But analysis can be an aid to judgment by defining issues and alternatives clearly: by providing responsible officials with a full, accurate, and meaningful summary of as many of the relevant facts as possible, an agreed-upon list of *dis*agreements and their underlying assumptions, and the probable cost of hedging against major uncertainties.[6]

Although the current study focused primarily on defense budget analyses and only peripherally on Vietnam, it is important to note Enthoven and Smith's argument that more systematic analysis of the Vietnam War was necessary.[7] They may not have championed the use of specific techniques or had any personal role in COIN assessment, but an OSA team conducted extensive systems analysis of the war. Thomas C. Thayer was the point man on Vietnam assessment for the Office of the Secretary of Defense. He served as the director of the Southeast Asia Intelligence and Force Effectiveness Division of the Southeast Asia Programs Office under the Assistant Secretary of Defense for Systems Analysis from March 1966 until June 1972. He applied Enthoven and Smith's description of systems analysis to COIN to produce a series of reports, *A Systems Analysis View of the Vietnam War.*[8] McNamara, a leading proponent of systems analyst in the U.S. government, also became embroiled in systems analyses of the Vietnam COIN campaign (see Chapter Six for a more detailed discussion). Both McNamara and Thayer also accepted the exacting challenges of COIN assessment, and like Enthoven and Smith, Thayer believed that systems analysis could penetrate complexity to find useful patterns even in the most challenging environments:

> The answer turned out to be finding the critical patterns of the war. . . . Any given action was seldom important by itself, and at first glance no patterns were seen. Analysis, however, revealed them. From these we, in Washington, were able to monitor the war surprisingly well by examining trends and patterns in the forces, military activities, casualties, and population security.[9]

---

[5]   In the chapter entitled, "Some Problems in Wartime Defense Management," Enthoven and Smith (1971/2005, pp. 307–308) argued that more systems analysis of the Vietnam problem would have better supported strategic decisionmaking. They did not address Thayer's work at OSA directly or in detail, but Enthoven hired Thayer, and Thayer conducted a thorough systems analysis of the Vietnam War for OSA.

[6]   Enthoven and Smith, 1971/2005, p. 63 (emphasis in original).

[7]   According to Enthoven and Smith (1971/2005, p. 307), "The problem was not overmanagement of the war from Washington; it was undermanagement. The problem was not too much analysis; it was too little."

[8]   See Thayer, 1975a, 1975b, and 1975c, for the reports referenced for the current research effort.

[9]   Thayer, 1985, p. 5.

Some of the efforts to build a centralized assessment process have been channeled into developing specific assessment tools, typically models of some sort. Figure 3.1 represents a widely publicized effort to capture and model the "system dynamic" of the COIN operating environment as described in FM 3-24, *Counterinsurgency*.[10] This model was developed as a systems analysis tool for COIN assessment. It is telling that the diagram is highly complex and all but unreadable in the absence of a magnifying glass. The purpose of including it here is to highlight the complexity of COIN modeling as it informs assessment, not to show the specific aspects of the model.[11]

Models like this are intended to help military staffs (not just assessment analysts) envision a COIN campaign or shape operations. A similar modeling study conducted, in part, to extend "the field of quantitative system dynamics analysis to explore modern counterinsurgency theory" provides a more detailed prescription for operations.[12] Other models have been specifically designed to apply systems analysis to COIN assessment. George T. Hodermarsky and Brian Kalamajka proposed applying a "systems approach

Figure 3.1
System Dynamics Model of Field Manual 3-24, *Counterinsurgency*

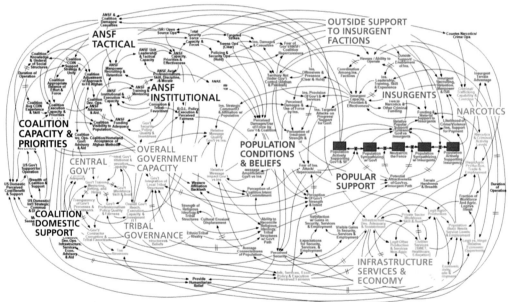

SOURCE: Mowery, 2009.
RAND *MG1086-3.1*

---

[10] HQDA, 2006c.

[11] The diagram originated in a briefing that was widely distributed within the U.S. military and, specifically, among ISAF staff (Mowery, 2009; see the bibliography for a link to a larger version of the model).

[12] See Anderson, 2011, p. 112.

to assessments" to deal with complexity. Their model closely parallels the EBA model described in Chapter Four and Appendix E.[13]

EBA is a centralized assessment methodology derived from EBO theory. It is similar to other centralized assessment in that it relies on an influx of reports from the field to a central repository, where the information is aggregated and analyzed. Unlike pattern and trend analysis, doctrinal EBA relies on the availability of highly accurate data. It interprets models of the COIN environment not as merely esoteric and informative, but as literal maps depicting people, groups, and their relationships with each other. Chapter Seven examines EBA as practiced in the field.

Differences Between Scientific Research and Campaign Assessment

In some cases in Vietnam and Afghanistan, centralized assessment methods have been likened to scientific research. Thus, it is important to draw a clear distinction between scientific research and campaign assessment and to explain why assessment findings are not comparable to research findings.

It is first necessary to draw a distinction between scientific methods and scientific research. Scientific methods are simply analytic tools that can be applied to problem-solving, while *scientific research* is a well-defined process with stringent guidelines and requirements. There have been many instances in which analysts have applied scientific methods to centralized campaign assessment with the intent of adding rigor to the process and output. These efforts include the application of social science methodology to population analysis, efforts to conduct regression analyses of multiple COIN data sets, and efforts to apply mathematical models to COIN assessment. While these analysts do not necessarily claim in explicit terms that their findings can be equated with research findings, nonanalysts in the military and policy community often take them as such. As a result, much greater faith is sometimes placed in these assessments than is warranted by the methods applied. For example, GEN William C. Westmoreland, the senior military officer in Vietnam from 1964 to 1968, stated that the HES was "scientifically developed" with "certain very precise criteria."[14] The difference between "scientifically developed" and scientific research can be unclear to nonscientists. While it is possible to apply scientific methods to some aspects of assessment, the application of scientific method alone does not constitute *scientific research*.

A number of scientific fields have developed or refined methods to analyze complex adaptive systems. These methods recognize the existence of data imperfections and gaps: They are designed to reduce uncertainty rather than produce absolute accuracy. The more rigorous the study, the more bounded the scope of the study, and the

---

[13] Hodermarsky and Kalamaja, 2008.

[14] Sheehan, 1988, p. 697.

more accurate and complete the data, the more uncertainty can be reduced. The purpose of rigorous criteria is to ensure that scientific findings can be judged according to generally agreed-upon standards. Indeed, this is the basis for peer review. Both research and review are highly structured and time-consuming processes that tend to produce moderate advances in knowledge in a focused area of interest. *The Belmont Report*, produced by the National Commission for the Protection of Human Subjects of Biomedical and Behavioral Research in 1978, provides a useful and widely recognized definition of scientific research:

> [T]he term "research" designates an activity designed to test a hypothesis, permit conclusions to be drawn, and thereby to develop or contribute to generalizable knowledge (expressed, for example, in theories, principles, and statements of relationships). Research is usually described in a formal protocol that sets forth an objective and a set of procedures designed to reach that objective.[15]

While it is possible to apply some scientific standards to COIN assessment, or to conduct small, isolated scientific research studies as a subset of COIN assessment, *holistic military assessment of a COIN campaign is neither general nor applied scientific research as it is defined in the scientific literature.*[16] Military staffs might conduct analysis, but this analysis tends to be far more informally structured than scientific analysis. For military purposes, assessment and analysis are often interposed, but these terms should not be confused with scientific research (the process of shaping a study, collecting data, analyzing data, and reporting findings) or scientific analysis (one aspect of research). Scientific research or experimentation depends on the internal and external validity of both method and information. The purpose of research is to improve general knowledge or general scientific knowledge, and research projects almost always have finite objectives. But holistic military campaign assessment is *by necessity* a loosely structured process that shifts according to conditions and the requirements of commanders and policymakers. It would be difficult to claim that a military campaign assessment produces generalizable scientific knowledge. The overall process of assessment is ongoing rather than finite, and assessment can continue even after a military withdrawal. It would be nearly impossible to determine the scientific validity of a military assessment that has been delivered to a policymaker and the public.

Scientific research also requires control of the data collection, data reporting, and data management process because these data are used to measure the attributes

---

[15] National Commission for the Protection of Human Subjects of Biomedical and Behavioral Research, 1978, p. 3.

[16] If one were to draw a comparison between military campaign assessment and scientific research, the assessment might be loosely categorized as applied evaluation research, but the structured requirements for this kind of research obviate the comparison.

of selected variables.[17] Data for research should be collected with a reliable measuring device (e.g., a structured survey or interview protocol) that has been created according to a set of strict standards. Lack of control over data collection and management throws a wrench in the scientific process. In COIN, the term *data* is used to describe any kind of information available to assessment analysts, commanders, or policymakers. The following are examples of the kinds of data that might be used to feed assessment:

- a one-time narrative report on a meeting with a key leader, filed by a civilian advisor
- a quantitative report showing the number of people treated at a hospital, filed by an NGO
- a radio report of an enemy attack, filed by a unit in theater
- a human intelligence report containing the observations and opinions of both the source and the reporter.

Each individual and unit reports information differently, depending on training, personal predilection, available resources (e.g., radio or email), and field pressures. Most initial field reports are filed under fairly trying conditions and often during combat, and in coalition COIN warfare, they are sometimes submitted in a number of different languages. There is no way to control or determine how these reports are filed in their original formats because almost all reporting is edited, collated, and aggregated in some way for incorporation into a quantitative assessment. People who are not trained scientific researchers are responsible for data management, so the way in which field reports are inducted into databases is neither controlled nor scientific (as discussed later in this monograph). Even data in a specified category, produced to meet a directed collection requirement, are not necessarily suitable for scientific comparison. As shown in the case studies presented here, this lack of control affects data validity.

These factors do not preclude the use of scientific *methods* of analysis in the production of holistic COIN campaign assessment. They simply draw a distinction between formal research and the less formal application of methods to feed what will eventually be subjective estimation. It is still possible for assessment staffs to conduct pattern or trend analysis using time-series data, but these analyses should be clearly defined as nonscientific and derived from a loosely controlled data set. Any effort to apply correlation analysis to differing sets of aggregated quantitative data should proceed with the utmost caution, since the correlation of uncontrolled data sets in a nonscientific environment tends to exaggerate or hide the impact of errors, gaps, and unknowns. Consumers of assessment reporting should not mistake a holistic COIN campaign

---

[17] Validity itself is a complex and multifarious concept that has many different meanings and applications in science. It is commonly used to refer to the construct validity of a study, but this section refers specifically to data validity.

assessment for scientific research, nor should they assume that results or individual elements of the report are scientifically valid without explicit substantiation.

Applying Scientific Rigor to Assessment

The seven best-practice standards for COIN assessment introduced in Chapter One do not constitute an exclusive list that precludes the application of scientific standards to assessment and analysis. If a staff determines that it will conduct some form of centralized campaign assessment, it might draw from scientific standards to try to improve these nonscientific assessment efforts. Noted cultural anthropologist H. Russell Bernard has stated that "nothing in research is as important as validity."[18] While validity and reliability are important criteria, there are other means of evaluating scientific studies that are equally important to gauging the integrity of a scientific study. Standards from many disciplines could be applicable to any one study or process:[19]

- *Replicability:* Can the measure produce commensurate findings when applied by different research groups (or, in this case, assessment staffs)?
- *Generalizability:* Can findings from the measurement be generalized across a broad population? This is used to check validity.
- *Understandability:* Can the intended audience understand not only the findings but also the process used to generate the findings? This feeds transparency.
- *Congruence:* Are the research methods used and the subject to be studied compatible (e.g., using both polls and macroeconomic data to measure opinion)?[20]
- *Credibility:* Is the study objective, or is bias present?
- *Independence:* Are the data and findings redundant or unique?
- *Confirmability:* Is there a way to independently confirm the data or findings?
- *Reflexivity:* Have all possible shortcomings in the method or measure been assessed, and are they apparent to outsiders (transparent)?

---

[18] Bernard, 2006, p. 53. Bernard goes on to say,

> Ultimately, the validity of any concept . . . depends on two things: (1) the utility of the device that measures it; and (2) the collective judgment of the scientific community that a concept and its measure are valid. In the end, we are left to deal with the effects of our judgments, which is just as it should be. Valid measurement makes valid data, but validity itself depends on the collective opinion of researchers. (Bernard, 2006, p. 60)

[19] Other criteria from a range of disciplines could be applied to metrics selection. This list is drawn from a survey of operations research, social science, political science, behavioral science, and other scientific literature. ISAF identifies two of the criteria—transparency and credibility—and U.S. military doctrine adds in several others: measurable (can be gauged either quantitatively or qualitatively), collectable (data associated with the indicator can be obtained more easily), and relevant (provides insight into a supported measure of performance or effectiveness).

[20] This differs from congruence of doctrine and assessment method.

- *Precision:* How precise is the metric? What does it portend to show in quantitative terms? This term might also be used to qualify data.
- *Accuracy:* Is the measuring device properly calibrated, or, in assessment terms, is the collector capable of proper data collection and reporting? This term might also be applied to data.

Some of these terms overlap or have similar meanings, but this list is far from exhaustive when considering all possible scientific disciplines. Assessment staffs might consider these standards when developing their processes. The rigor of scientific thinking is very useful when it comes to brainstorming the process of centralized assessment. However, it is not possible to prescribe the use of any one of these criteria for assessment because holistic assessment should not be held to scientific standards.

How can military assessment staffs make use of the concepts found in scientific research to improve the transparency and credibility of their work? Skeptical consumers might take the application of scientific terminology to military assessment as a veneer, so a good first step would be to clearly define holistic COIN campaign assessment as nonscientific. The best time to start shaping assessment methods and processes would be prior to deployment. It would not be necessary or even possible to apply each of the aforementioned criteria to assessment, but the list could be used to gauge each step from direction to presentation. If analysts accept that COIN assessment is a nonscientific process, thinking through these criteria would at least help define caveats.[21] Appropriate questions might include the following:

- *Direction:* Are the commander's requirements for information and subordinate assessment logical and reasonable? For example, if the unit is using centralized assessment, are the core metrics generalizable, reliable, replicable, precise, relevant, measurable, parsimonious, and understandable to subordinate commands and higher-level consumers? If not, how could these gaps and inaccuracies be reconciled or explained?
- *Collection:* Are collectors trained to collect information consistently and in the right format? Is the information required actually collectable at the local level? If not, is it because it is simply not available (e.g., determining the cost of bananas when no bananas are sold) or because it would require significant risk to obtain? Additionally, is the collection method accurate? In other words, is it likely to produce valid and consistent results?
- *Analysis:* Is it possible to determine whether there are nonrandom patterns that might show progress or setbacks? Is the information received confirmable, or are assessment analysts working with information that they cannot confirm as accu-

---

[21] In this case, the step-by-step process reflects the observations of actual assessment staffs and not the doctrinal process defined in FM 5-0 (HQDA, 2010).

rate? In this case, how would they add caveats to their analyses? Is there congruence among the methods used to collect information (e.g., polls), the original CCIRs (commander's critical information requirements), and the actual information needed by the analysts? Are the sources of data independent or circular? In other words, is the same information from a single source being reported multiple times through separate channels and therefore being interpreted as corroboration rather than redundancy?

- *Production and presentation:* Is the assessment clear and understandable to the consumer? Is it transparent and sourced? Can the consumer "see" through the summary to find original reporting in order to check validity? Are all the decisions and methods behind the mathematical formulas, weights, values, thresholds, metric selection decisions, and other subjective inputs to the system visible or easily obtainable?

It would be impossible to apply precisely structured scientific research to ongoing combat decisionmaking because combat decisions require timely and necessarily imperfect input and output. While campaign assessment has more extended timelines, the pressures on assessment staffs are also constant and their control over their own processes is minimal. Access to data is always limited, and data—the essence of valid research—are often flawed and inconsistent to the point that they are not scientifically reliable. Some scientific rigor can help improve informal, nonscientific pattern and trend analysis, but these studies should contain appropriate caveats.

Time-Series Analysis

> [T]he issue is not so much qualitative versus quantitative approaches to measurement as the degree of complexity one is willing to admit into the measurement theory. But complexity quickly leads to perplexity, vagueness, and defeatism unless systematic and rational grounds can be established for handling this complexity.
>
> —*Hubert M. Blalock*[22]

Centralized assessment often relies on the analysis of aggregated data or information over time. This section examines the basic concept of time-series analysis as applied to military assessments, as well as some of the specific concerns that might apply to the use of time-series analysis for assessment.

Unlike relatively high-tempo conventional campaigns, COIN operations tend to develop and end slowly, over a number of years. Progress in COIN tends to be incremental and vague rather than dramatic and obvious. The United States spent more

---

[22] Blalock, 1974, p. 3.

than ten years in Vietnam, shifting from an advisory approach to direct combat and, eventually, back to advisors as the south gradually crumbled. U.S. forces in Afghanistan overthrew the Taliban in direct combat in a matter of months but then became embroiled in what has become the longest continuous war in American history. Similarly, in 2003, U.S. forces quickly decimated the Iraqi Army and then engaged in a fight against an amorphous insurgency that is still ongoing at low levels in 2011. When counterinsurgents win modern COIN campaigns, they tend to win slowly over time, with violence abating gradually.[23] To track incremental progress toward strategic objectives in this kind of campaign, operational staffs try to develop a baseline of information and then track and compare data over time. Reliance on time-series graphs by military assessment staffs reinforces the notions that COIN campaigns tend to be lengthy and difficult to assess.

Time-series analyses depend on data acquired over time, baseline data, and (in some cases) a threshold to assess progress. Staffs set a baseline by identifying the best available sources at the time and then tying information from those sources to a visual measurement standard that can be used on the horizontal and vertical axes of a graph.[24] For example, attack reporting might show that there are typically 50 attacks per day across a country. In this case, the baseline might be set at 50. The staff then determines how the indicator might show progress toward the linked objective. To assess progress, the staff must decide how often data will be collected and reported to feed the analysis. The more often this cycle occurs, the easier it should be to show progress. However, frequent collection of volatile data sets can show variability that is simply "noise" or random. For example, if the indicator is the price of grain and the price fluctuates wildly from day to day, frequent reporting may show tremendous variability without long-term effect. Staffs can address this problem by shifting reporting increments: They can collect data frequently but report it only once per month or quarter. It is important to note, however, that assessment staffs usually do not control the reporting process or timetable.

To show definitive progress (at least for EBA or pattern and trend analysis), assessment staffs tend to select a time-series threshold or milestone.[25] Thresholds were prolific in COIN assessment in Vietnam (and remain so, to a lesser extent, in Afghani-

---

[23] Connable and Libicki, 2010, p. 15.

[24] A time-series analysis might involve *determinate* data or (typically quantitative) data that are clearly tied to a measure of effectiveness (MOE), measure of performance (MOP) or objective, or *indeterminate* data. Indeterminate data are informative but not necessarily definitive. A determinate data set might be "number of children attending school," while an indeterminate data set might contain public opinion polling. In practice, staffs do not ordinarily draw a distinction between determinate and indeterminate data, and indeterminate data are often used to track definitive MOE targets (e.g., "*x* percentage of the population supports the coalition").

[25] The difference between a threshold and a milestone is that a threshold measures progress on the y-axis against a scale (e.g., number of attacks) while a milestone measures whether a threshold has been reached by a certain period. Therefore, a milestone is a threshold on the y-axis coincident to a point on an x-axis timeline.

stan). Depending on the type of data reported, the threshold will be either higher or lower than the baseline. For example, if the indicator is violence and the goal (MOE) is a reduction in violence, then the threshold will be lower than the baseline. Conversely, if the MOE calls for increased electricity production, the threshold will be higher than the baseline. Threshold selection is almost always a wholly subjective process, so the use of thresholds adds a layer of subjectivity and opacity to the assessment.[26]

Time-series analysis is affected by both data collection and the dynamics of complex adaptive systems marked by chaos. For example, economists deal with leading, lagging, and coincident indicators to assess macroeconomic trends. These kinds of indicators also appear in COIN time-series analyses:

- *Leading indicator:* Information that helps predict future events. Intelligence professionals refer to these simply as "indicators" as part of the "indications and warning" process. For example, the massing of Iraqi troops on the border of Kuwait in 1990 was a leading indicator of possible invasion.
- *Lagging indicator:* Information that surfaces after an event occurs but helps explain or confirm that event in retrospect.
- *Coincident indicator:* Events or information that parallels other indicators to help analysts understand the meaning of either or both. For example, high winds coupled with dark skies may indicate a coming storm; the indicators are reinforcing.

Leading indicators are used to support predictive analysis, but this analysis typically requires context of some sort. Lagging indicators can considerably reshape the assessment of an event after it occurs. An example of a lagging indicator in COIN would be a shift in popular perception in the days, weeks, and months after a U.S. operation or, perhaps, an election; "atmospheric" data like these come in from various sources over time. The effect on the population might emerge slowly over time, or reports of popular perception might not be available until well after the event. It might not be apparent whether the operation or election succeeded or failed until weeks or months later. In this case, any initial assessment might have to be retroactively adjusted.

Time-Series Graphs in Military Assessment

Military staffs tend to report time-series assessment through graphs to help strategic commanders and policymakers see progress, or lack of progress, over time. To build

---

[26] It adds opacity because the consumer may have no way of determining how or why a threshold was set. Therefore, achieving a threshold objective may have little meaning. If the policymaker sets the threshold, this is not a concern of the military assessment staff. However, it may affect the credibility and transparency of the report when it is issued to the general public.

one of these time-series graphs and use it in accordance with doctrine, the staff theoretically requires the following:[27]

- clear strategic objectives and, in the case of EBO/EBA, effects and indicators
- a baseline of information provided by credible sources
- thresholds or milestones (typically for determinative assessments)
- systematic, consistent information over time that can be compared with the baseline
- license to retroactively adjust data and timelines to reflect new knowledge.

Optimally, graphs help the consumer understand one idea or, at most, a very small set of interrelated ideas. For example, a time-series graph might show the number of friendly deaths compared to enemy deaths over time. Time-series graphs are intended to present a relatively narrow range of understanding: the meaning of the data becomes clear as the data change or (according to U.S. doctrine) show variance—"a difference between the actual situation during an operation and what the plan forecasted the situation would be at that time or event."[28] Overly complex graphs are distracting and ultimately of little use in explaining data. Time-series graphs can be used to represent small-scale events, but they are more often used to describe large-scale trends over long periods. Edward R. Tufte, the author of several popular manuals on the art of visual data display, states that time-series graphs are "at their best for big data sets with real variability."[29]

DoD used time-series graphs extensively throughout the Vietnam era to show progress against both the Viet Cong (VC) and North Vietnamese Army (NVA) and to show progress in pacification efforts. Both military and policy staffs use time-series graphs to help describe and analyze the COIN campaigns in Afghanistan and Iraq. Based on a review of a broad sample of assessment products from these two theaters, it appears that time-series analysis is the most commonly used method of operational assessment in contemporary Western-led COIN campaigns. Figure 3.2 depicts attack incidents reported in Iraq from 2004 to 2007 in an aggregated time-series graph.

Time-series graphs are so prolific in contemporary operations that policymakers at all levels expect to see them with every assessment and in some cases demand that the assessment consist solely of maps and time-series graphs.[30] DoD reports to Congress

---

[27] This list of requirements is based on an analysis of time-series methodology, basic requirements for COIN assessment according to RAND research, and U.S. military doctrine, most of which is examined in this monograph.

[28] HQDA, 2010, p. 5-6.

[29] Tufte, 2006, p. 30.

[30] Author interview with an ISAF assessment analyst, Kabul, Afghanistan, May 6, 2010. This conclusion is corroborated by my personal observations during a May 2–8, 2010, visit to ISAF headquarters in Kabul and during three successive tours on division-level staffs in Iraq, as well as time spent with a service-level headquarters staff in Washington, D.C., during Operations Enduring Freedom and Iraqi Freedom.

Figure 3.2
Attack Incidents Reported in Iraq, 2004–2007

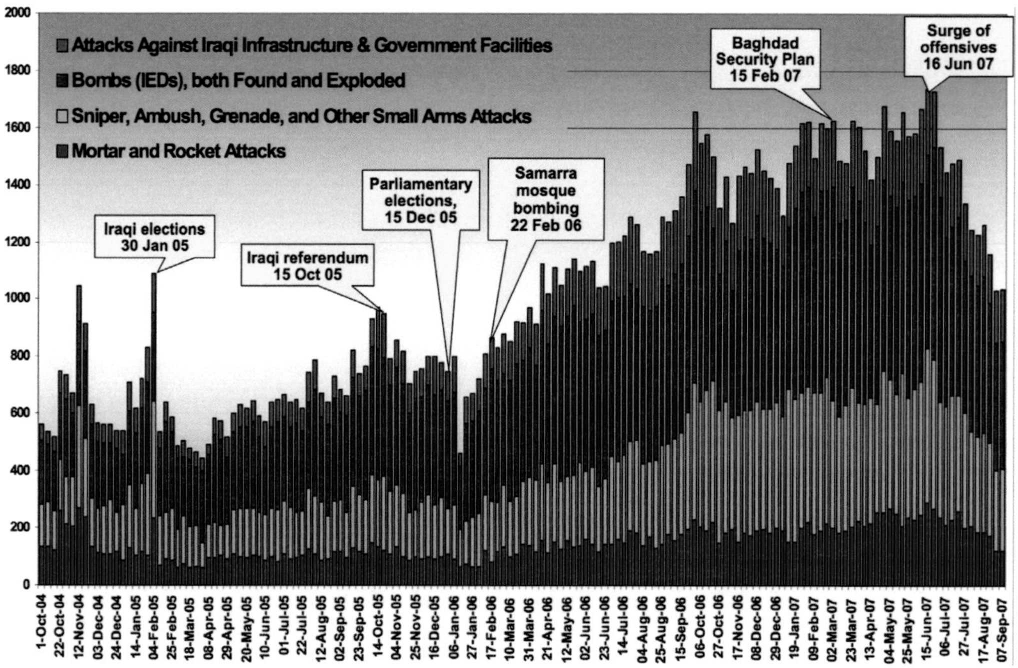

SOURCE: Multi-National Forces–Iraq, 2007, slide 2.
RAND *MG1086-3.2*

on Iraq and Afghanistan convey many key points in time-series displays. For example, ISAF's 2009 "Unclassified Metrics" briefing consists almost entirely of graphic polling data, maps, and time-series graphs.[31] Time-series COIN assessment is an unofficial standard in step 6 (delivering the report) in the assessment process as detailed in FM 5-0.[32] Time-series charts are often used as centerpieces in oral briefings. These briefings are intended to assist military officers or civilian defense officials in their efforts to characterize data for policymakers, the press, and the public. For example, GEN David H. Petraeus's 2007 congressional testimony on Iraq was framed by time-series charts.[33] Figure 3.3 shows the commanding general of U.S. Forces–Iraq, GEN Raymond G. Odierno, presenting a briefing at the Pentagon on February 22, 2010, on the then-upcoming Iraqi elections. The graph to his right shows attack incidents over time (i.e., significant activities, or SIGACTs), and the graph to his left shows attacks by explosives over time (a subset of SIGACTs). The photograph shows common practice for both briefings and more formal assessments in the Vietnam, Iraq, and Afghanistan cases.

---

[31] ISAF Headquarters Strategic Advisory Group, 2009.

[32] HQDA, 2010.

[33] See Multi-National Forces–Iraq, 2007.

Figure 3.3
Iraq Briefing with Time-Series Graphs

SOURCE: DoD photo by Robert D. Ward.
RAND *MG1086-3.3*

Retroactively Adjusting Time-Series Analysis and Assessment

Lagging indicators, data corrections, and late reports are commonplace in COIN. To be precise, then, time-series analyses require regular review and possibly retroactive adjustment in response to the discovery of errors or new information.[34] Because operational COIN assessment involves the analysis of many interrelated types of data (e.g., attack data, popular opinion, information about friendly operations), any retroactive change would also change the holistic theater assessment. Command decisions can also affect the holistic theater assessment. If a commander or staff decides that the definition of "attack" needs adjustment halfway through a campaign, all information recorded up to that point must be adjusted to meet the new definition. For example, if IED "finds" are no longer counted as attacks, the official time-series analysis must be retroactively adjusted to remove reference to all IED finds as attacks. If the staff fails to adjust the graph or report, the assessment becomes inaccurate or suspect. It may appear that there was a sudden, unexplained reduction of violence when in fact a simple change in definition altered the entire assessment.

When operational assessment staffs do try to sustain accuracy and credibility by making retroactive adjustments, they should not expect to be rewarded for their efforts. Wicked problem theory again rears its head: Senior officers and commanders

---

[34] Staffs will have to define "regular" and determine how often they should review time-series data based on reporting increments, available manpower, and consumer requirements.

who have been presenting and explaining the existing graphs to policymakers and the media for months, if not years, tend to be nonplussed when the accuracy of their entire data set is thrown into question by their own assessment staffs. The suspicions of civilian leadership and the international media regarding U.S. operational assessments may lead them to see any retroactive change in data as a potential whitewash. Enthoven and Smith describe this effect on a late 1960s effort to retroactively adjust inaccurate pacification data prior to the development of the HES:

> The reluctance to correct the pacification numbers is not surprising. The reason most widely given was that the public would never understand why past numbers being used now differed from those used in the past. Considering the "credibility gap" problems of officials in Saigon and Washington, this was understandable, even though the gap was widening in part *because* the corrections were not made.[35]

Further, if time-series analyses are presented at face value and not carefully explained and contextualized over time, the negative reaction to retroactive adjustments can be exacerbated. The effort to simplify assessments through time-series graphs can backfire precisely because the graphs are deceptively simple: Consumers will not expect the data or definitions to change if military staffs have not helped them appreciate the complexities behind the colored lines and bars. Staffs also should assume that any time-series graph they publish will at some point be separated from the assessment brief or report and presented without context; this has occurred routinely in both Iraq and Afghanistan. In other words, any time-series graph should be "self-contextualizing": It should be very difficult to misinterpret the value and meaning of the data even when viewing the graph as a stand-alone product. Because time-series graphs show only one narrow aspect of a larger data set and are hard to contextualize, this is a pressing challenge.

Pattern and Trend Analysis

Pattern and trend analysis is one of the two existing approaches to COIN campaign assessment discussed in this monograph. Modified with some elements of EBA, it is currently the most commonly used assessment approach in Iraq and Afghanistan, and it was the most commonly used approach in Vietnam. Pattern and trend analysis is centralized assessment that leverages a number of statistical methods to show the direction of a campaign over time. While pattern and trend analysis can be presented through narrative, tables, or maps, the most common means of displaying such data are time-series graphs. The previous section explained many of the central methods and challenges in time-series analysis and thus, by default, pattern and trend analysis.

---

[35] Enthoven and Smith, 1971/2005, pp. 302–303 (emphasis in original).

This section describes the overall methodology, delineates the difference between patterns and trends, and explains the general theory behind the application of pattern and trend analysis to COIN assessment.

One of the Vietnam-era OSA's unofficial mottos was, "It is better to be roughly right than precisely wrong." Enthoven and Smith state that this motto was "a reminder to analysts to concentrate, not on pinpoint accuracy on a part of the problem, but on approximate accuracy over the total problem."[36] They add, "Even when uncertainties are present, it is better to use numbers than to avoid them. Quantitative analysis is possible even if there are uncertainties."[37] This approach frames the effort to find what might be described as *reasonable* patterns and trends in COIN information to support policy decisions. This implies that the precision required for EBA is not obtainable, and it calls into question the use of baselines and thresholds, both of which are associated with and depend on precise and accurate data. However, in many cases, both baselines and thresholds are used in pattern and trend analysis despite the lack of precise and accurate data. For the purposes of this research effort, I draw the following distinction between pattern and trend:[38]

- *Pattern:* a consistent arrangement or behavior, not necessarily over time
- *Trend:* prevailing tendency over time.

A pattern is an identified consistency in any kind of data. Patterns can be found over time, but for the purposes of this monograph, they are not *anchored* in time. It would be possible to find patterns of movement by looking at a single aerial photograph of a crowded city street; one could find densities of people at crosswalks and on sidewalks. In another example, birds migrate during winter seasons; this pattern is observed and revealed only over a number of seasons. The pattern is in the consistency of the behavior over time (multiple winters) and not in one specific winter season. However, birds may or may not migrate in any one specific season; the pattern could change for some unpredicted reason. While patterns are often used for predictive analysis, applying pattern analysis to complex adaptive environments like COIN carries risks that will become evident in the discussion of the Vietnam case study in Chapter Six.

Trends indicate a consistency or shift in behavior or activity over a *specific* period of time. For example, a military assessment graph might show violence dropping steadily from month to month over a one-year period in Iraq (see Figures 3.2 and 3.3,

---

[36] Enthoven and Smith, 1971/2005, p. 68.

[37] Enthoven and Smith, 1971/2005, p. 70.

[38] These definitions are aggregated from various dictionary definitions and reference publications on time-series analysis. They will not meet any one single standard, and many criticisms of these definitions will probably be valid. However, these definitions also generally reflect the working definitions of pattern and trend in the military and policy circles most interested in the information presented here. They also loosely reflect various assessment analysts' interpretations of pattern and trend analysis.

for example). Such a graph might be used to show causation: Violence has been drop-ping over this period because, e.g., troop presence increased or troops were withdrawn. Assessment staffs might conduct centralized *trend analysis* to determine whether an input is having an effect.[39] FM Interim 5-0.1, since superseded, described this process:

> A *measure* is a data point that depicts the degree to which an entity possesses an attribute. . . . Once two or more measures are taken, they can be plotted to deter-mine patterns and trends. These reveal whether an attribute is more or less preva-lent at different times. Commanders and staffs also develop a standard or baseline against which they compare measures and trends. Once established, this baseline remains a fixed reference point. From this information and analysis of why a trend is up or down, staffs can identify trouble spots and plan operations to reverse negative trends. They can also capitalize on positive trends by determining what is causing the positive increase and apply those tactics, techniques, and procedures more broadly.[40]

Trend analysis is part and parcel of centralized assessment.[41] It is closely tied to the concepts of variability and correlation. This monograph argues that, in COIN, the identification of a trend will rarely, if ever, show clear causation. Nevertheless, it is possible to find patterns or trends in inaccurate and estimated data. For example, most U.S. economic reporting is based on data with inaccuracies and gaps. The U.S. Bureau of Labor Statistics (BLS) makes it clear that data are estimates based in large part on polling and do not reflect accurate counts.[42] Yet, few credible experts would dispute the idea that BLS unemployment statistics reflect genuine trends in unemploy-ment. Why are the BLS statistics different from statistics collected and analyzed in places like Vietnam, Iraq, or Afghanistan? The most obvious difference is the environ-ment. COIN data are collected in a hostile, chaotic, and unpredictable environment. A mix of trained and untrained people, each of whom may be risking his or her life in the process, collect, and report these data. In contrast, BLS can generally rely on a well-structured government bureaucracy that operates at the submunicipal level across a very stable country. It uses a cadre of trained and experienced pollsters and statisti-cians to capture and analyze the data. Unlike people in COIN environments, who sometimes shape their answers to manipulate the questioner, Americans have no sig-nificant incentive to throw off BLS statistics with false responses to polling. And, if the

---

[39] Perry, 2011.

[40] HQDA, 2006a, p. 5-6. Although this interim manual is no longer in effect, it provided more detail on some aspects of assessment than the updated 2010 version, FM 5-0, *The Operations Process*.

[41] In practice, the Afghanistan case shows that there is little structured analysis of time-series data for holistic assessment.

[42] U.S. Bureau of Labor Statistics, 2009. BLS uses the Current Population Survey of approximately 60,000 households to help determine unemployment statistics.

BLS data inaccurately represent 15 percent of the U.S. population, for example, that error might not turn out to be significant in terms of national-level policy. In COIN, failing to address 15 percent of the population in a national-level assessment could be disastrous.[43]

Assessment analysts can try to identify trends through structured techniques, and they can also use these techniques to correlate trends with other inputs. To determine meaningful correlation with both partially known variables (e.g., number of violent incidents) and unknown variables (e.g., precise locations of insurgents), it is possible to use a range of approaches backed by any number of statistical or modeling techniques from scientific studies. For example, an assessment staff could compare two or more trends, or a pattern and a trend, through correlation analysis. This analysis might show statistical correlation (e.g., variable A went up when variable B went down, consistently over time). But because there is no control over data collection and no means of ascertaining data validity, there is no way to prove that these correlations are meaningful. In other words, there is no way to prove causation through pattern and trend analysis using common COIN data sets available to assessment staffs. Therefore, correlation analysis of time-series trend data does not meet the standards for effective campaign assessment as defined here.

Pattern or trend analysis in COIN is possible, but these techniques are risky for both collectors and consumers of the analysis. These approaches tend to produce a series of narrow windows into specific issues without providing a means of holistic analysis. The Vietnam case shows how OSA produced these "windows" in the form of dissociated data sets and charts without producing an overarching analysis. This process would later be replicated in Afghanistan. This brief introduction to pattern and trend analysis is intended to frame a more detailed discussion of the process in Chapter Four. The Vietnam War provides a well-documented case of pattern and trend analyses, as discussed in greater detail in Chapter Six.

Chapter Summary

All analytic methods are designed to simplify complex problems to reduce uncertainty and foster better understanding. Centralized quantitative methods, such as systems analysis, EBA, and pattern and trend analysis, seem tailor-made to address a complex challenge like campaign assessment: They provide commanders with ostensibly objective evidence to feed requirements from senior military leaders and policymakers. These centralized analytic methods rely to varying extents on the idea that the COIN operational environment is a system, or it is made up of interconnected and interdepen-

---

[43] In *On Guerrilla War*, Mao Tse-Tung states that in order to win, insurgents may need the support of only 15–25 percent of the population (Mao, 2000, p. 27). Data or "patterns" that do not address populations in hostile areas can easily miss a key element of the population.

dent systems. Analysts relying on effects-based theories contend that complexity can be overcome with technology and intuitive methodology. Some analysts associate their work with a more traditional understanding of complex problems; they believe that, at best, they can provide reasonable pattern and trend analysis of these systems.

No matter which approach is taken, practitioners and consumers of centralized assessment should assume that military campaign assessment cannot reflect the kind of control, precision, and accuracy offered by scientific research. And while some military commanders and staffs might prefer to deliver assessments that are grounded in strict scientific method, they are prevented from doing so by the realities of both military operations and the COIN environment. Most of the operational reasons are practical. Policymakers can and should shift strategic objectives as needed. Commanders will always have control over assessment and they can—and should—be able to adjust the process to meet their needs or adapt to changing circumstances as they see fit. No tactical commander will ever respond in lockstep to assessment information requirements; these requirements will be modified and sometimes ignored due to the vagaries of combat. Even a team of highly trained scientists attempting to produce a valid centralized analysis of a COIN campaign would be prevented from doing so by the lack of control over data collection and data reporting, the decentralized application of policy, shifts in strategic objectives, and routine adjustments to core metrics. It might be possible to conduct limited scientific studies in a COIN theater, but these studies would merely serve as inputs to the holistic campaign assessment process.

A broadly accepted understanding that the military's assessment of COIN campaigns is not scientific research might actually improve the way in which assessments are built, reported, and consumed. Centralized assessment methods are heavily reliant on the capabilities of a single staff section (or small group of top-level staffs) to produce a holistic campaign assessment. These staffs are, in turn, almost wholly dependent on the quality and flow of data from the periphery to the center, but they have little actual control over the collection and reporting of these data. Military assessment staffs should not be burdened with the requirement to find a scientific method that might explain aggregated quantitative data; they could instead create a more reasonable set of analytic methods (preferably, all-source intelligence analysis methods) that would be applicable to the realities of their work. Policymakers should expect any methods applied to assessment to be somewhat ad hoc, but they can use this understanding of centralized assessment to decide whether to rely on determinative or impressionistic decisionmaking.

All consumers of centralized assessment reports should be aware that quantitative time-series graphs and quantitative assessment findings are necessarily imprecise and potentially misleading. They sometimes show consistency over time that is precise but not accurate. Without context, time-series analysis conducted according to OSA's standard for "reasonable" accuracy will, by definition, provide only reasonably accurate results, with "reasonable" lacking clear definition or boundaries. With this standard so loosely defined, it would be unreasonable to establish and rely on precise time-series thresholds and milestones to ascertain campaign progress over time. Finally, correlation does not infer causation, so correlation of loosely structured COIN data should not form the basis for holistic campaign assessment.

CHAPTER FOUR

The Effects-Based Approach to Assessment

Chapter Three described the general approach to centralized assessment with a focus on one of the two assessment methods used by the U.S. military (pattern and trend analysis). This chapter describes the other approach, effects-based assessment, or EBA. EBA is the U.S. military's official assessment process, according to doctrine as of early 2011. EBA is derived from EBO, a process rooted in conventional air operations but intended to be applied across the full spectrum of military operations, from humanitarian relief to conventional warfare to COIN. To understand EBA, it is first necessary to understand some of the fundamental theories and language of EBO. Effects-based theories are spelled out in official U.S. military publications and are woven throughout standing doctrine. But any military process (e.g., logistics, operations, maintenance) is rarely spelled out in a single doctrinal publication or manual. Assessment is no exception. This chapter draws on what are often dissociated segments of various military documents to summarize and explain both EBO and EBA.

EBA is a staff process designed to assess a military campaign by gauging progress toward objectives by examining *effects* (MOEs) or *performance* (MOPs) by tracking *indicators*.[1] It is intended to help staffs envision the operating environment and to help commanders and staffs find clear pathways to achieve campaign objectives.

This chapter describes the centralized EBA process and the systems analysis theory that is central to this type of assessment model. It reveals some of the friction inherent in the application of precision measurement and analysis to complex and chaotic environments. It also shows some of the internal contradictions in U.S. doctrine that complicate the work of assessment analysts, commanders, and policymakers alike.

---

[1] The term *objectives* is used to generalize several different concepts in joint and service doctrine, such as end state, task, mission, commander's intent, or termination criteria. Joint, Army, and Marine Corps terminology vis-à-vis the steps and strata in the planning and operations process is inconsistent, and injecting these various terms into an examination of EBA would not serve to clarify the EBA process. Joint doctrine states, "Assessment uses measures of performance and measures of effectiveness to indicate progress towards achieving objectives" (U.S. Joint Chiefs of Staff, 2006, p. III-27). But Army assessment doctrine states, "A formal assessment plan has a hierarchical structure—known as the assessment framework—that begins with end state conditions, followed by [MOEs], and finally indicators." (HQDA, 2010, p. H-2). Army doctrine describes objectives as intermediate goals that might be used to help shape end state, but objectives are not necessarily integral to the assessment process.

Subsequent chapters argue that doctrinal EBA is not applicable to distributed COIN operations because EBA is a top-down, centralized assessment process, while COIN is a bottom-up, distributed type of operation.

What Is Effects-Based Operations Theory and How Does It Shape Assessment?

It is necessary to understand effects-based theories and EBA to understand the official U.S. military approach to assessment as published in doctrine. While assessment staffs do not necessarily apply EBA literally in real-world COIN campaigns, assessment processes in both Iraq and Afghanistan have been shaped by EBA doctrine. Because parts of U.S. doctrine (and nearly all assessment doctrine) are effects-based, the training and education on assessment that does exist in the U.S. military also tends to be effects-based.

Effects-based operations, or EBO, describes an approach to military planning and operations that focuses on the precise and accurate production and assessment of *effects* with the intent of improving military performance.[2] Arguably, it differs from traditional military theories such as *maneuver warfare*, which tend to focus explicitly (or more directly) on *objectives* or *missions*. Proponents of EBO contend that it is a means of helping staffs and commanders focus on those actions that will help them accomplish missions and achieve objectives. Doctrinal EBO appears to be systematic, precise, and focused on immediate and cyclical actions and reactions, while contemporary interpretations of EBO describe a less precise and prescriptive and more informative process. Critics of EBO tend to compare it to traditional approaches to warfare, or the more modern incarnation of *maneuver warfare*, which is nonsystematic, generally imprecise by design, and focused directly on a final end state and tasks that lead to that end state. The two sides of this debate are explored in greater detail in Appendix E.

Introduction to Effects-Based Operations

No single document summarizes all interpretations of EBO—and they are many and varied. However, the U.S. Joint Forces Command's *Commander's Handbook for an*

---

[2]   EBO is derived from the broader concept of network-centric warfare (NCW) and the theory of revolution in military affairs (RMA). The literature on RMA is expansive, but *Lifting the Fog of War* by ADM (ret.) Bill Owens (2000) and Stephen Biddle's *Military Power: Explaining Victory and Defeat in Modern Battle* (2004) might sum up the two sides of the debate. NCW proposes that technology offers contemporary Western militaries the opportunity to dominate battlefields by "lifting the fog of war." This section also describes system-of-systems analysis (SoSA) and (more briefly) operational net assessment (ONA), two components of NCW that are closely linked with effects-based theory. NCW, SoSA, ONA, and EBO shaped most of the military doctrine, tactics, techniques, and procedures promulgated in the late 1990s and early 2000s. As a result, at some point in their careers, most field-grade U.S. and NATO military officers have been trained and educated to study effects-based theory and to apply EBO to their plans, operations, and (to a lesser extent) assessments.

Effects-Based Approach to Joint Operations and the U.S. Army FM 5-0, *The Operations Process*, are the most comprehensive and applicable to this discussion. Both describe not only the theory and process of EBO but also EBA.[3]

Figure 4.1 shows the flow of EBO from action to cascading effect. In the first step, an action is taken, resulting in an effect. This effect then ripples outward, creating new effects with each ripple. While the chart shows three levels of effects with arrows suggesting interconnected reactions between other effects, it gives a disarming appearance of simplicity. In COIN, this would be a much more complex, three-dimensional model made up of thousands of nodes and links, most of which would not be fully understood.

In this model, the United States or its allies would take an action or inject an input into the system to achieve an effect or cascading series of effects. For example, in Vietnam, the United States attempted to force the North Vietnamese to the bargaining table by strangling their ground lines of communication with deep interdiction air strikes. Analysts calculated the amount of supplies needed to support operations in the south, how many trucks it would take to move these supplies, and how much disruption in the supply lines (the effect) could be achieved through a specified amount of bombing.[4] Air crews then dropped bombs on the targets, and intelligence analysts

Figure 4.1
Effects-Based Operations and Cascading Effects

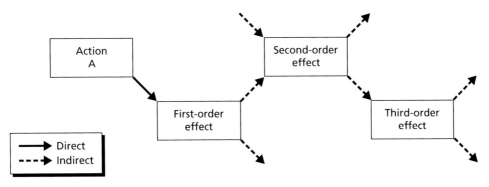

SOURCE: Mann, Endersby, and Searle, 2002, p. 33, Figure 3.
RAND *MG1086-4.1*

---

[3] See Joint Warfighting Center, 2006, and HQDA, 2010. FM 5-0 incorporates some elements of EBO but clearly depicts EBA. The joint manual provides a more comprehensive discussion of EBO.

[4] This approach failed and also proved very difficult to assess. A military intelligence analyst at MACV Headquarters in Vietnam stated,

> I have always, in my own mind, used my pi factor [failure rate adjustment factor] on any battle damage assessment. But even if you go down to 50 percent factor, it's still staggering [the number of North Vietnamese trucks destroyed]. We're getting 200 trucks a night. We're going to reach 20,000 trucks. And even if you take it down to 50 percent, I still can't figure out where the hell all the trucks are coming from. (quoted in Sorley, 2004, pp. 594–595)

attempted to gauge the effect of the bombing on supply lines and NVA and VC military operations through analyses of reports and data on the operation.[5] In this case, effects-based theory was applied in a very practical sense: Force is used to achieve an effect that will lead to (or contribute to) victory. Modern effects-based theories originate from U.S. Air Force models designed to shape military campaigns through an analysis of battle damage assessment and the higher-order effects that might ensue from air-delivered attacks.

Chapter Two showed that, in U.S. service doctrine, both strategic and operational objectives tend to be broadly prescriptive and vague to give leeway to subordinate commanders. Subordinate units interpret the commander's intent and higher-order missions, deriving from them specified and implied missions (what it is they need to achieve to accomplish their objective) and then *tasks* (what it is they need to do to accomplish their mission).[6] Instead of supplanting traditional military theory, EBO is intended to enhance performance by bridging what some perceive to be a conceptual gap between objectives and tasks. Essentially, "Effects are derived from objectives. They help bridge the gap between objectives and tasks by describing the conditions that need to be established or avoided within the [operating environment] to achieve the desired end state."[7]

The next section describes the SoSA model of the interconnected and interdependent warfare environment, a central tenet of EBO. SoSA explains how effects are used to shift behavior within a complex system like a COIN operating environment. The subsequent section presents the debate over the applicability of EBO to modern warfare.

Elements of Effects-Based Theory: Definitions and System-of-Systems Analysis

To some extent, all centralized assessment techniques incorporate elements of systems analysis, and EBO is a process designed to exploit systems. At least in some inter-

---

[5] See Rehm, 1985, Part II, first section. There are no page numbers in this section, but it shows the handwritten calculations that went into this effects-based process. Many of books on Vietnam strategy also describe this process. For example, Sorley (2004) documents how these operations were tracked at the theater level for a period of two years. The analysis of effects was conducted in part by intelligence staffs and in part by assessment staffs. It was generally acknowledged to have been a failed strategy, but this failure is not necessarily an indictment of the effects-based assessment process.

[6] This is a very brief interpretation of a thoroughly documented process. For more detail on the command and control process, see HQDA, 2008. For example,

> Commanders analyze a mission in terms of specified tasks, implied tasks, and the commander's intent two echelons up. They also consider the missions of adjacent units to understand their relative contributions to the decisive operation. Results of that analysis yield the essential tasks that—with the purpose of the operation— clearly specify the actions required. This analysis also produces the unit's mission statement—a short description of the task and purpose that clearly indicates the action to be taken and the reason for doing so. (HQDA, 2008, p. 6-8)

[7] Joint Warfighting Center, 2006, p. III-5.

pretations (e.g., the *Commander's Handbook*), SoSA is a theoretical subset of EBO. It describes the operating environment as an interconnected set of systems, each of which is made up of interdependent nodes, key nodes, and links. Nodes are people or things that can be targeted, while links represent relationships that can be triggered or changed through effects. The *Commander's Handbook* defines the elements of SoSA:[8]

- An *effect* is the physical and/or behavioral state of a system that results from an action, set of actions, or another effect. It is also a change to a condition, behavior, or degree of freedom.
- A *system* is a functionally, physically, or behaviorally related group of regularly interacting or interdependent elements that form a unified whole. Systems associated with national security include political, military, economic, social, infrastructure, information, and others.
- A *node* is an element of a system that represents a person, place, or thing.
- A *key node* is a node that is related to a strategic or operational effect or center of gravity.[9]
- A *link* is an element of a system that represents a behavioral, physical, or functional relationship between nodes.

The SoSA concept is used to define, shape, and explain EBO and EBA processes and methods, and these models can be found in a number of joint and service doctrinal publications. Figure 4.2 presents a graphical depiction of SoSA. This kind of graphical node-and-link depiction is common in modern U.S. COIN operations. It is intended to highlight "decisive points against which the joint force can act to render the terrorist [or other] system unable or unwilling to fulfill its mission."[10] It shows smaller systems (e.g., the "military" system) and how they connect to other small systems (e.g., the "social system") within a system-of-systems.

SoSA drives operational concepts for both air and ground combat, and it is used to shape the ways in which analytic methods (typically statistical analyses) are applied to COIN assessment. It fosters a belief that not only can systems analysis be used to discern quantifiable progress but that it can also provide a *clear* understanding of the operating environment. Because nearly all the literature on warfare—including capstone U.S. military doctrine—rejects this possibility, the ways in which SoSA is applied to warfighting and assessment are problematic. The authors of the *Commander's Handbook*, which was published by U.S. Joint Forces Command, rejected

---

[8]   Joint Warfighting Center, 2006, p. I-3. The list very closely reflects the wording used in that document.

[9]   As used here, "center of gravity" is a military term referring to a center of power—military, political, or other—that is crucial to the targeted entity.

[10]   Joint Warfighting Center, 2006, p. II-5.

Figure 4.2
System-of-Systems Analysis Perspective of the Operational Environment

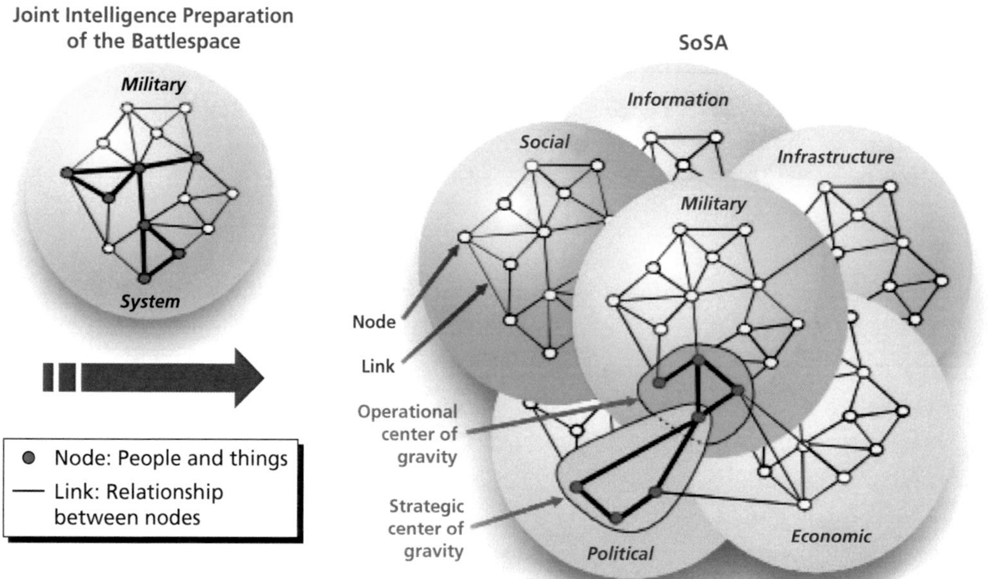

SOURCE: Joint Warfighting Center, 2006, p. II-2.
RAND MG1086-4.2

the notion that the battlefield or COIN operating environment is a complex adaptive system:

> [T]he effects-based concept was founded on "General Systems Theory," not "Chaos Theory" or "Complex Adaptive Systems" methods addressed in the mathematical sciences. In other words, [we] view the real world as a set of systems composed of tangible elements (nodes) and their relationships (links) to each other. The nodes represent discrete elements (people, material, facilities, and information) and the links portray the physical, functional and/or behavioral relations that can exist between and among nodes and systems. Both nodes and links are *only* symbolic. They are "icons" meant to simplify the complexity of the real world.[11]

This description is self-contradictory: It presents SoSA as both a concrete and abstract depiction of the COIN environment or system.[12] This excerpt also presents

---

[11] Joint Warfighting Center, 2006, p. V-1 (emphasis in original). While the handbook relies on general system theory (GST), other EBO literature clearly accepts the existence of chaos and the nature of complex adaptive systems as applied to COIN. Specifically, Edward Allen Smith (2002, pp. 231–352) makes an effort to both account for and exploit complexity. Complexity theory, which is tied to complex adaptive systems, is not entirely distinct from GST; the models used to explain complexity rely on the idea that there is a "universal pattern of life," as Neil F. Johnson (2009, p. 56) states in his lay introduction to complexity.

[12] See Ackoff, 1971, p. 662, for definitions of abstract and concrete systems in systems analysis literature.

GST as an alternative to complex adaptive system theory. GST applies mathematical formulas and modeling to systems with the goal of clarifying and simplifying the complex. Indeed, it is one of the foundational theories behind systems analysis.[13] GST assumes that all systems (e.g., mechanical, political, social) have a basic underlying structure that can be observed and defined.[14] Through the lens of GST, a SoSA schematic of the COIN environment portrays a system that is similar to other, familiar systems, such as a car or government bureaucracy. The scientific intent behind both chaos theory and complex adaptive system theory also is to simplify, or reduce, the complexity of the real world, but the real world they describe does not necessarily possess an underlying system that is generally analogous.[15]

The more salient distinction between SoSA and complexity theories is that (arguably) SoSA and EBO claim to offer a practical and readily achievable way to see through complexity to achieve desired effects, while chaos and complex adaptive system theories offer only tentative formulas and limited results.[16] The claim that "nodes and links are *only* symbolic" in the *Commander's Handbook* is disingenuous; the handbook contradicts itself not only here but also on page II-2, where it states, "System nodes are the tangible elements within a system that can be 'targeted' for action, such as people, materiel, and facilities."[17] This depiction of EBO clearly tries to directly portray real people, places, and things on a schematic SoSA map. Indeed, some analytic software used by the U.S. military to help understand insurgencies is predicated on concrete node and link analysis. At least in this accounting, the system-of-systems approach relies on the belief that it is possible to see, understand, and predict all *relevant* (as ascertained by the military staff) nodes and links in the operating environment.[18] As discussed later, *at least in COIN*, this is a faulty assumption. Appendix E presents

---

[13] For a full explanation of GST, see Bertalanffy, 1974. Bertalanffy states, "A consequence of the existence of general system properties is the appearance of structural similarities or isomorphisms in different fields. There are correspondences in the principles that govern the behavior of entities that are, intrinsically, widely different" (p. 33).

[14] Yehezkel Dror of Hebrew University of Jerusalem, working at RAND in 1969, stated that GST can "be used to better analyze and explain behavior and to provide a unifying and general theoretic framework for comprehending in common terms a large number of more heterogeneous phenomena" (Dror, 1969, p. 1).

[15] *Commander's Appreciation and Campaign Design* recognizes this disparity and cautions against reductionism when analyzing an interactively complex system (Joint Warfighting Center, 2006, p. 6).

[16] It is open to debate where Enthoven et al. would come out on a scale of certainty between these two positions, but the *Commander's Handbook* is fairly definitive.

[17] Joint Warfighting Center, 2006, p. II-2.

[18] The *Commander's Handbook* makes brief mention of uncertainty and describes the role of intelligence in ferreting out unseen nodes and links using both quantitative and qualitative methods. It does not clearly offer the possibility that the operating environment could obscure nodes and links to the point that EBO would be ineffective or inaccurate. This concern is reflected in efforts to apply agent-based models to COIN problem sets. See Gilbert (2008) for a lay introduction to such models. JP 2-01.3, *Joint Intelligence Preparation of the Operational Environment*, states,

a counterargument to this analysis of EBO, contending that SoSA and its component elements are not to be taken literally and that warfighters have misread and improperly applied EBO doctrine.

Effects-Based Assessment

The previous section described effects-based and systems analysis theories that shape some U.S. military doctrine on COIN operations. This section examines the official assessment method derived from effects-based theory. Like EBO, EBA depends on a system-of-systems understanding of the COIN environment. It also uses trend analysis and therefore incorporates some of the concepts described in Chapter Three. This discussion introduces distinct terminology and theories that form the framework for the EBA process.

Measures and Indicators

EBA is a system of quantitative measurement. It measures effects and performance by collecting and then sorting indicators. A measure of effect, or MOE, is a standard designed to gauge progress against a military end state; a measure of performance, or MOP, measures friendly performance; and indicators are the categories of data used in the measurement process. MOEs "assess changes in system behavior, capability, or operational environment. MOEs measure the attainment of an end state, achievement of an objective, or creation of an effect; they do not measure task performance."[19] FM 5-0 states, "MOEs help measure changes in conditions, both positive and negative. MOEs help to answer the question 'Are we doing the right things?' MOEs are commonly found and tracked in formal assessment plans."[20] On the other hand, MOPs

> measure task performance. MOPs are generally quantitative, but also can apply qualitative attributes to task accomplishment. They are used in most aspects of combat assessment, since it typically seeks specific, quantitative data or a direct

---

Analyzing all possible nodes and links in the operational environment would be an insurmountable task. However, not all nodes and links are relevant to the [Joint Forces Command's] mission. [Joint intelligence preparation of the operational environment] analysts should develop their understanding in sufficient detail to identify relevant systems, subsystems, nodes, and potential key nodes. (U.S. Joint Chiefs of Staff, 2009b, p. II-44)

This assumes that a staff can (1) know which nodes and links are relevant, (2) find these nodes and links and describe them in sufficient detail over time to create a realistic SoSA map, and (3) determine all the ways in which friendly actions, nonfriendly actions, and happenstance might shape or reshape nodes and links and make unseen nodes relevant while making other nodes less relevant.

[19] U.S. Joint Chiefs of Staff, 2010, p. IV-33.

[20] HQDA, 2010, p. 6-2.

observation of an event to determine accomplishment of tactical tasks, but have relevance for noncombat operations as well.[21]

An indicator is "an item of information that provides insight into a measure of effectiveness or measure of performance."[22] Table 4.1 shows the relationships between MOEs, MOPs, and indicators.[23]

Table 4.1
Effects-Based Assessment Terminology in U.S. Army Doctrine

MOE	MOP	Indicator
Answers the question: Are we doing the right things?	Answers the question: Are we doing things right?	Answers the question: What is the status of this MOE or MOP?
Measures purpose accomplishment	Measures task completion	Measures raw data inputs to inform MOEs and MOPs
Measures *why* in the mission statement	Measures *what* in the mission statement	Information used to make measuring *what* or *why* possible
No hierarchical relationship to MOPs	No hierarchical relationship to MOEs	Subordinate to MOEs and MOPs
Often formally tracked in formal assessment plans	Often formally tracked in execution matrixes	Often formally tracked in formal assessment plans
Typically challenging to choose the correct ones	Typically simple to choose the correct ones	Typically as challenging to select correctly as the supported MOE or MOP

SOURCE: HQDA, 2010, p. 6-3, Table 6-1.

NOTE: FM 6-0, *Mission Command: Command and Control of Army Forces*, offers up "criteria of success" instead of MOPs and the idea of a "running estimate" that can be used to supplement the standard assessment (HQDA, 2003, p. 6-17).

Doctrinal Effects-Based Assessment Process: Overview

Building on MOE, MOP, and indicators, FM 5-0 breaks the effects-based assessment process down into six steps:

1. Gather tools and assessment data.

2. Understand current and desired conditions.

3. Develop assessment measures and potential indicators.

4. Develop the collection plan.

---

[21] U.S. Joint Chiefs of Staff, 2010, p. IV-33.

[22] HQDA, 2010, p. 6-3.

[23] HQDA, 2010, p. 6-3.

5. Assign responsibilities for conducting analysis and generating recommendations.

6. Identify feedback mechanisms.[24]

Each of these steps can be taken prior to deployment. EBA starts to shape operations during this predeployment phase. The assessment staff section or, doctrinally, the Assessment Working Group, helps draft MOEs derived from the commander's objectives.[25] Depending on the approach this may take the form of a complex timeline with milestones and measurable effects, or it may simply be a more detailed restatement of the commander's intent.

The ultimate step in the doctrinal assessment cycle is producing the assessment report (or reports). While doctrine suggests ways of delivering assessments, there was no joint U.S. assessment report format as of early 2011; the instructions in FM 5-0 are generic and, as is typical of any military doctrine, suggestive rather than prescriptive. Assessment staffs develop their own reporting mechanisms or, more often, tailor reports to meet consumer (commander or policymaker) preferences.

Joint intelligence doctrine provides a schematic of the effects-based assessment process. Figure 4.3 shows how effects (MOEs, MOPs) are integrated between military tasks and military objectives at each level of focus—national strategic (policy), theater strategic, operational, and tactical. The national strategic level might be equated with policymaking, the theater strategic level with a theater command like MACV or ISAF, the operational with a regional or division command (~15,000 people), and the tactical with a battalion command (~1,000 people). In this model, tactical assessment focuses on task performance reporting aligned with MOPs and also "combat assessment," such as battle damage assessment from air or artillery attacks.[26] Tactical units (e.g., battalions) are responsible for describing how they are affecting enemy forces and performing military tasks.[27] This process is tailored for conventional military operations.

---

[24] HQDA, 2010, p. H-1.

[25] According to FM 5-0,

> The assessment working group is cross-functional by design and includes membership from across the staff, liaison personnel, and other partners outside the headquarters. Commanders direct the chief of staff, executive officer, or a staff section leader to run the assessment working group. Typically, the operations officer, plans officer, or senior [operations research/systems analysis] staff section serves as the staff lead for the assessment working group. (HQDA, 2010, p. 6-9)

[26] JP 2-0, *Joint Intelligence*,

> Tactical-level assessment typically uses MOPs to evaluate task accomplishment. The results of tactical tasks are often physical in nature, but also can reflect the impact on specific functions and systems. Tactical-level assessment may include assessing progress by phase lines; neutralization of enemy forces; control of key terrain, people, or resources; and security or reconstruction tasks. Combat assessment is an example of a tactical-level assessment and is a term that can encompass many tactical-level assessment actions. Combat assessment typically focuses on determining the results of weapons engagement. (U.S. Joint Chiefs of Staff, 2007, p. IV-22)

[27] U.S. Joint Chiefs of Staff, 2007, IV-20.

Figure 4.3
The Effects-Based Assessment Process

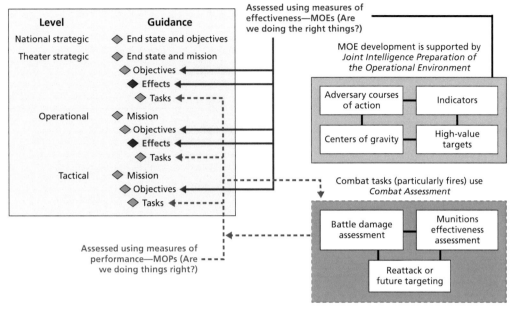

SOURCE: U.S. Joint Chiefs of Staff, 2007, p. IV-20, Figure IV-6.
RAND *MG1086-4.3*

Selecting Measures and Indicators

The *Commander's Handbook* describes how staff should gauge effects at the operational and strategic levels by tracking friendly MOPs and other indicators. It offers the preferred EBA approach to developing MOEs and MOPs during the planning process:

> MOEs and MOPs can be qualitative or quantitative. Whenever possible, quantitative measurements are preferred because they are less susceptible to interpretation—subjective judgment. They demand more rigor (or proof) and can be replicated over time even if the analysts and the users—the commanders—change.[28]

The following text box presents a set of recommended "progress indicators" for COIN. This example is drawn directly from FM 3-24.

Table 4.2 is an example of an MOE spreadsheet used to track progress against an MOE using specific indicators.[29] This example is drawn from FM 3-05.40, *Civil Affairs Operations*. It shows an objective, then subordinate effects with an associated timeline on the right.

---

[28] Joint Warfighting Center, 2006, p. IV-15.

[29] HQDA, 2006b.

The literature on COIN assessment tends to focus on which MOE, MOP, and indicators are appropriate for a specific campaign or for COIN in general. Chapter Five extends this brief introduction, addressing the selection of EBA metrics and the complications associated with the application of core metrics lists, such as the one in Table 4.2.

Box 4.1
Example Progress Indicators in Counterinsurgency

- **Acts of violence** (numbers of attacks, friendly/host-nation casualties).

- **Dislocated civilians.** The number, population, and demographics of dislocated civilian camps or the lack thereof are a resultant indicator of overall security and stability. A drop in the number of people in the camps indicates an increasing return to normalcy. People and families exciled from or fleeing their homes and property and people returning to them are measurable and revealing.

- **Human movement and religious attendance.** In societies where the culture is dominated by religion, activities related to the predominant faith may indicate the ease of movement and confidence in security, people's use of free will and volition, and the presence of freedom of religion. Possible indicators include the following:
  –Flow of religious pilgrims or lack thereof.
  –Development and active use of places of worship.
  –Number of temples and churches closed by a government.

- **Presence and activity of small- and medium-sized businesses.** When danger or insecure conditions exist, these businesses close. Patrols can report on the number of businesses that are open and how many customers they have. Tax collections may indicate the overall amount of sales activity.

- **Level of agriculture activity.**
  –Is a region or nation self-sustaining, or must life-support type foodstuffs be imported?
  –How many acres are in cultivation? Are the fields well maintained and watered?
  –Are agricultural goods getting to market? Has the annual need increased or decreased?

- **Presence or absence of associations.** The formation and presence of multiple political parties indicates more involvement of the people in government. Meetings of independent professional associations demonstrate the viability of the middle class and professions. Trade union activity indicates worker involvement in the economy and politics.

- **Participation in elections, especially when insurgents publicly threaten violence against participants.**

- **Government services available.** Examples include the following:
  –Police stations operational and police officers present throughout the area.
  –Clinics and hospitals in full operation, and whether new facilities sponsored by the private sector are open and operational.
  –Schools and universities open and functioning.

- **Freedom of movement of people, goods, and communications.** This is a classic measure to determine if an insurgency has denied areas in the physical, electronic, or print domains.

- **Tax revenue.** If people are paying taxes, this can be an indicator of host-nation government influence and subsequent civil stability.

- **Industry exports.**

- **Employment/unemployment rate.**

- **Availability of electricity.**

- **Specific attacks on infrastructure.**

SOURCE: HQDA, 2006c, p. 5-28, Table 5-7.

Table 4.2
Example of Measures of Effectiveness Spreadsheet with Indicators

MOE Spreadsheet			

Objective 1: Gain public support for U.S./coalition military forces and interim Iraqi government.

Effect A: General populace supports U.S./coalition efforts.

Measures	October	November	December
Number of offensive gestures directed at U.S./coalition patrols by Iraqi civilians	10	12	9
Number of instances involving anti-U.S./coalition graffiti	9	11	8
Number of anti-U.S./coalition demonstrations	12	11	5
Number of pure Iraqi events U.S./coalition respresentatives are invited to attend	4	3	5

Effect B: Civil leadership at district and local levels supports U.S./coalition efforts

Measures	October	November	December
Number of civil or religious leaders actively supporting U.S./coalition initiatives	20	20	25
Number of civil or religious activities U.S./coalition representatives are invited to attend	8	10	12

	Baseline		Neutral
	Positive		Negative

SOURCE: HQDA, 2006b, p. 4-9, Table 4-3.

Weighting the Assessment

Both the *Commander's Handbook* and FM 5-0, *The Operations Process*, encourage the use of quantitative indicators because EBO theory generally sees quantitative data as objective and qualitative reports as subjective opinion.[30] Thus, data are used to populate weighted mathematical models that theater-level staffs then use to determine overall progress.[31] Figure 4.4 shows the mathematical assessment model suggested by FM 5-0. It is broken down into levels from end state to "condition" (another term that is not thoroughly explored in doctrine but is equivalent in some ways to end state), MOE, and subordinate indicators.

---

[30] While FM 5-0 recognizes the need for both types of indicators, it states, "Quantitative indicators prove less biased than qualitative indicators. In general, numbers based on observations are impartial" (HQDA, 2010, p. 6-8).

[31] See HQDA, 2010, Appendix H.

Figure 4.4
Field Manual 5-0 Mathematical Assessment Model

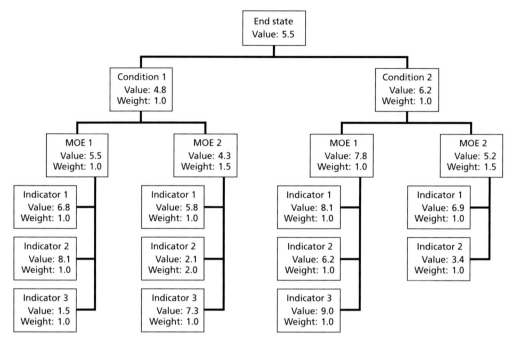

SOURCE: HQDA, 2010, p. H-6, Figure H-3.
NOTE: "Value" is also a subjectively assigned number in the weighting process.
RAND MG1086-4.4

By weighting, the staff gives each metric a mathematical percentage value in relation to the other metrics. FM 5-0 both describes and recommends the implementation of a comprehensive weighting scheme to establish an overall single measurement of progress:

> A weight is a number that expresses relative significance. Some indicators may be more significant than others for informing a given MOE. They count for more in the real world and should literally count for more in the mathematical assessment framework. Weights are used as multipliers for MOEs and indicators. The standard weight of 1.0 implies equal significance. A weight of 2.0 for an MOE (or indicator) implies that MOE carries twice the significance.[32]

For example, in a simplified case, a staff would assign a percentage weight out of 100 percent to each indicator and MOE. First, they would have to determine a weighting scheme and record the justification for the scheme to ensure transparency (this

---

[32] HQDA, 2010, p. H-5.

latter step rarely happens in practice). Next, the staff would have to assign a weight to each MOE:

- Violence: 50 percent
- Public opinion: 20 percent
- Government staffing: 15 percent[33]
- New construction: 15 percent.

Once the weights were assigned, the staff could then develop a process to track percentage changes over time, precisely compare those changes, and show overall progress toward an MOE, or even an overall campaign objective, using weighted indicators. However, the doctrinal model displayed in Figure 4.4 also incorporates "value," another subjective term tied to a 1–10 rating scale.[34] If weights show the importance of an indicator to the consumer (commander or policymaker), value purports to show whether a reported indicator is "good" or "bad." The notional example presented in FM 5-0 shows that more overall ransom money paid for hostages in a month is bad, while less would be good. This scheme interlaces two subjective rating scales and then combines them in a mathematical formula. The inherent subjectivity of this mathematical model is both compounded and opaque.

Determining Thresholds for Indicators

Not every assessment incorporates weighting, but some effects-based products derived from core metrics lists require the identification of quantitative time-series thresholds. To select a threshold, the commander or staff selects a point on a graph that will identify when an effect has been achieved. The *Commander's Handbook* states that

> quantitative measurement is not complete until the metric is compared against a specific criteria [sic] (standard or threshold). These thresholds can be minimums, maximums, or both. Comparison of measures against established criteria gives commanders a sense of whether they are making progress in accomplishing their objectives, effects, or tasks.[35]

---

[33] This is referred to in Afghanistan as *tashkil* fills.

[34] The FM offers the following explanation for value:

Standardization means that each component is expressed as a number on a common scale such as 1 to 5 or 1 to 10. Setting a common scale aids understanding and comparing as well as running the mathematical model. For example, Indicator 1 for MOE 1 for Condition 1 in figure H-3 [reprinted as Figure 4.4 here] could be monthly reported dollars in ransom paid as a result of kidnapping operations. For the month of June, that number is $250,000. That number is normalized to a scale of 1 to 10, with 1 being bad and 10 being good. The value of that indicator within the framework is 6.8. (HQDA, 2010, p. H-5)

[35] Joint Warfighting Center, 2006, p. IV-15.

Figure 4.5 depicts this process, showing the number of attacks over time in a notional location. It shows a measure as a single point on the scale, a metric as the difference between two points (or the change), and the threshold as the acceptable number of hostile actions (about 12).

FM 5-0 offers a similar definition and ties thresholds to the use of color-coding for assessment reporting:

> A threshold is a value above which one category is in effect and below which another category is in effect. Thresholds answer the question for a given indicator or MOE of what good and bad is. The categories can be whatever the commander finds useful, such as colors or numbers.[36]

The *Commander's Handbook* does not suggest any systematic method (or any method) for setting thresholds, but FM 5-0 states, "A significant amount of human judgment goes into designing an assessment framework. Choosing MOEs and indicators that accurately measure progress toward each desired condition is an art."[37] If this is the case, how should one choose the number 12 as an acceptable number of attacks, as depicted in Figure 4.5? Would this number be selected through a careful modeling

Figure 4.5
Example of Thresholds in Quantitative Assessment

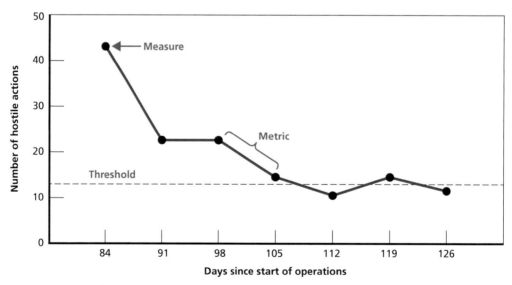

SOURCE: Joint Warfighting Center, 2006, p. IV-15, Figure IV-7.
RAND *MG1086-4.5*

---

36 HQDA, 2010, p. H-6.

37 HQDA, 2010, p. H-5.

process that would show how many attacks host-nation forces could handle per reporting period, or would it simply be chosen because it "seemed about right" to a staff or commander? If the latter method is used—it seems to be the most common choice—then the selection of thresholds is a highly subjective process that exposes the entire mathematical assessment scheme to legitimate critique.

Examples of Color-Coded Reports

Staffs at all levels tend to deliver EBA in both graphic and narrative format. Because graphics are easier to digest, such reports are more commonly produced than narrative reports. The *Commander's Handbook* offers an example of a graphic color-coded assessment report. Figure 4.6 shows color-coded ratings for a number of desired effects. Color codes reflect input from indicators: red is bad, yellow middling, and green good.

Figure 4.7 depicts a modification of the color-coded assessment as developed by staffs in Iraq and Afghanistan in the 2000s.[38] This version also relies on a sliding color scale from red to green, with red depicting poor performance or results and green depicting positive performance or results. In the figure, the first example (specific assessment) attempts to display pinpoint accuracy, the second (ranged assessment)

Figure 4.6
Example of a Color-Coded Effects-Based Assessment Report

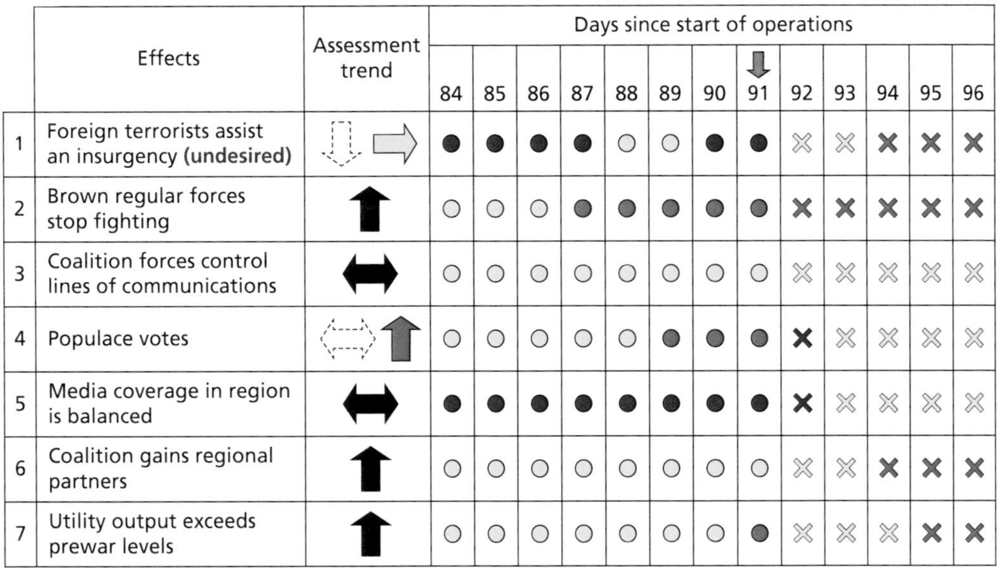

	Effects	Assessment trend	Days since start of operations												
			84	85	86	87	88	89	90	91	92	93	94	95	96
1	Foreign terrorists assist an insurgency (undesired)	⇨	●	●	●	●	○	○	●	●	⊠	⊠	✕	✕	✕
2	Brown regular forces stop fighting	⬆	○	○	○	●	●	●	●	●	✕	✕	✕	✕	✕
3	Coalition forces control lines of communications	⬌	○	○	○	○	○	○	○	○	⊠	⊠	⊠	⊠	⊠
4	Populace votes	⬆	○	○	○	○	○	●	●	●	✕	⊠	⊠	⊠	⊠
5	Media coverage in region is balanced	⬌	●	●	●	●	●	●	●	●	✕	⊠	⊠	⊠	⊠
6	Coalition gains regional partners	⬆	○	○	○	○	○	○	○	○	⊠	⊠	✕	✕	✕
7	Utility output exceeds prewar levels	⬆	○	○	○	○	○	○	○	●	⊠	⊠	⊠	✕	✕

SOURCE: Joint Warfighting Center, 2006, p. IV-14, Figure IV-6.
RAND MG1086-4.6

---

[38] The examples in Figure 4.7 were compiled based on observation of and participation in assessment report development groups. Data in the figure are notional.

Figure 4.7
Example of Color-Coded Assessment Reports on a
Sliding Color Scale

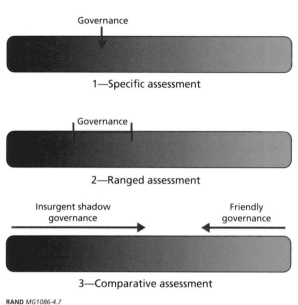

RAND *MG1086-4.7*

is intended to account for the relative inaccuracy of COIN data, and the third (comparative assessment) shows a comparative assessment of progress and setback for both insurgents and friendly forces. These notional examples use "governance" as the MOE (Is governance better or worse?), but this technique could be applied to any MOE or line of operation.

Chapter Summary

The original design of EBO has been expanded over the past two decades and is now central to joint U.S. military theory, planning, operations, and assessment. While there is ongoing debate over the relevance and efficacy of EBO (see Appendix E), effects-based and related systems analysis concepts continue to shape standing doctrine and operations. Because EBA is derived immediately from EBO, it is also a process originally intended to address conventional warfare. It is predicated on the need to interpret a cohesive enemy force or national threat through the application of systems analysis; identify centers of gravity, critical capabilities, and critical vulnerabilities of that threat; and subsequently predict and assess second- and third-order effects of friendly actions designed to shape the threat. To be successful, EBA depends on the availability of a system-of-systems understanding of the environment and the threat, a steady flow of complete and accurate data, and the ability to manage and analyze these data with mathematical precision.

Doctrine describes some ways in which EBA might be interpreted for COIN operations, but it does not clearly describe how to modify EBA so that it can better account for the complexity, chaos, and mosaic nature of the COIN environment. There does not seem to be a way to reconcile the centralized and precise nature of EBA with the distributed and imprecise nature of COIN. Efforts to explain how EBA might contend with complexity, chaos, and the absence of clear and comprehensive data seem to highlight rather than alleviate these stark inconsistencies. Some of the literature on EBO is more articulate in attempting to address these inconsistencies, but no document reviewed for this study seems to resolve them.

While this very brief introduction to the inner mechanics of EBA is not comprehensive, it does address all the essential elements that might go into an EBA product. Issues only hinted at here—subjectivity, complexity, opacity—come to life in the case studies (Chapters Six and Seven). The underlying inconsistency between complexity and the near-perfect accuracy needed to produce an EBA assessment requires the use of field modifications designed to make EBA work in the COIN environment. These modifications tend to result in a scattershot and sometimes inconsistent assessment process that lacks transparency and credibility. The following chapter goes into greater detail on how assessments are actually developed and presented.

CHAPTER FIVE

Choosing Core Metrics

> In today's increasingly dynamic operating environment, [commanders] can only gain sufficient situational awareness and adapt their current operations, future operations, and future plans if the staff is assessing "the right things" [the right metrics] in the operational assessment.
>
> —*Commander's Handbook for an Effects-Based Approach to Joint Operations*[1]

> Our metrics suck.
>
> —*Director, ISAF Afghan Assessment Group, 2009*[2]

All centralized assessments, including both EBA pattern and trend analysis, use a set of "core metrics" to assess COIN campaigns. Core metrics serve a dual purpose: They direct the collection of what is deemed to be relevant information for the assessment, and they shape the analysis of information and data once they are acquired. Selecting core metrics for centralized quantitative assessment is tricky. What matters most and why? Is information available, or will the commander have to induce risk to get new information? How should a core metric be defined for military units in the field and for policymakers? Selecting and using a set of core metrics is difficult in practice.

The indicators described in this chapter have been used in Afghanistan, and some were used in Vietnam. The purpose of providing these examples is not only to present real-world metrics but also to show, in detail, the friction points between theory and reality. These examples also help illuminate and inform other considerations: Can one consider specific metrics to be intrinsically good or bad for campaign assessment? Which set of metrics should be incorporated into doctrine and used in the field? What can one learn from looking at metrics like first-to-fire ratios? The first-to-fire ratio is a

---

[1] Joint Warfighting Center, 2006, p. IX.

[2] Quoted in Soeters, 2009, p. 11.

particularly timely example because this metric is in vogue among both experts and operational staff officers in Iraq and Afghanistan.[3]

Other considerations shape the analysis of the core metric selection process. Is the current process of selecting core metrics adequate to provide good centralized effects-based assessments, pattern analysis, or trend analysis? How do commanders and staffs usually select metrics, and is this process adequately rigorous? This chapter answers these questions and raises others in an effort to plumb the depths of complexity faced by policymakers, commanders, and military assessment staffs in COIN. It shows that there are no inherently good or bad metrics because no core metric is equally applicable to all situations. It also shows that the process of selecting metrics is highly subjective and often poorly understood. This analysis reveals the hidden "what-if" questions that erode the underlying assumptions behind centralized assessment and preclude the ability to apply core metrics to distributed operations overlaid on a mosaic environment.

Choosing "Metrics" and Determining Values

This section describes both the process of selecting individual criteria for centralized assessment and the challenges in doing so. It presents subject-matter expert positions on the selection of core metrics and examples of specific core metrics that highlight selection challenges. The purpose is not to recommend any one process over another, but instead to explore the topic that tends to dominate contemporary assessment discussion: finding the right things to count. Many experts working to help improve centralized assessment focus not only on what to count ("good metrics") but also on what not to count ("bad metrics"). They tend to refer to *metrics* in the vernacular; they use the term in the same way FM 5-0 would describe MOE, MOP, or, more commonly, indicators. This section uses the term *metrics* to describe EBA-related concepts, such as MOEs, MOPs, and indicators, as appropriate, to reflect various expert positions. However, core metrics are also used in pattern and trend analysis and could probably be used for any other kind of centralized assessment not identified here.

Policymakers, commanders, and staffs face considerable difficulty in selecting metrics for centralized assessment. Military staffs are tasked by doctrine with finding the "right things to count" but typically cannot settle on what those things might be for a variety of reasons. Each of the following considerations shapes the selection and analysis of core metrics to some degree:

---

[3]   I have personally observed contentious debate over the definition of "first to fire" in both theaters. The first-to-fire ratio was often highlighted as a key metric at several levels of assessment.

- Any single core metric must be generally applicable everywhere at all times, because core metrics are intended to show progress across the entire theater.[4] However, finding a metric that meets this requirement has proven to be very difficult, if not impossible.
- Core metrics demand the generalized and recurring collection of information, but this is not possible in distributed COIN operations, because the information varies greatly in consistency, quality, and availability from village to village (or hamlet to hamlet).
- There is no agreed-upon set of necessary qualities for a core metric, even in U.S. doctrine. Experts tend to agree that metrics should show variability (their words) and that metrics lists should be parsimonious, but they otherwise disagree or fail to provide clear guidance.
- There are no agreed-upon definitions for individual metrics, including the ones listed as examples in this monograph. Therefore, it is unclear to both staffs and reporting units what they should be looking for and why.
- Definitions that do exist not only are contested but also change with policy or commander preference. When this happens, the metric and associated collection requirements change, and any efforts to maintain reliability or track variability fall apart. It is very difficult to predict how definitions might change when selecting metrics to avoid this outcome.
- It is not clear how individual core metrics should be tied to end states or how they might show progress toward those end states (or, to use a more common term, objectives). Doctrine provides insufficient guidance as to how metrics should show progress in assessment reports.

Intense debates over "good" and "bad" metrics are commonplace. The *Commander's Handbook* clearly states that staffs should be able to find a list of inherently good metrics, or "the right things to assess."[5] David Kilcullen lists what he describes as good and bad core metrics in a December 2009 paper on Afghanistan metrics. First, he addresses metrics (or MOE, MOP, and indicators) that he believes can be deceptive and should be avoided. These include counts of enemy dead, reported number of enemy attacks (reported as SIGACTs in both Iraq and Afghanistan), any metric that produces a single number out of context, and any metric that measures inputs, such as the number of schools built. He builds a case describing why each of these should be avoided. He then provides an extensive list of things that should be considered metrics. This is a sample from Kilcullen's list: progress of NGO construction projects,

---

[4]   This consideration applies even when core metric indicators are broken into provincial or district-level blocks of data. Indicators can show different levels of activity or effect from area to area, but core metrics are designed to provide reliable measurement using a like standard across all areas.

[5]   Joint Warfighting Center, 2006, p. IV-18.

transportation prices, the influence of Taliban versus government courts, kill ratios, kill-versus-capture ratios, civilian casualties, first-to-fire ratios, and several others along five primary lines of operation.[6] Kill ratios measure the number of casualties inflicted versus the number of casualties suffered, while the first-to-fire ratio measures which side fired first to determine who has the tactical initiative in combat.[7]

Considerations for Selecting Core Metrics

Kilcullen's intent is for operational staffs to select appropriate metrics from his list and elsewhere, shape them to fit specific needs, and adjust them over time as necessary. He argues that it is better to develop a common set of core metrics at the theater level (a point at which the operational and strategic meet) that can be readily tracked and adjusted, rather than creating long lists of metrics that may be cumbersome and narrowly applicable; he recommends parsimony. Core metrics lists are typically designed and maintained at the theater level, but in both Iraq and Afghanistan, several such lists (each somewhat different from the next) existed or continue to exist simultaneously at various levels. In theory, subordinate lists would be derived from a centralized list, but in practice, core metrics lists reflect a mix of specified, implied, and ad hoc items. Since the term *metrics* is nondoctrinal and rather vague, a core metrics list for EBA can contain MOEs, MOPs, output indicators, and input indicators. Some staffs have found these lists difficult to interpret, particularly since indicators should be subordinate to MOEs and MOPs in accordance with FM 5-0.[8]

In their various works on Afghanistan assessments between 2008 and 2010, both Kilcullen and Anthony H. Cordesman of the Center for Strategic and International Studies argue that selecting the right criteria is essential to the overall assessment process. Cordesman also argues for a holistic approach to objectives-based assessments and calls for the careful selection of core metrics.[9] Jonathan J. Schroden of the Center for Naval Analyses offers a detailed examination of security metrics in the *Journal of Strategic Studies*. He accepts the wicked problem construct and has produced a body of work on adapting various methodologies to try to compensate for data flaws. In his article, he tries to steer the counterinsurgent away from traditional metrics, like casu-

---

[6] Security, development, governance, rule of law, and essential services (Kilcullen, 2009b).

[7] Kilcullen, 2009b.

[8] Anthony H. Cordesman of the Center for Strategic and International Studies concurs with Kilcullen on core metrics but also believes that there is a lack of coordination in developing core metrics lists:

> Various elements of the US government—civil and military—often generate metrics and analytic models without adequate efforts to ensure comparability. Net assessment and fusion are still the exception and not the rule. The end result in war after war has been that different value systems are applied to some degree to US, allied, host country, and threat forces; and kinetic measures of tactical progress have more emphasis than the equally important data on ideology, politics, and governance, economics and the perceptions of the population. (Cordesman, 2008, p. 3)

[9] Cordesman, 2010.

alty counts and body counts, and toward a complex analysis of attack reporting to analyze security. He states that these more complex metrics

> have a direct relation to several key aspects of counterinsurgency operations: removing the insurgent's advantage of initiative; developing local forces to fight and generate intelligence; and gaining the support of and securing the local population. The same cannot be said about many other metrics currently in use.[10]

Parsimony is a central theme in many professional discussions of COIN assessment.[11] There is general agreement that a core metrics list should be kept brief and broadly focused to ensure that subordinate commands are afforded maximum flexibility. A detailed and extensive list of metrics might not be equally applicable across an entire country (an idea that will be explored later), and the micromanagement of the assessment process even at the operational level of command—in this case, the theater operational level—would deter full participation from units in the field. This, in turn, would prevent the acquisition of the information necessary to feed the centralized assessment process. U.S. Army assessment doctrine states, "Excessive reporting requirements can render an otherwise valid assessment plan onerous and untenable."[12]

Campbell, O'Hanlon, and Shapiro discuss the idea of parsimony in metrics lists but argue that it is sometimes useful to have large quantities of data available to various staffs and researchers. Their Brookings Index on Iraq consists of about 50 indicators, while the index on Afghanistan is slightly more condensed due to the lack of available data. In their report, *Assessing Counterinsurgency and Stabilization Missions*, they state,

> If we were confident about which 10 or 15 or 20 metrics could best tell the story of the efforts in Iraq or Afghanistan, and had reliable data for those categories, we might have focused more narrowly on them. The truth is that the wars in both Afghanistan and Iraq have usually demonstrated an ability to confound short lists of metrics.[13]

---

[10] Schroden, 2009, p. 741.

[11] Hriar Cabayan of the Office of the Secretary of Defense produced a list of MOEs and associated metrics for the ISAF Joint Command staff in Afghanistan. His working group consisted of approximately 70 assessment, modeling, and COIN experts from government, academia, and the military. Most of the MOEs that the group proposed were shaped to measure popular perception. The group generally concurred that if the population is the center of gravity in COIN, and winning popular support is critical to winning the campaign, then measuring popular support is the most important method in campaign assessment (Cabayan, 2010a, 2010b).

[12] HQDA, 2010, p. H-3.

[13] Campbell et al., 2009a, p. 4. They add, "In retrospect, some key metrics seem to emerge with greater clarity." This trend away from parsimony and toward providing a detailed, or at least specific, list of metrics is partially attributable to an effort to meet specific requirements from the field, but it also may reflect a natural inclination among both academic and military experts to seek and present detail in assessments.

Parsimony was one of the key axioms identified by the participants of an assessment workshop held by the Military Operations Research Society. The results of the workshop reflected the combined viewpoints of many of the leading proponents and practitioners of operations research in the U.S. government and in academia. Their list of axioms included the following:

- *Parsimony:* Use the simplest tool that will get the job done.
- *Sensitivity:* A good measure should support change detection within the review time frame.
- *Validity:* The indicator must measure what we need to measure.[14]
- *Reliability:* If collected again by a different person or a short time later, it should yield the same answer.
- Not everything that can be counted counts, and not everything counts that can be counted (citing Albert Einstein).[15]

It may be possible to identify desirable traits for metrics, but deciding on individual metrics is challenging. The varying subjective views of policymakers, commanders, and staff officers can generate debate that muddies definitions and reduces reporting incentive. If the description is too vague, each tactical unit will define the metric differently. For example, a battalion commander might count IED attacks as enemy fire and found IEDs (i.e., discovered before they are detonated against a friendly target) as a friendly success, while another might not consider an IED to be a type of fire at all. Since IEDs make up a large portion of incidents in modern COIN, this kind of differentiation would skew the metric to the point that it would be misleading when aggregated only one level above that at which the attacks occurred.

Schroden conducted a detailed analysis of the "ratio of incident initiation" between May 2005 and August 2007 in Anbar Province, Iraq.[16] While he views this as a useful metric when properly analyzed, he elaborates on the problem of definitions:

> [W]e realized it was necessary to have a clear understanding of what it meant to "initiate" an incident. . . . However, in Iraq the situation is not always this clear.

---

[14] There are several types of validity, including construct, internal, and conclusion validity. For the sake of simplifying a complex subject, this report takes a broad approach to the concept of validity, drawn primarily from social science literature. Also, occasionally the term *valid* will be used to describe the relative value of data.

[15] Grier et al., 2010, p. 9. Some of these bullets are paraphrased.

[16] Schroden differentiates between the more simplistic first-to-fire ratio and the ratio of incident initiation. He states that "the latter allows one to define an actual categorization scheme that makes operational sense for incidents like IEDs." Schroden also believes that while it is not possible to obtain total accuracy with any indicator, it is possible to analyze large data sets like this one and produce operationally useful conclusions (Jonathan J. Schroden, email exchange with the author, September 29, 2010b).

. . . Thus, a more rigorous definition was needed for what it meant to initiate an incident.[17]

Each staff from the tactical to strategic level must agree on definitions, or the values of the data will change (possibly more than once) as they are assimilated up the chain of command. If too much detail is provided in the description of the indicator, however, tactical commanders might feel that they are being micromanaged or might report too narrowly. Finding a middle ground is tricky, and the more contested the definition the more likely it is to be changed over time. Each change in definition sets off a ripple of retroactive changes in assessment or, worse, distorts current assessments.

The examples in the remaining sections of this chapter are intended to demonstrate the complexity encountered by military officers as they attempt to select core metrics by which to measure a COIN campaign.

Example of a Core Metric: First-to-Fire Indicator

This indicator is designed to show which side—counterinsurgent or insurgent—fired first in a specific engagement to determine which side generally has the initiative in combat. The idea behind first-to-fire is that relative initiative shows either momentum or lack of momentum. Kilcullen recommends this as a useful metric because it shows which side controls the casualty rate. Further,

> The first-to-fire ratio is a key indicator of which side controls the initiation of firefights, and is a useful surrogate metric to determine which side possesses the tactical initiative. If our side fires first in most firefights, this likely indicates that we are ambushing the enemy (or mounting pre-planned attacks) more frequently than we are being ambushed. This in turn may indicate that our side has better situational awareness and access to intelligence on enemy movements than the insurgents, and it certainly indicates that we have the initiative and the enemy may be reacting to us.[18]

Because it is presented as a ratio, it is a quantitative metric (or indicator) requiring hard, quantitative information from the field. Combat units would have to report which side fired first for each specific combat incident. Once the metric is defined and reporting requirements (i.e., how often and in what format) are established, the data start to flow in from subordinate units. The theater assessment staffs will then have a data set that, in theory, shows friendly to enemy fire ratios over time. The total number of incidents would then be added up to show a comparison between friendly

---

[17] Schroden, 2009, pp. 724–725.

[18] Kilcullen, 2009b, p. 17.

and enemy-initiated attacks. First, though, the staff has to define and incentivize the capture of data to ensure the successful use of the indicator:

- What is an "incident"? This may be rather simple to define for a single ambush. However, if there is a running, interconnected firefight between multiple units over the course of a day, should the whole battle be considered one incident or should each fight be considered an incident? This would have to be precisely defined for tactical commanders.
- What is "fire"? Does this metric track direct-fire attacks only, or does it also include IED attacks? What about indirect fire? Mine strikes? A single sniper shot from an unseen foe? Does an assassination of a key government or military official count, or would that fall under another category? Does a friendly airstrike or artillery strike not called in by infantry units—perhaps a missile strike based on intelligence—count?
- Do these definitions match the definitions of other metrics? In other words, is a first-to-fire incident in the SIGACT database exactly the same as a first-to-fire incident in other databases, or in the original definition of the metric? If not, how will this affect cross-comparison of information or data?
- Why does this metric matter? The staff should provide a detailed explanation to subordinate commands to ensure that they are incentivized to report accurately. If they do not believe that the metric is important, they may fail to report the data or simply fill in forms with little thought.

How should an assessment staff interpret this ratio? Is it true that, as Kilcullen states, "If our side fires first in most firefights, this likely indicates that we are ambushing the enemy," or might other complexities be hiding under the surface? Assessment staffs would need to seek parity on the following considerations, among others:

- What if the rules of engagement require positive identification before a friendly unit can initiate fire? If this is the case, how often do counterinsurgents identify insurgents—i.e., friendly spots enemy first—but have to wait to be fired upon before they are comfortable engaging? What if they are engaged while waiting for approval to fire on an enemy that they observed first? In these cases, the incident report might be misleading.
- There is little incentive for tactical units to report that they are routinely caught off-guard by insurgents. There may be great personal motivation for the combat leader to report that his side had the initiative when the matter is in doubt. *This is particularly true if the command has emphasized firing first by creating a first-to-fire reporting requirement.*
- When opposing units chance upon each other (in conventional terms, a meeting engagement), the issue of who fired the first shot may be irrelevant or misleading.

In instances like this, in which both sides are surprised, the first shots are often wildly inaccurate or ill-considered; it may be more tactically sound to hold fire until a clear shot is offered.

- One side may initiate most of the fighting but also cause fewer casualties and, overall, be less effective.[19] Undisciplined units—particularly poorly trained host-nation security forces on which counterinsurgents might have to rely in the early stages of a campaign—often discharge their weapons at the slightest hint of danger. U.S. advisors in Iraq referred to this reaction as the "death blossom." Behavior like this is a sign of incompetence. These units are also the ones most likely to make false claims of initiative when delivering unsupervised combat reports. In Vietnam, even U.S. combat activity reports were often conflicting and misleading for this reason.[20]
- Conversely, insurgents may be incompetent or relatively less competent than the security forces. They may get off the first shot because they can blend in with the population, but they may lose—badly—once the tactical combat is under way. Sometimes, the act of attacking is sufficient to achieve a propaganda victory, but what about situations in which insurgents tend to fire first and are then defeated, weakening their structure and possibly undermining the psychological value of their attacks? Who controls the loss rate in these situations? Is firing first always "good," or can it sometimes backfire? While it may be possible to balance this indicator against other indicators, doing so with aggregated data would make accurate analysis difficult.
- Is it possible to retain analytical context as first-to-fire information or data are aggregated up to the policy level?

Schroden does not specifically address all of these complexities, but he believes that trained analysts can break through some of the chaos of combat reporting to provide useful analysis. He argues that a shift from enemy- to friendly-initiated attacks in mid-2007 in Anbar Province was a clear indicator of success for the coalition: "Since insurgents rely on initiative to compensate for being overmatched in personnel, technology, and firepower, the shift of the incident ratio to favor friendly forces was a clear and quantitative indication that significant progress was being made against the insur-

---

[19] Analysis of core metrics might try to compare these data with another data set to show whether firing first also produced more control or more casualties. However, it is not clear how this might work in the absence of a clearer definition of control or with body count reporting (see Chapter Six for additional discussion of this point). At any rate, comparison would have to be conducted at a relatively low level of aggregation to ensure accuracy; this kind of low-level assessment might require extensive effort at the tactical level and also obviate the need for core metric assessment.

[20] Thayer, 1985, p. 56. Thayer describes the problems with "operational days of contact" statistics.

gency in Al Anbar."[21] This explanation closely matches Kilcullen's description of the first-to-fire ratio.

So, assuming that good and bad metrics or indicators do, in fact, exist, is first-to-fire inherently good? To what degree do unaddressed complexities shape the data? Can one description of first-to-fire be used in every case, or even within one campaign?

Differences in the meaning of data from place to place call into question not only the reliability and generalizability but also the validity of the first-to-fire metric. A measuring device (e.g., an MOE or indicator) is valid if "it does what it was intended to do."[22] First-to-fire might not show what it portends to show. Taken at face value, the device might be considered valid if it showed which side fired first. However, for the purposes of operational assessment, this may not be sufficient. If the first-to-fire metric claims to show an *advantage* in initiative for one side or the other, it must show actual advantage. For example, direct-fire attacks in a notional "Sector A" might show that the insurgents have found a way to initiate attacks against friendly forces while taking few casualties in return. This might show both a tactical and psychological advantage for the insurgents in Sector A. In a notional Sector B, however, the use of sniper fire (a type of direct fire that allows insurgents to fire first in most cases) might show that the insurgents have been pushed from operating in strong, semiconventional units to relying on asymmetric attacks. This might be especially relevant if the attacks in Sector B have been ineffective in that they have not caused many casualties.

In this case, friendly casualty rates would be only one of many coincidence indicators or confounding variables that are not incorporated into the first-to-fire assessment.[23] A confounding variable is a variable that questions the assumption drawn from a dependent variable (indicator) or places the dependent variable in context. For example, if insurgents fired first more often than counterinsurgents but also caused very few casualties, would the first-to-fire data have the same meaning as it might when viewed independently? What would it mean for the overall progress of the campaign if, at the same time, 30 stores opened in the market, insurgents bombed five schools, ten government jobs were filled, eight new roads were built, and so on? This kind of complexity can quickly overwhelm an analyst attempting to correlate incomplete and aggregated quantitative data sets.

For the first-to-fire indicator, confounding variables could include both friendly casualty rates and civilian casualty rates. What if coalition (friendly) forces in a specific area are getting the jump on the enemy but then attack so aggressively that they cause increasing numbers of civilian casualties? This could result in tactical improvement but also a strategic setback. Considering these two variables in conjunction would require

---

[21] Schroden, 2009, p. 726.

[22] Carmines and Zeller, 1979, p. 12.

[23] This metric and the SIGACT metric are also particularly vulnerable to the kinds of data paradoxes that are described later in this chapter.

the kind of context that can be found only in disaggregated, primary sources. Schro-den notes that another admitted drawback of the first-to-fire metric is "that we have counted all incidents equally, so for example a suicide bomber who kills 100 people in a single explosion is counted the same as a single bullet from a sniper. However, thus far we have been unable to identify a weighting scheme that is both clear and non–ad hoc."[24] Lack of centralized control over the collection and reporting of data in dis-tributed operations like COIN degrades the reliability of measurement and the validity of findings.

Many questions about such individual metrics remain unanswered in the peer-reviewed literature, doctrine, and (particularly) in the field in Iraq and Afghanistan. The debate over the definition of first-to-fire has been subjective and generally unre-lated to reliability or validity, although Schroden has done some empirical analysis on this specific metric for ISAF.[25] The value of first-to-fire as a stand-alone metric, or even as part of a broader operational assessment, appears to be uncertain but Schroden concurs with Kilcullen: "These drawbacks aside, we maintain that the incident ratio is a better measure of progress for a counterinsurgency than many others in common use."[26]

Example of Input Measures of Performance or Indicators

Some indicators in use in Afghanistan and Iraq track inputs rather than outputs. In other words, they count or describe things that coalition forces or government agen-cies deliver (input) into the environment with the goal of achieving a certain result. For example, a government agency might input money into the economy to achieve the effect of improved economic opportunity, and then attempt to determine the success of this effort by measuring the amount of money input into the economy and not the effect of the action (or the degree to which it furthered objectives). Kilcullen places input indicators in the "bad" column. He states that they "are indicators based on our own level of effort, as distinct from the effects of our efforts. . . . These indicators tell us what we are doing, but not the effect we are having."[27] A typical input metric in Afghanistan is *tashkil* fills. A tashkil is a government staff in a particular office or sta-tion. This indicator (or MOP) shows how many government jobs in a particular district were filled by civil servants. In theory, this could show how successful the government has been in placing its civil servants and, in turn, how responsive the government was to the population. In practice, there may be flaws:

---

[24] Schroden, 2009, p. 731.

[25] Jonathan Schroden, email exchange with the author, October 27, 2010d.

[26] Schroden, 2009, p. 731.

[27] Kilcullen, 2009b, p. 7.

- Are these civil servants actually doing any work of value, or are they adding to the anger and frustration of the population by being visibly ineffective?
- Have the civil servants been through any kind of training, or are they untrained and therefore inadequately prepared to do their jobs?
- This input does not track corruption. If the civil servants are corrupt by Afghan standards and perpetuate the perception of the Afghan government as corrupt, their presence is counterproductive.
- Are they showing up for work or staying at home? Are they actual civil servants or "ghost" employees whose paychecks are sent directly to a corrupt official in Kabul? This input does not measure physical presence, which, itself, is also not necessarily valuable.
- Are the civil servants local or from out of the area, and how might this bear on their ability to survive and operate in their official positions?

A counter to the criticisms of this input metric might claim that the data would not be analyzed in isolation but would be compared to other data, such as corruption levels, popular opinion of the government in that area, whether or not the officials live near their places of work, and other factors. If a systematic and thorough analysis of this kind were being carried out with reasonably sound data and kept in context, then this might be a valid defense of the *tashkil* fill metric. However, this would require an individual analyst to match corruption reporting, polling data, or some like-opinion data to the performance of an individual Afghan government official. This is not the kind of all-source analysis that can be easily aggregated for assessments. Nonetheless, even aggregated and out-of-context input metrics like this are used to decide resource allocation at top levels of command: A commander might assume that if a district has "100-percent *tashkil* fills" then it must not need any more civil servants. Stephen Downes-Martin describes how even the basic arithmetic used to describe *tashkil* fills in a police unit—before the data were aggregated—can be misleading:

> Another observed example of junk arithmetic (this time leading to an overly optimistic claim) was an Afghan National Police (ANP) assessment that claimed nearly 100% *tashkil* filled. The underlying data, however, was that the patrolmen were overfilled and NCOs [noncommissioned officers] and officers under-filled by significant amounts.[28]

Tashkil fills is a useful input indicator only if it is used to measure input and not as a campaign assessment tool. In other words, it can only describe how many people were *assigned* to a specific office. This kind of number is useful to budgeting officers

---

[28] Downes-Martin, 2010b, p. 4. Downes-Martin delivered his draft report after being embedded with the Regional Command Southwest Assessment Section both prior to its deployment and then for three months in Helmand Province, Afghanistan.

and government planners, but only for very limited purposes. It cannot describe how many of those people showed up for work, how many were competent, or how many were corrupt, effective, responsive, or tied to the insurgency. In the absence of additional information, such as performance reports, it would seem unwise to draw any operational (as opposed to logistical or bureaucratic) conclusions from this indicator.

Example of Output Measures of Effectiveness or Indicators

A popular output MOE or indicator counts the number of new businesses opened in a district or village.[29] The metric report would state, "District A—10 new businesses opened—increase of 20 percent from last reporting period." At the operational level (battalion, regiment, brigade, or perhaps region), the assessment staff would see that not only had ten new businesses opened but that these data reflect a positive trend: More new businesses are opening in this area than was the case in the previous period. At first glance, this would seem to show that security is improving, business is returning, people are finding ways to get their goods to market, and so on. However, if "new businesses" are reported as a core metric, the following issues would have to be taken into consideration:

- Like the *tashkil* fill problem, there is no method to show direct correlation between opening markets and improvement in security outside of individual intelligence analysis, which is not suitable for data aggregation in operational assessment.
- Further, only intelligence reporting or the local commander's analysis could show whether these new stores reflected civil stability or whether insurgents owned them, whether they were fronts for criminal gangs, or whether these businesses otherwise represented an increase in corruption or hostile control.
- Is it common for businesses to open and then close quickly in this specific area, perhaps according to the seasonal harvests? What are the prospects for longevity? If new businesses open and then fail in rapid succession, the population might lose faith (perhaps unjustifiably) in the local or national economy.
- In the absence of detailed contextual narrative or very clear measuring criteria, both the increase in businesses and the percentage increase over time have little meaning. What is the optimal number (or threshold) of businesses in this district or village?[30] Is there any way to show that this "optimal" number, if it can be identified, could be tied to improved security or an improved perception of government legitimacy? This is unlikely, and there appears to be no evidence that

---

[29] Business and commercial availability metrics have been used at various times in both Afghanistan and Iraq and were used to track performance in both post–World War II Germany and Bosnia-Herzegovina. Sources for this information are the author's observations, interviews conducted for this study, and Grier et al., 2010, p. 8.

[30] See the discussion of thresholds in Chapter Four.

this level of analysis has occurred in practice—at least not successfully—at the operational assessment level in Vietnam, Iraq, or Afghanistan.

Every core metric used in Vietnam, Iraq, and Afghanistan contains internal contradictions like these; they are unavoidable because context shapes information in complex and chaotic environments. Efforts to compensate for these problems tend to focus on finding ways to compare metrics to each other to determine nonscientific correlation (e.g., Does it seem like one thing is making another thing happen?). This kind of correlation assessment then leads the analyst to bring in more sources and focus at lower and lower levels in an effort to obtain context, which, in turn, makes centralized and aggregated quantitative assessment all but impossible. Chapter Six describes OSA cycled through this process; concerns with centralized core metrics in Afghanistan led to the development of a contextual analytic report called District Deep Dive.[31]

Example of Measures of Effectiveness and Indicators from Doctrine
The following text box presents an example list of MOEs and indicators for stability operations as provided in FM 5-0. Although it is merely an example using notional data, the list shows many of the inconsistencies inherent in the selection of core metrics. None of the MOEs are worded similarly; most identify general, undefined standards (e.g., "improved"), while MOE 2 for condition 2 is simply a report and not an effect at all. Indicators for some MOEs are quantitative data categories that are ostensibly tied to MOEs (e.g., monthly number of reported kidnappings), while others are polling questions and are not strictly indicators. There is no explanation for why any of these MOEs are important, and the "effects" are vague. The list does not explain what "disrupted" means for condition 1, MOE 1, and it does not define "improved" for condition 1, MOE 2. Nothing explains how these measures might be related to each other, how indicators for one MOE might show a change in another MOE, or how any of these criteria should be analyzed holistically.

While this doctrinal example may not have been intended to show such detail, it does set a standard for assessment. This is typically how MOEs and indicators are presented to operational units in U.S. COIN operations. Any subordinate staff (e.g., a brigade headquarters) attempting to put such a list into effect would be forced into developing highly subjective interpretations of each MOE and indicator. It would be possible to interpret field reporting for nearly all of these indicators as either positive or negative in the absence of clear and contextual criteria for evaluation.

The MOEs and indicators suggested by FM 5-0 are shown in the following text box.

---

[31] Because the District Deep Dive is a relatively new process, an examination of its efficacy or applicability to the overall assessment process was outside the bounds of the current research effort.

Box 5.1
Measures of Effectiveness and Indicators from Field Manual 5-0

Condition 1: Enemy defeated in the brigade area of operations.

MOE 1: Enemy kidnapping activity in the brigade area of operations disrupted.

- Indicator 1: Monthly reported dollars in ransom paid as a result of kidnapping operations.
- Indicator 2: Monthly number of reported attempted kidnappings.
- Indicator 3: Monthly poll question #23: "Have any kidnappings occurred in your neighborhood in the past 30 days?" Results for provinces ABC only.

MOE 2: Public perception of security in the brigade area of operations improved.

- Indicator 1: Monthly poll question #34: "Have you changed your normal activities in the past month because of concerns about your safety and that of your family?" Results for provinces ABC only.
- Indicator 2: Montly K–12 school attendance in provinces ABC as reported by the host-nation ministry of education.
- Indicator 3: Monthly number of tips from local nationals reported to the brigade terrorism tips hotline.

MOE 3: Sniper events in the brigade area of operations disrupted.

- Indicator 1: Monthly decrease in reported sniper events in the brigade area of operations. (Note: It is acceptable to have only one indicator that directly answers a given MOE. Avoid complicating the assessment needlessly when a simple construct suffices.)

Condition 2: Role 1 medical care available to the population in city X.

MOE 1: Public perception of medical care availability improved in city X.

- Indicator 1: Monthly poll question #42: "Are you and your family able to visit the hospital when you need to?" Results for provinces ABC only.
- Indicator 2: Monthly poll question #8: "Do you and your family have important health needs that are not being met?" Results for provinces ABC only.
- Indicator 3: Monthly decrease in the number of requests for medical care availability from local nationals by the brigade.

MOE 2: Battalion commander estimated monthly host-nation medical care availability in battalion area of operations.

- Indicator 1: Monthly average of reported battalion commander's estimates (scale of 1–5) of host-nation medical care availability in the battalion area of operations.

SOURCE: HQDA, 2010, p. H-4, Figure H-2.

Chapter Summary

This chapter asked the question, "How hard could it be to select a practical core metrics list?" Policymakers and military staffs have discovered over the past ten years or so that the answer to this question is, "Quite hard if not impossible." That there was no agreed-upon single list of metrics for Afghanistan as of early 2011 drives this point home. There is a clear lack of consensus among experts and practitioners, and *between* experts and practitioners, over how to tackle a seemingly simple task like selecting core metrics. No assessment group applies a clear set of broadly agreed-upon standards for academic, scientific, or even military analytic rigor in the selection and valuation of metrics. Doctrine provides little clarity and, in most cases, fails to address the realities of COIN assessment. Ultimately, this failure to settle on a core metrics list stems from the failure of centralized COIN assessment theory to account for and address the reality of COIN. In other words, policymakers, commanders, and staffs cannot settle on a unified core metrics list because core metrics are inherently impracticable for COIN campaign assessment.

CHAPTER SIX

Vietnam-Era Assessment

This chapter describes key elements of the Vietnam-era assessment process, examines the use and misuse of data in the resulting assessments and in policy decisionmaking, and ties this historical example to contemporary centralized assessment. It begins with an explanation of the assessment processes used during the Vietnam War era and then explores the details of assessment requirements, assessment methods, field collection, and reporting. The intention is to show how assessments and assessment data were used in the military and civilian strategic decisionmaking processes of the period and why lessons from Vietnam are relevant to Iraq and Afghanistan. Assessment efforts in Iraq and Afghanistan bear an irrefutable resemblance to Vietnam-era assessments, and there is a clear linear connection between U.S. assessment practices in these two periods.[1]

The Vietnam War offers a breadth of resources on assessment that are unavailable from either the Iraq or Afghanistan campaigns. Arguably, the assessment processes used during the Vietnam War were the most complex and comprehensive in the history of modern COIN. Assessment analysts from Saigon to Washington, D.C., employed what were then cutting-edge computer programs to tabulate millions of reports of all kinds, including attack data, hamlet pacification data, and operational data on U.S. forces; the sheer amount of data collected in Vietnam is probably unparalleled in the history of warfare.[2] Many if not most of these reports are now available to the public online or in hard copy—often measured in thousands of linear feet of paper—at the U.S. National Archives or in any one of the other Vietnam databases and archives across the country (e.g., Texas Tech University, Cornell University). Not only are the

---

[1] This does not imply that there is a direct linear connection between the Vietnam, Iraq, and Afghanistan COIN cases. Each COIN case is necessarily unique, and there are distinct differences from one case to another. This chapter focuses on the similarities in the assessment process employed by the United States and its allies and the similarities in the COIN environment.

[2] Part of the reason for this volume of data was the sheer number of military personnel deployed in Vietnam across the peak of operations (between 1967 and 1969)—approximately five times the number of U.S. forces in Afghanistan as of late 2010. The National Archives alone house more than 9 million Vietnam War reports of various types (including official cables and memos), and these collections represent only a fraction of the overall data reported.

raw data available and relatively easy to find, but the scholarship on Vietnam-era assessment is also extensive, detailed, and typically of very high quality. Research for this study uncovered not only detailed analyses of the assessment process by key participants (e.g., Thayer, Brigham, McNamara, Komer, Race) and contemporary observers (e.g., Karnow, Lewy) but also a series of lesser known but equally insightful analyses commissioned by the U.S. government both during and after the war (e.g., Cooper et al., U.S. Defense Logistics Agency).

The Vietnam case is rich with examples and revealing analyses, but it is also a complex analytic minefield. Any research on the Vietnam War runs the risk of being either too narrow (thereby missing critical strategic context) or too shallow (missing details that explain critical decisions or complex combat environments). It is impossible to find a perfect balance or to produce a comprehensive analysis of Vietnam-era assessment in a single chapter. This chapter is therefore admittedly incomplete. It does not address all the reasons behind key decisions at any individual level, nor does it explain all the challenges—analytic, physical, or bureaucratic—faced by those tasked with reporting and assessing the war. It also does not directly address one of the most contentious assessment issues of the war: estimates of enemy strength. This issue is too complex and contested to be adequately examined here.[3]

What this chapter does show is the depth of the COIN assessment challenge. It profiles the well-meaning, at times intellectually courageous, and exhaustive efforts made to collect and analyze data and why it was so difficult to collect and report those data accurately. It describes an effort to apply mathematical processes and (to some extent) scientific rigor to centralized assessment. Indeed, there is a two-sided struggle in the centralized assessment cycle: On one side, analysts fight to obtain, collate, and understand vast reams of decontextualized data while under intense pressure from policymakers and senior military leaders to show progress; on the other side, troops in the field are tasked with reporting data that often do not exist, in formats that make little sense, for objectives they do not understand or believe in, while also under intense pressure to show progress. This chapter provides clear evidence that combat data were often erroneous or fabricated during that period and shows how efforts to analyze these data with pattern and trend analysis fell short of expectations and were ultimately not effective in supporting policy. It also examines Thomas Thayer's assertion that the data quality was "reasonable" and therefore sufficient to find "definite patterns" with which to support decisionmaking. Finally, because the Vietnam system is so similar to the Iraq and Afghanistan systems, this chapter presents a case study analysis on what was effectively modified EBA.

---

[3]   Instead, I refer to specific elements of data and strength estimates as part of broader examinations. For example, this chapter addresses estimates of VC infrastructure (VCI, the intelligence and political wing of the VC).

Orientation to Vietnam, Circa 1967–1973

A July 26, 1970, U.S. military intelligence briefing listed 44 provinces, 257 districts, 2,464 villages, 11,729 hamlets, and 1,500 miles of coastline from the demilitarized zone with North Vietnam to the border with Cambodia in the Gulf of Siam.[4] Figure 6.1 is an undated map of Vietnam showing the demilitarized zone between North and South Vietnam, surrounding geography, and the U.S. military corps sectors under MACV. Figure 6.2 depicts most of the key figures in Vietnam policy and assessment that are cited or referred to in this chapter.

**Figure 6.1
Map of Vietnam, Including Military Sector
Delineation**

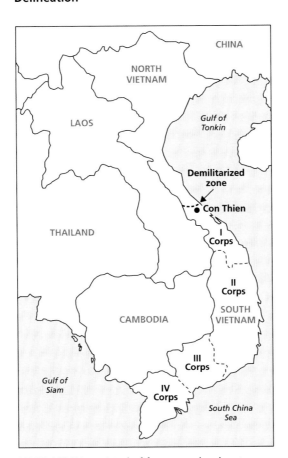

NOTE: MACV consisted of four corps-level sectors.
RAND *MG1086-6.1*

---

[4]   Sorley, 2004, p. 454. The number of villages and hamlets should have been presented as an estimated number, since it was in flux throughout the war, depending on the information available.

Figure 6.2
Photographs of Key Figures Cited or Referenced in This Chapter

SOURCE: U.S. government archival photos.
NOTE: From left to right, top row: William E. Colby, Robert W. Komer,
Creighton W. Abrams, McGeorge Bundy; bottom row: Robert S. McNamara,
William C. Westmoreland, Julian J. Ewell, Paul D. Harkins. (Thayer is not pictured.)
RAND *MG1086-6.2*

Overview of Vietnam-Era Assessment Processes

> [T]he more statistics you assemble the greater your appetite becomes for even *more* statistics. Somehow—somehow you feel that, in the *end*, you can solve the whole goddamn problem if you just had *enough* statistics.
>
> —*GEN Creighton W. Abrams*[5]

This section briefly describes the various assessment processes in use during the Vietnam War; subsequent sections explore some of these processes in detail.

Assessments of the Vietnam War varied in type, purpose, and intended consumer. The entire process changed and grew between the early 1960s and the early 1970s; there is no single "Vietnam War assessment." As in Iraq and Afghanistan, various Vietnam War assessments were published by a number of different organizations, all in different formats and based on different sources. While no single organization or entity was entirely responsible for providing campaign assessments to the President

---

[5]   Quoted in Sorley, 2004, p. 195 (emphasis in original). Abrams appears to have made this comment with some intended irony at a May 24, 1969, intelligence briefing in Saigon. However, he went on to state that the statistics were helpful.

and his advisors, two groups played key roles in the process. The Military Assistance Command, Vietnam, or MACV, was the theater-level military command in Saigon, South Vietnam. MACV was the neck of the funnel for nearly all field reports on operations, intelligence, pacification, and other data categories. MACV commanding generals including GEN Paul D. Harkins, GEN William C. Westmoreland, and GEN Creighton W. Abrams wrote formal assessments for the President based on the work of their staffs, accumulated data, and their personal perspectives on the war.[6] MACV both collated and filtered data for submission to the Office of the Secretary of Defense (OSD) and wrote its own reports, including detailed accounts on the development of the Army of the Republic of Vietnam (ARVN).[7]

The other key assessment group outside the intelligence analysis community was the Office of Systems Analysis (OSA) in the Pentagon. Thomas C. Thayer, whose work is examined later in this chapter, was hired by Alain C. Enthoven to work on Vietnam issues in OSA.[8] In an effort to improve insight into force deployments, Thayer took it upon himself to put together a comprehensive scientific review of the available data and publish reports based on his analyses.[9] His office became an assessment center that applied systems analysis methodology to COIN campaign assessment. Thayer points out that he never had the opportunity to task data collection to the field, but others in OSD, including Secretary of Defense Robert McNamara, had that authority and exercised it often. Thayer's work was published separately in what were labeled "unofficial" reports, while the data he amassed were used to feed other reports in various sections of OSD.

Some of the MACV reports and nearly all of the OSD reports relied heavily on aggregated quantitative data and pattern and trend analysis. In some cases, data collection requirements were developed to meet perceived operational or strategic needs; in other cases, they were specifically designed to provide data that would show some kind of progress without context. For example, in 1968, MACV reported the number of cakes of soap it had issued to Vietnamese villagers in 1967 (572,121), an irrelevant input metric.[10] Data flowed up from the hundreds of thousands of troops on the ground, province advisors, military advisors to Republic of Vietnam units, U.S. civilian officials, and U.S. intelligence officers, as well as Republic of Vietnam military units, government agencies, and civilian development teams. These data were then fed into catalogs and computer databases, including the Hamlet Evaluation System (HES), the

---

[6] See, for example, Westmoreland, 1967a. The *Measurement of Progress* report was the most consistently published MACV theater-level holistic assessment.

[7] MACV, 1969c.

[8] Rehm, 1985, p. 4-3.

[9] Rehm, 1985, p. 4-3.

[10] Lewy, 1978, p. 93. It was irrelevant to campaign assessment, though it might have been relevant to the officer in charge of the budget.

Terrorist Incident Reporting System, the Territorial Forces Effectiveness System, the Pacification Attitude Analysis System (PAAS), the Situation Reports Army File, and many others.[11] "Data" could mean anything from a simple number (e.g., rounds fired in a single artillery attack) to a more complex set of ostensibly correlated survey data (e.g., hamlet evaluation data).

Often overlooked are the thousands of detailed narrative assessments written by military and civilian officers working in the field.[12] These assessment reports, containing quantitative and qualitative data as well as analysis, are probably the best sources of contextual reporting on the Vietnam War. They describe in depth the successes and failures of the Americans and the Vietnamese (both North and South), potentially provide rich context, and serve as de facto histories of the war at the district and province levels. These reports suffered from the same kind of subjectivity as any field report, and most were necessarily focused in scope; thus, they do not provide comprehensive, theater-wide assessments of the entire U.S. COIN campaign in Vietnam. Despite the fact that MACV relied on narrative province reports as one of two "primary documents upon which the periodic theater-wide assessment of overall progress in the pacification effort is made" (a claim that is not reflected in the campaign assessment), a 1969 MACV order limited province reports to four pages for easier reading by senior leaders.[13]

The challenges of assessment in Vietnam are the same challenges faced in any COIN campaign, including Iraq and Afghanistan: How should policymakers determine progress and decide strategy if all they have to choose from is inaccurate, decontextualized, and aggregated numbers or thousands of pages of lengthy narrative? In Vietnam, theater-level reports produced by military and military intelligence officers tried to bridge the gap between these two options, as did those produced by various government and intelligence agencies in Washington, D.C. While some of these

---

[11] Some of these are reporting categories, others are databases, and not all were necessarily fully computerized. Thayer describes some of these databases and systems, but many others were available to him during his service with OSA. He did not necessarily control all of these databases, and some may have been compiled after his departure. The Cornell University Library and the U.S. National Archives now hold many of these reports. *Introduction to the Pacification Data Bank*, a government pamphlet released in 1960, lists and describes the automated pacification databases that existed in November 1969. They included HES, the Territorial Forces Effectiveness System, the Terrorist Incident Reporting System, the Revolutionary Development Cadre System, the Assistance in Kind System, the Refugee System, the Village and Hamlet Radio System, the People's Self Defense Force System, and the VCI Neutralization System, the last of which was crossed out in the document. The Pacification Data Bank itself was designed to provide "the ability to construct reports to meet the management and analysis requirements of individual users" (*Introduction to the Pacification Data Bank*, 1969, p. 1). The list is not exhaustive; there were a number of other automated databases in use that were not related to pacification but also fed campaign analysis at OSA. The Southeast Asia Province File, Project Corona Harvest, the System for Evaluating the Effectiveness of [South Vietnamese] Subsystems, and the Air Summary Data Base are all available through the Cornell University Library in Collection Number 4406.

[12] See, for example, MACV, 1968a.

[13] MACV, 1969b, Annex A, p. 7.

reports were useful and many were well written, they often contained the flaws by virtue of their overaggregated data analyses or lengthy narratives.

Douglas Kinnard's 1974 survey of 110 U.S. general officers who had been involved with the Vietnam War showed that only 2 percent felt that the system used to measure the war was valid.[14] This is a resounding dismissal from those who used and fed the assessment system throughout the war. Their viewpoint is by no means an empirical refutation of the MACV or OSD assessments, but it shows that Vietnam-era COIN assessment was not adequate. As the remainder of this chapter shows, the flaws in the system were pervasive. The application of this flawed system had dire consequences for the personnel in the field and for U.S. national security.

Thomas C. Thayer and the Southeast Asia Analysis Reports

An examination of the Vietnam-era assessment process should ideally begin with an overview of the assessment process at OSA. As director of OSA's Southeast Asia Intelligence and Force Effectiveness Division, Thomas C. Thayer led the most comprehensive analyses of COIN metrics to date. This section explains Thayer's role in the Vietnam-era assessment process, the problems Thayer faced in terms of data accuracy, and Thayer's proposition that "reasonably accurate" data are sufficient to find definite patterns and trends in a COIN campaign.

Thayer's reports, and his analysis of his own work in *War Without Fronts*, are perhaps the best and most accessible insights into the nuts and bolts of COIN assessment. Upon joining OSA, Thayer already had credibility as an analyst, having served three years in the field in Vietnam as a civilian (albeit on specific technical projects and operations research initiatives).[15] During the war, he published his analyses in more than 50 editions of the *Southeast Asia Analysis Report*, later republished in a 12-volume series titled *A Systems Analysis View of the Vietnam War 1965–1972*.[16] Although his reports were ostensibly unofficial, there is little doubt that they were taken by most readers as official reports issued by OSD.[17] Thayer himself lists the accolades he received from consumers of his reports, including the White House, DoS, and MACV. Because Thayer's reports were then the only serious effort to aggregate all the available and hard-to-obtain war reporting, and because Thayer's analyses were well written and interesting, the Southeast Asia Analysis Reports may have been as influential as those issued by MACV or the Central Intelligence Agency (CIA)—and perhaps even

---

[14] See Kinnard, 1977, p. 172.

[15] Rehm, 1985, p. 4-1.

[16] See Thayer, 1975a, 1975b, and 1975c, for volumes referenced for this study.

[17] Thayer (1985, p. 261) acknowledges this concern.

more so.[18] With their COIN assessments, Thayer and his analysts shaped decision-making on the Vietnam War, and Thayer himself could easily be considered the most influential operational and strategic assessment analyst in the history of U.S. COIN.

Limitations of Thayer's Data

To understand how Thayer used the data he collected, it is important to first understand the quality of the data in question. Throughout *War Without Fronts* and in his other publications and public discussions, he reaffirms that the data were often of very poor quality. While he has some biases as the senior analyst and he can never know the full degree of data inaccuracies or gaps, he is remarkably forthright in showing the errors that he knew existed. It is possible that the actual inaccuracies in these data were much more or less problematic than Thayer's reports. The following quotes from *War Without Fronts* address the quality of specific data types encountered by Thayer and his team of analysts.

Counting Insurgents. "The estimators [of enemy force data] were aware of the great uncertainties in their estimates and did their best to furnish a reasonably accurate picture of the communist forces" (p. 30). The postulated combined estimate of total forces ranged between 395,025 and 482,452 (p. 31). Palmer concurs that estimates of combined enemy force data—NVA and VC—were generally "best guesses" (Palmer, 1984, p. 79)

Communist Attacks. "The most serious problem in dealing with the official U.S. figures is that they do not include communist actions reported by the Vietnamese National Police and other civilian authorities" (p. 44). Regarding ground assaults: "The figures probably understate the actual rate at which ground attacks took place because communist attacks in reaction to allied operations were seldom, if ever, included in the data" (p. 46).

Anti-Aircraft Incidents. "[W]hile the number of air sorties tripled the number of anti-aircraft incidents fell from 6,800 in 1971 to 800 in 1972. Such a drastic change in the face of intense combat suggests that the pilots simply stopped reporting anti-aircraft fire, not that the fire itself had stopped" (p. 51). Thayer's speculation may be accurate, but in the absence of field investigation, there is no way to determine why these numbers dropped or whether they were accurate in the first place.

ARVN Leadership. "American advisors' assessments gave the impression that all of the ARVN division commanders were capable. Experienced observers disagreed" (p. 62). If this was commonly the case, as Thayer asserts, then these specific data as used to determine host-nation military readiness for the period in question are flawed.

---

[18] Thayer is quoted describing how he managed to obtain the data for his reports in Rehm, 1985, pp. 4-2–4-3. He essentially found the data with the assistance of his analysts; there was no systematic process to funnel data to Thayer's office, no order from on high granting him access to hidden sources, and no comprehensive data collection plan.

Civilian Casualties. "[N]o official U.S. estimate of civilian casualties exists for the Vietnam War. [A]n estimate is developed here and compared with another made by the U.S. Senate Subcommittee on Refugees but both estimates are unofficial guesses" (p. 125).

Hamlet Security and Development Data (pre-HES) 1964–1967. "The data support the notion of an optimistic bias" (p. 138). "The method of reporting also suggests that the population in communist controlled areas was understated because the pre-1967 joint system counted only the hamlets planned for pacification. It ignored the ones not in the pacification plan, most of which were probably communist hamlets" (p. 143). Therefore, the 1964–1967 data described only hamlets in friendly control. This statement also assumes a great deal about those hamlets not covered by this process.

Hamlet Security and Development Data (HES). "In terms of percentages, the gain in [GVN control over the population] is from 42 percent of the total population in 1964 to 93 percent in 1972" (p. 141). This shows that in the year in which the United States essentially declared defeat and withdrew from Vietnam, the notoriously ineffective and corrupt GVN *controlled* 93 percent of the South Vietnamese population to some degree. This unlikely statistic is based on what are questionable data (as discussed later in this chapter), and, as Race points out, control is irrelevant in the absence of willing support. Thayer also quotes a hamlet-level poll showing that only "54% of the respondents generally agreed with the HES description of their hamlet" (p. 151). On January 16, 1971, a senior officer at a MACV intelligence briefing referred to this disparity between HES and the polling data, stating that this "means our HES rating is optimistic—in the view of the people" (quoted in Sorley, 2004, p. 523). By the end of 1972, there were only about 35,000 U.S. troops in Vietnam, so most of this information must have come from GVN and ARVN reporting, which was typically less reliable than U.S. reporting.[19]

The People's Self-Defense Forces Rosters. "The figures are notoriously unreliable" (p. 170). The data purported to show how many members of the People's Self-Defense Forces, a local civil defense organization, were on the active roll on a yearly basis.

PAAS (a survey of popular sentiment). "The PAAS attempted to portray urban and rural South Vietnamese attitudes toward security, politics and economic development. . . . Any systematic effort to portray attitudes and beliefs is subject to error . . . and the conditions in South Vietnam further limited the accuracy." Because it relied on semistructured interviews and not standard polling methods, "the results must be viewed as being much less precise" than those of opinion polls conducted in the United States. "Quota rather than probability sampling techniques were used to select the hamlets and the individual respondents, so the sample from which interviews were drawn was not necessarily an accurate representation of the South Vietnamese Popu-

---

[19] The purpose of these quotes is to identify data flaws only.

lation" (pp. 174–176). Thayer uses the remainder of this chapter of his book to draw fairly comprehensive conclusions from the PAAS data. In March 1970, MACV's Civil Operations and Revolutionary Development Support (CORDS) director William Colby stated, "There is no fully reliable system for assessing overall public attitudes in Vietnam so most judgments are only estimates" (U.S. Senate Foreign Relations Committee, 1970, p. 417).

Estimating VC Infrastructure (clandestine agents). American officials "attempted to estimate the numbers and types of clandestine communists by adopting the techniques used by police everywhere to compile lists of persons wanted for crimes . . . [and] communist subversive presence." This information "can be obtained from questions found in the HES and the PAAS, although these estimates are also tenuous" (pp. 205–206). Estimates of VCI were tenuous and used methods that were not clearly articulated. Furthermore, it is not clear that every estimate of VCI from all U.S. sources relied on standard police techniques; poorly informed guesswork probably fed these statistics to a significant degree. These estimates, however, were used to measure success based on a body count formula or, according to Thayer, "VCI neutralization." A September 12, 1970, estimate by MACV put the number of VCI personnel at 65,000, of which 37,000 had been "identified by name and position."[20] This number did not include "guerrilla or low-level supporters." So, assuming that analysts were actually able to identify 37,000 individuals as clandestine agents—in and of itself a staggering intelligence coup—the total number was still a gross estimate of only part of the VCI. This did not stop MACV from setting VCI "neutralization" objectives; in one case, the objective was to kill or capture 1,800 VCI personnel per month. It is not clear how this number was agreed upon.[21]

Refugee Statistics. The statistics were designed to "identify numbers of individuals to whom payments were due, not to count all refugees and war victims in South Vietnam." However, "outside observers believed they represented the total number of refugees. . . . If [the refugees] didn't register [with the government] they didn't show up in the statistics" (pp. 221, 225). "Despite their lack of precision, the refugee figures are fairly reliable in indicating the magnitude of the problem and they did fluctuate with the tempo of combat. When combat increased, so did the number of refugees. When it decreased the number of refugees did too" (p. 222). Thayer's caveats are straightforward and helpful in understanding the statistics. However, it may or may not be true that the *actual* number of refugees rose and fell with combat. While this statistic may reflect a broad trend, it might instead reflect increased efforts by GVN and U.S. refugee assistance program officers to travel to areas that had recently experienced fighting

---

[20] Sorley, 2004, p. 478. This estimate is taken from the MACV Weekly Intelligence Estimate Update.

[21] Sorley, 2004, p. 303. Colby states that the primary objective was to capture, not kill, VCI personnel (Sorley, 2004, p. 381).

to register new refugees.[22] This would reflect an artificial bump in the data created by proactive registration and would not reflect the existence of refugees in areas that had no registration officials. Because refugees could not register where there were no GVN officials, NGOs (e.g., Red Cross), or U.S. officials, no refugees in communist-held or in hotly contested areas could be counted.[23]

Refugees also may flee when faced with threats or intimidation. How many refugees were created when U.S. forces ceded areas to communist control, as often happened? How many refugees did the communists displace in their own areas, thereby losing popular support? How many people registered for refugee payments because they needed money and were not, in fact, refugees?[24] Were GVN statistics—the primary source of refugee data, according to former Ambassador Robert W. Komer—accurate? Were the United States and GVN defining refugees the same way? These relevant inputs are unknown, but Komer doubted the accuracy of the GVN data.[25] The CIA described a ghost population of unregistered refugees who did not register "to circumvent local government policies which exclude them from certain areas."[26] William Colby described the magnitude of the data gap at a briefing with the MACV senior staff: "We estimate something between three and four million people have been in refugee status in this country sometime in the past three years."[27] Further, it seems incongruous to state that the statistic was not designed to identify all refugees in the war zone and then to state that the level of combat had a clear impact on the rise and fall of the "number of refugees." It would be safe to say that *more fighting probably causes more refugees in a specific area for a specific period of time*, but the data do not support any broader conclusions.

South Vietnamese Military Casualties After 1972. According to Thayer, the low number of battle deaths reported by the South Vietnamese was used by Congress "as

---

[22] Field reporting shows that this is exactly what transpired in at least some cases. For example, in Phong Dinh Province in early 1968, the GVN surged refugee services to meet an increase in refugees after intense fighting in the provincial capital. The official numbers did not include an estimate of an additional 4,000 refugees who had not registered (MACV, 1968a, pp. 12–13).

[23] While there may be exceptions to this assumption, they are probably few and statistically insignificant.

[24] Refugee and damage claims could be lucrative. A Vietnamese citizen could claim *solatia*, or compensation payments of thousands of dollars, and could pursue claims against both the United States and GVN simultaneously for the same incident. The U.S. government paid more than $4.7 million in claims as of March 1970 (U.S. Senate Foreign Relations Committee, 1970, p. 420).

[25] Komer, 1967, p. 12.

[26] CIA, 1966, p. VII-2. This declassified report also states that there may have been between 30,000 and 40,000 refugees living in only one of nine Saigon precincts, while GVN statistics listed only 1,518 in the entire city.

[27] Sorley, 2004, p. 459. Colby presented this 3–4 million figure on July 26, 1970. But in his congressional testimony in March of that year, he stated, "About 3,500,000 people have been uprooted during the past six years" (U.S. Senate Foreign Relations Committee, 1970, p. 420). The latter estimate seems much more accurate than the former by virtue of its presentation as a more precise number. This disparity exemplifies the casual way in which statistics are misrepresented—even unintentionally—in official briefings.

part of the rationale for slashing aid to South Vietnamese forces during the summer of 1974. The problem was that the official South Vietnamese figures for battle deaths [for 1974] turned out to be twice as high as the figures reported to Washington in the operational messages" (p. 256). This comment highlights the danger of depending on host-nation reporting to determine policy.

The Concept of "Reasonable" Accuracy

Thayer believed that field reporting in Vietnam was significantly inaccurate, and that many of the aggregated data sets that he worked with were also significantly inaccurate. Yet, like some of his contemporaries and many modern assessment analysts, he believed that the data were "reasonably" accurate and could be used to show useful patterns and trends. He presented these patterns and trends in his reports, along with the data he collected from MACV and other sources. The idea that useful patterns and trends can be gleaned from aggregated data that are known to be both significantly inaccurate and incomplete drives to the heart of the debate over centralized assessment.

It is helpful to understand the depth of the chasm between "accurate" and "reasonable" before examining the "good-enough" proposition, or the idea that useful patterns and trends can be identified through centralized assessment. The following two sections of this chapter build upon Thayer's descriptions of data quality to provide in-depth analyses of two specific data sets used by Thayer and other Vietnam analysts: body counts and hamlet pacification evaluations. The subsequent section, "Pattern and Trend Analysis of Vietnam War Data," addresses the idea of pattern and trend analysis in holistic assessment. The chapter concludes with a discussion of how trend analysis was used to guide strategy and to try to shape popular opinion on the war and whether lessons from Vietnam are relevant to modern COIN campaign assessment in Iraq and Afghanistan.

Body Counts

> [They were the] bane of my existence and just about got me fired as a division commander. They were gross exaggerations by many units primarily because of the incredible interest shown by people like McNamara and Westmoreland. I shudder to think how many of our soldiers were killed on a body-counting mission—what a waste.
>
> The immensity of the false reporting is a blot on the honor of the Army.
>
> A fake—totally worthless.
>
>   *—U.S. general officers' opinions of body-count data*[28]

---

[28] Quotes from Kinnard, 1977, p. 75. The first comment was in response to Kinnard's survey of U.S. general officers in 1974.

This section addresses the body count metric in the context of the Vietnam War. The purpose of this section is to (1) reveal flaws in this set of aggregated data, (2) provide a detailed example of data collection and reporting problems in COIN, and (3) build a foundation for the later discussion of data analysis and strategic decisionmaking. *Body count* was a term for the officially mandated tallying of the bodies of enemy combatants killed by U.S. and GVN military forces. This assessment method had been used in conventional contests, including World War II and Korea; in Vietnam, its implementation is attributed to Secretary of Defense McNamara, General Westmoreland, and Westmoreland's predecessor, GEN Paul D. Harkins.[29]

The previous section described Thomas C. Thayer's caveats of Vietnam-era aggregated data. In *War Without Fronts*, Thayer also expressed doubt regarding the veracity of estimates of enemy dead coming from the field: "It is doubtful whether anybody, including Hanoi [Democratic Republic of Vietnam], really knows how many communist troops died."[30] Thayer goes on to detail why body count estimates from the field were so inaccurate: the communists' emphasis on reclaiming bodies, difficult terrain, duplicate reporting, and the temptation to exaggerate, among other reasons. He then describes in detail the unsuccessful methods that his assessment staff used to try to compensate for the problems with body count data and time-series analysis. "Considerable effort was made to check the validity of the communist loss estimates but the results were not conclusive."[31] Just how accurate were the body-count statistics? How were these data used? There is both anecdotal and official evidence from the tactical to the strategic level that body-count reporting was taken seriously and that it shaped both strategy and assessment, for better or worse. LTC Douglas S. Smith, a mechanized infantry commander in Vietnam, had his reservations about body counts and weapon seizure statistics but thought them accurate enough to be a useful indicator:

> [Before I arrived, one of our units'] body count for VC . . . was 103 in a six-month period. The U.S. losses during this time frame was 36 which was about a 3:1 ratio. . . . [In] the past almost six months now that I've had the battalion, we see almost a reversal. We see our body count in the same period of time nearing 750 and our losses down in the vicinity of 25 killed. So our casualties have gone down and the number of VC eliminated has gone up tremendously.[32]

It is impossible to know whether or not Smith's information was accurate, as he claims, down to single digits; he does not describe how he checked the accuracy of information from his subordinate commanders, and there is no record to show whether

---

[29] The process started with Harkins (Sheehan, 1988, p. 287).

[30] Thayer, 1985, p. 101.

[31] Thayer, 1985, p. 102.

[32] Smith, 1969.

his data were checked by a higher-level unit. Body counts figured prominently in official unit reporting and were typically used as the primary gauge of success in both combat operations promotions.[33] Most anecdotal and much of the official reporting, however, seems to show that body count data were inaccurate or created from whole cloth. Official Army history shows that the MACV intelligence staff filed a series of formal complaints to MACV that claimed military staffs were underestimating enemy strength and overstating enemy losses.[34] It is not clear that senior staff and officers really had any idea how accurate the numbers were from any source at any one time.[35] A transcript of an intelligence briefing delivered to GEN Creighton W. Abrams (Westmoreland's replacement as the senior officer in Vietnam) on March 5, 1971, reveals this uncertainty at the top of the military intelligence and operations hierarchy in Vietnam. In response to Abrams's question regarding the veracity of ARVN body count reporting, the briefer states that "it is probably a physical impossibility for them to tabulate the bodies, especially after B-52 strikes. . . . But at the same time, we feel that the number from air and artillery and so forth that are never uncovered probably make up the difference."[36] When asked whether the ARVN estimates were accurate to *within an order of magnitude*, he responded, "I feel so."[37] An official CIA analysis of the body count provided by U.S. and GVN military forces in the wake of the Tet Offensive called the estimate "exceedingly difficult to accept."[38]

Guenter Lewy describes a survey of former ground commanders at the Army War College in 1968. He states that 60 percent of the officers thought that body counts were based on a "combination of body counting and estimates" and that "body counts were usually 'upped,' sometimes honestly and sometimes with great license." Some of these officers reported pressure to turn in higher counts and that "this pressure was

---

[33] See, for example, Sykes, undated.

[34] Cosmos, 2006, p. 86. As stated earlier in this chapter, estimates of enemy strength were the most contested elements of campaign and intelligence assessments throughout the war. Westmoreland sued CBS News for broadcasting a report that accused him and MACV staff of falsifying estimates to sustain the impression that the United States was succeeding in its attrition warfare strategy. Testimony and evidence from the case are revealing and provide good insight into the issue of estimates. The RITZ and CORRAL programs, the testimony of former Marine and then-Congressman Paul N. McCloskey, and the letters of MACV intelligence officer James A. Meacham are particularly informative. See *Westmoreland v. CBS*, 1984.

[35] There is considerable evidence that the numbers were sometimes intentionally falsified, or known errors were papered over. Sorley (2004, pp. 221–222) includes an example of a military intelligence briefer preparing to deliver a brief on enemy strength who talks about how to "explain away" a gap of 72,000 enemy forces between varying estimates. Another briefer states that the number is "too big to cry, and it hurts too much to laugh."

[36] Sorley, 2004, p. 555 (emphasis added).

[37] Sorley, 2004, p. 555.

[38] CIA, 1968, p. 4. The report presented a detailed analysis of available reporting and compared the body count to estimates of enemy forces, which were also suspect. The more salient points in this report address the ratio of civilian casualties to those of VC irregular or cadre forces. It asserts that many of the bodies that were physically counted may have been civilians.

especially pronounced when the actual score was in the enemy's favor."[39] One division commander, Julian J. Ewell, put pressure on his subordinate units to report high body counts. He allegedly set up a quota system and threatened commanders with relief of command if they did not meet their quotas.[40] This pressure to rack up high body counts may have resulted in the deaths of civilians or at least encouraged the counting of dead civilians to achieve higher "scores" during Operation Speedy Express in the Mekong Delta in early 1969.[41] Ewell described his firm belief in the body count system in a discussion in Saigon with General Abrams, commanding general of MACV, on June 21, 1969. In the quote below, he is responding to Abrams' statement that "going out there and killing a few of them is not, in my opinion, going to have the effect [sic]. I think they're [the VC combatants] all sort of fatalists."[42]

> Well, I don't agree with you, General. You can get a sapper [combat engineer] unit mining the road, and you kill two or three and they'll knock it off. It may be that a month later they'll come back. These people can *count*. And, boy, when you line them up [bodies] and they count one, two, three, four, their enthusiasm is highly reduced. That's the way we opened up Highway 4—just killing them. It doesn't take many.[43]

Other anecdotal evidence calls into question the accuracy of Vietnam-era body count statistics. A U.S. Army advisor to the ARVN in Phong Dinh Province in 1967 observed an ARVN unit to which he was assigned fabricating a body count and probably a weapon count after engaging with a VC unit in rough terrain. While this single anecdote is not reflective of all body count reporting in Vietnam, it is similar to most other anecdotal evidence on the veracity of body counts:

> We got into a fight, boxed in this group of VC and they tried to duck out through the perimeter. [We] killed a few of them. The senior advisor said, "Verify the body count." No way! The fight was all over the place in rough terrain; we just couldn't count them. The best gauge [of success] was the weapons that we had recovered, but not all VC carried weapons. Some were specialists that carried ammunition or

---

[39] Lewy, 1978, p. 81.

[40] Lewy, 1978, p. 81.

[41] These accusations remain contested, and Ewell denied them until he died in 2009. Evidence supporting the accusations is anecdotal and circumstantial. For example, the unit reported 10,899 enemy dead and 748 weapons seized during the same period. It is possible that inflated body counts provided during this operation backfired in that they exaggerated the discrepancy between the body count and recovered weapons.

[42] Sorley, 2004, p. 213. Abrams responded, "One last thing you ought to consider when you handle this is how it's going to look when the *New York Times* and *Newsweek* and some of those describe it to the American people. I'm not quite ready to take that one on. They might flavor it. They might lose their objectivity."

[43] Sorley, 2004, p. 213 (emphasis in original).

other materiel. And the ARVN had no interest in body counts. They were paid to recover weapons, so they were motivated by different reporting criteria.[44]

Gibson relates several anecdotes that describe the falsification and extreme inflation (ten times higher in one case) of body counts by combat units.[45] For example, one officer claimed, "If you come across dead bodies, you count the dead bodies. You re-sweep the area, recount the numbers, *double it*, and call it on in."[46] Alain Enthoven and K. Wayne Smith eviscerate the body count process: "Errors could and did frequently creep in through double counting, counting civilians . . . or counting graves, or through ignoring the rules because of the pressures to exaggerate enemy losses or the hazards of trying to count bodies while the enemy was still in the area."[47] They explain how the body count system encouraged inflation, exacerbating an already difficult problem:

> The incentives for field commanders clearly lay in the direction of claiming a high body count. Padded claims kept everyone happy; there were no penalties for overstating enemy losses, but an understatement could lead to sharp questions as to why U.S. casualties were so high compared with the results achieved. Few commanders were bold enough to volunteer the information that they had lost as many men in an engagement as the enemy—or more. The net result of all this was that *statistics regarding body counts were notoriously unreliable.*[48]

Enthoven and Smith go on to state that General Westmoreland (the senior officer in Vietnam at the time) claimed to know the body count to within 1.8-percent accuracy based on captured enemy documents but that OSA thought the same documents were 30-percent understated.[49] Contemporary COIN doctrine has resolved this question somewhat, uncovering the critical flaw in the use of aggregated body count

---

[44] Former U.S. Army psychological operations advisor to Phong Dinh Province, Vietnam, interview with the author, Washington, D.C., April 19, 2010. Note that he states that the ARVN unit still provided a body count.

[45] Gibson (2000, pp. 125, 127) quotes first-hand accounts from the field: "I know of one unit that lost 18 men killed in an ambush and reported 131 enemy body count. I was on the ground at the tail end of the battle and I saw five enemy bodies. I doubt if there were many more." And "in counting, a weapon captured is counted as five bodies. In other words, if you shoot a guy who's got a gun and you get that gun, you've shot six people."

[46] Gibson, 2000, p. 127 (emphasis in original).

[47] Enthoven and Smith, 1971/2005, pp. 295–296. They add,

> Off-the-record interviews with officers who had been a part of the process revealed a consistent, almost universal pattern: in a representative case, battalions raised the figures coming from the companies, and brigades raised the figures coming in from the battalions. In addition, something had to be (and was) put in for all the artillery and air support, which the men on the ground could not check out, to give the supporting arms their share of the "kill." (p. 298)

[48] Enthoven and Smith, 1971/2005, p. 295 (emphasis added).

[49] Enthoven and Smith, 1971/2005, p. 297.

metrics. FM 3-24 first states, "Body count can be a partial, effective indicator only when adversaries and their identities can be verified." This might be true if the data were not aggregated or if they were used to track the decimation of a specific, known unit (usually unlikely in COIN). More importantly, it points out that if body counts are used to measure attrition of an enemy, one must know the total enemy strength for the numbers to have any meaning.[50] Thayer makes it clear that estimates of total enemy strength were estimates and that estimates of irregular forces were not accurate. Without accurate data in both data sets, the entire premise of employing an aggregated body count metric to determine attrition unravels. This approach would also assume that the North Vietnamese had a casualty threshold, an idea that is explored later in this chapter.

GEN Bruce Palmer, Jr., had extensive experience in Vietnam dating back to the 1950s and served as General Westmoreland's deputy for part of the war. He believed that counting enemy dead encouraged exaggerated claims but that "higher headquarters, nonetheless, have ways to judge the validity of unit claims."[51] While some reports were undoubtedly questioned from above, there is no indication in Palmer's book, in Thayer's work, or in any of the official records reviewed for this study that there was any comprehensive or systematic effort to correct inflated body counts. As a retired general with extensive Vietnam experience, Palmer's voice is relevant. However, 61 percent of the generals who responded to Kinnard's 1974 survey believed that body count reports were "often inflated."[52] And, as stated earlier, Thayer admitted that his analytic efforts to rectify body counts were inconclusive.

The body count metric shaped behavior and put the lives of U.S. personnel at risk.[53] Body counts in several U.S. military units, including Ewell's, were informally but implicitly tied to promotions and assignments through the use of published scoring systems. For example, Ewell described a "highly skilled" U.S. military unit as having a 50:1 ratio of enemy to friendly casualties and an "average" unit having a 10:1 ratio.

---

[50] HQDA, 2006c, p. 5-27. My professional experience, both in Iraq as a senior intelligence analyst and working with senior analytic staffs in the United States tasked with estimating the number of insurgents in Iraq, suggests that these numbers either were pure guesses or reflected inappropriate, haphazard, or debunked methodology. In one case, a staff attempted to count insurgents using a method that U.S. park rangers employ to count wild deer. There is no method that can be employed to accurately count something that is almost impossible to define at the individual level (in the absence of uniformed cadre) and usually only reveals itself by choice. Intelligence activities can uncover specific insurgents and elements of an insurgent cadre, but intelligence estimates of overall insurgent strength are rarely (if ever) derived from comprehensive and accurate data. This finding also stems from previous RAND research conducted for *How Insurgencies End* (Connable and Libicki, 2010). Order-of-battle assessments of conventional units are a separate matter, and certainly a complicating factor in a case like Vietnam. For an example of a declassified Vietnam-era intelligence order-of-battle summary, see MACV, 1972.

[51] Palmer, 1984, p. 165.

[52] Kinnard, 1977, p. 172.

[53] In some cases, the use of published metrics lists may have encouraged risk aversion. According to Gibson, points were deducted for friendly casualties in the 25th Infantry Division's "Best of the Pack" contest.

The 25th Infantry Division held a running "Best of the Pack" contest that awarded points for "possible body counts," and the 503rd Infantry Regiment had a military performance indicator chart that awarded points for body counts and captured prisoners, and deducted points for various problems (e.g., cases of malaria).[54] Because body counts were informally but clearly tied to promotions, the desire to raise the body count may have led some officers to conduct unnecessary missions. There is no clear way to tell whether any specific mission was conducted to attrite the enemy—a valid if misguided objective in an attrition warfare strategy—or to simply "up the score." This is a subtle yet critical difference between these two types of missions: The former might be conducted thoughtfully and within the scope of larger strategic objectives, while the latter are often conducted haphazardly and carry risk that far outweighs any potential military benefit. Gibson presents a number of anecdotes describing how troops were wounded and killed on missions designed to raise body counts or simply to count existing bodies.[55] In a self-critical evaluation of its tactics, the U.S. Army published a primer on lessons learned written by veteran war historian and retired Army BG S. L. A. Marshall. In it, Marshall and his co-author, David Hackworth, describe several incidents in which lives were needlessly endangered or lost in the hunt for enemy bodies. For example,

> A U.S. rifle company in a good defensive position atop a ridge is taking steady toll of an NVA force attacking up hill. The skipper [company commander] sends a four-man patrol to police weapons and count bodies. Three men return bearing the fourth, who was wounded before the job was well started. Another patrol is sent. The same thing happens. The skipper says, "Oh, to hell with it!"[56]

It probably was not General Westmoreland's, Secretary McNamara's, nor President Johnson's intention to cause additional or unnecessary risk to U.S. troops to count enemy dead. But their undue emphasis on this assessment information *requirement*—like many other requirements for information in COIN—became a military objective unto itself. In the end, it failed to provide accurate data, failed to accurately or usefully inform the decisionmaking process, and cost the lives of an untold number of troops. The Vietnam-era body count metric is a cautionary tale not only for those who might apply attrition warfare to COIN but also to those who might apply centralized metrics without consideration for the risk and the unwanted behavior they might induce in the field. Putting aside the ultimate sacrifice that some men may have paid to feed the body count system, it certainly is not conducive to military good order and discipline—

---

[54] Gibson, 2000, pp. 113–115. Gibson's chart on p. 115 indicates the 503rd Infantry Division, but it should probably read 503rd Infantry Regiment.

[55] Gibson, 2000, pp. 112–122.

[56] Marshall, 1966, p. 33.

or to civil-military relations—to establish an assessment system that encourages institutionalized lying.

This section called into question the use of directed collection requirements for data that are difficult or dangerous to collect and likely to be tied to performance ratings. The following section presents an examination of hamlet pacification metrics.

The Hamlet Evaluation System

This section considers the Vietnam-era HES and places it in context with modern assessment methods. This examination is important for a number of reasons. The HES is arguably the de facto gold standard for assessment, at least in the community of experts interested in assessment. It is referenced as the only truly comprehensive and structured effort to assess population security in a COIN campaign.[57] It was also intended as a structured endeavor that employed polling techniques, computer-assisted analysis, time-series analysis, and correlation analysis. Therefore, it offers excellent insight into the application of scientific process in COIN assessment.

Hamlet Evaluation System Concept and Execution

> Perhaps the best single measure of pacification is the extent to which the population has been brought under Government control and protection. To provide a more valid standard of measurement, we have developed a device called the hamlet evaluation system.
>
> *—Secretary of Defense Robert S. McNamara*[58]

> We've been using [HES] and defending it, over the years. We've emphasized that we don't think it's a precise thermometer for the situation, but it's been a *very* handy tool. It's given us an idea of differences over time and it's given us an idea of differences over space.
>
> *—William E. Colby, CORDS director*[59]

While the HES may be the gold standard for quantitative COIN assessment, it is also emblematic of the failures of centralized assessment in COIN (despite Colby's arguments to the contrary). The CIA developed the HES in 1966 and DoD implemented it in 1967 as part of the Pacification Evaluation System (PACES), a "fully automated

---

[57] See, for example, Gayvert, 2010.

[58] McNamara, 1968, p. 266.

[59] Sorley, 2004, p. 367 (emphasis in original). Colby made these comments at a February 11, 1970, briefing to then–Secretary of Defense Melvin R. Laird and then–Chairman of the Joint Chiefs of Staff Earl G. Wheeler.

system" designed to determine who controlled the population of Vietnam.[60] The CIA and DoD created PACES to measure pacification programs under MACV's CORDS program.

The establishment of the HES mandated the collection of standardized security and development survey data on approximately 11,000–13,000 Vietnamese hamlets. A hamlet is a community that is smaller than a village; many hamlets made up one village in Vietnam, a district contained many villages, and every province contained several districts. Hamlets varied in size dramatically, from approximately 50 people to as many as 20,000.[61] The HES aggregated the survey data describing these hamlets into centralized computer databases and then presented them as quantitative outputs. These outputs were then used to generate a number of different types of analyses and analytic reports.[62] As mentioned earlier, the system is perhaps the single most structured and thoroughly implemented assessment collection system in the history of COIN, but it is also remarkable in that it closely reflects contemporary effects-based theory and EBA. Indeed, the HES has been suggested as a possible model for Afghanistan,[63] and the Hamlet Evaluation Worksheet (survey sheet) guidelines are remarkably similar to the rating guidelines used to develop the ISAF Joint Command (IJC) district assessments. Furthermore, HES metrics are very similar to metrics used in both Iraq and Afghanistan. Other systems used in Afghanistan and suggested for incorporation into contemporary COIN doctrine, including TCAPF (to be replaced by the District Stability Framework) and MPICE, are similar to the HES or incorporate some elements of HES methodology.

The HES was a controversial program and was the subject of intense congressional and media scrutiny from its inception in 1967. Archival information on the system is expansive and includes reams of data, maps, reports, scorecards, press conference transcripts, handbooks, briefings, and official testimony. While this section provides a review of the published and archival data, it cannot and does not offer a thorough examination of the entire issue of hamlet pacification.[64] Instead, the purpose of this section is to use the HES as a window into the complexities of centralized analysis for COIN assessment.

Concluding his chapter on the HES in *War Without Fronts*, Thayer states, "In the absence of an absolute criterion of truth, the data can be interpreted in many ways

---

[60] Kinnard, 1977, p. 107; Brigham, 1968, p. 22.

[61] CORDS director Robert Komer (1968a, p. 5) described this size disparity as comparing "apples to grapefruit."

[62] The HES produced six published reports in both graphic and tabular format. Brigham (1968, pp. 12–13) provides a list of these reports.

[63] See Gayvert, 2010.

[64] See Komer, 1970, Race, 1972, and Cooper et al., 1972a, 1972b, for a more in-depth examination of the pacification program. This monograph does not address the inherent validity of the pacification approach in the context of Vietnam or COIN strategy.

and at various levels of aggregation. Indeed, this chapter has done so and has shown the futility of assuming that the data represent a completely accurate statement at any point in time." But, he adds, "The trends seem reasonable."[65] In 2009, noted civil violence expert Stathis N. Kalyvas and Matthew Adam Kocher wrote, "The HES is likely to remain for some time the only systematic micro-level database of civil war dynamics covering such a large territory and time period." In their analysis of the HES, they address criticisms of the data but find the data set to be a "unique resource for the study of the dynamics of civil wars in terms of its scope, detail, and level of analysis."[66] Anders Sweetland conducted a statistical item analysis of December 1967 HES data and found the system to be meaningful and mathematically stable.[67]

HES survey data were drawn from a series of worksheets filled out by officials at the district and province levels (by provincial advisory teams, which may have had as many as six people).[68] The HES worksheets are too large to reprint here, but they included a range of questions in a format similar to a modern bubble sheet. Some of the forms had spaces for narrative input.[69] The questionnaire rated six major "factors"—perhaps the equivalent of MOEs or MOPs—each of which had three associated indicators (analogous to the indicators in FM 5-0). District or province advisors

---

[65] Thayer, 1985, pp. 151–152.

[66] Kalyvas and Kocher, 2009, pp. 343 and 342, respectively.

[67] Sweetland, 1968, p. ii. Sweetland's study is discussed in greater detail later in this chapter.

[68] There are disparities in the description of the HES process even among MACV staff members. The MAC-CORDS Field Reporting System order of July 1, 1969, stated that province advisors were responsible for the PACES reports but that district advisors were responsible for HES reporting (Mission Coordinator, U.S. Embassy, Saigon, Vietnam, Annex C). Brigham (1968) indicates that data collection occurred primarily at the province level, while Komer (1970), MACV (1971a), Roush (1969), and Cooper et al. (1972a, 1972b) state that the data were collected and recorded by more than 200 district advisors. Anecdotal field reports are also conflicting; for instance, the U.S. Army psychological operations advisor to Phong Dinh Province filled out the report himself at the province level, while Donovan describes a district-level process (Former U.S. Army psychological operations advisor to Phong Dinh Province, Vietnam, interview with the author, Washington, D.C., April 19, 2010; Donovan, 1985). Nearly all sources show that GVN and ARVN officials continued to have input in even the HES 70 and 71 reports. Clark and Wyman (1967, p. 26) summarize the original OSD plan for the HES, published in late 1966:

> The objective was to obtain from each U.S. subsector (district) advisor a detailed evaluation of each hamlet with some degree of GVN control in the district to which he was assigned. There are on the average about 50 hamlets per district, the range extending from about 25 to a maximum of more than 100. The basic element of the proposed system was a questionnaire containing 191 separate evaluations to be filled out for each hamlet with some vestige of GVN control (about 70 percent of all the hamlets in the country). Responses to the questions were to be recorded on a one-page checklist as a yes/no response. After completing the evaluation checklist the advisor was to be asked to compute from his responses an overall evaluation index for the hamlet. These forms were to be submitted to sector (province) headquarters where the number of hamlets in five evaluation levels were to be tabulated and the summary data from them forwarded through division and corps headquarters to RDSD [Revolutionary Development Support Directorate] and to the Office of the Deputy Ambassador.

[69] HES statistics can be found in Thayer (1975b) and Sweetland (1968); for worksheets, see Thayer (1975b) and Brigham (1968).

tasked with filling out the forms would mark each of the categories with grades ranging from A to E. This is no different than a color-code scheme: like colors, the letters are directly equated to numbers, 1–5. The numerical responses to these questions were then averaged. In later versions of the system (HES 70 and 71, for the years 1970 and 1971), respondents filled out slightly more qualitative questionnaires, and analysts at some level above them would assign codes to the data.[70]

The HES exemplifies most of the criteria used to define centralized and heavily quantitative assessment. Every version of HES, including the 70 and 71 versions, aggregated subjective input from a range of officials, each of whom had different priorities, faced different challenges, and had very different degrees of access to the hamlets assessed; their individual experiences reflected the mosaic nature of COIN. These officials were then channeled into providing information that they may or may not have had, within narrow categorizations that may or may not have reflected the range of possible conditions from hamlet to hamlet. These subjective inputs were then transformed into narrow numerical codes that were devoid of context once aggregated and processed. (It is impossible to retain context when aggregating more than 8,000 reports.) These aggregated data were used to produce a countrywide perspective on hamlet security and development. Thayer's data on GVN control of the countryside is derived primarily from the HES (and also from PAAS and the Territorial Forces Effectiveness System).

Narrow, standardized guidelines for the rating system forced the respondents to provide answers for each indicator in one of five categories. For example, the 1968 version of the HES rated VC military activities, focusing on the activity of "VC guerrilla units" to help the respondent rate the category.[71] The precoded options were as follows:

A. Guerrillas (Gs) remain driven out. No threat of harassment or intimidation from Gs in adjacent villages.

B. G control reduced to 1–2 hamlets on village periphery or 2–3 hours travel to hamlet: could make a desperation raid. Activities of Gs from adjacent villages limited by no havens or by friendly defenses.

C. Military control of village broken, most Gs identified, 50 percent losses, havens destroyed, activity below platoon level; can harass but not prevent GVN activities in hamlet.

---

[70] Revision recommendations can be found in the Simulmatics report, described later in this chapter, and in a 1973 report by the Department of Applied Science and Technology at the Bendix Aerospace Systems Division, *Analysis of Vietnamization: Hamlet Evaluation System Revisions* (Prince and Adkins, 1973). The latter provides a more in-depth analysis of the questions and data than the former. A December 27, 1969, MACV briefing in Saigon reported that when HES was modified to HES 70, 42 percent of hamlets changed category (e.g., from B to C). See Sorley, 2004, p. 331.

[71] Each indicator on the worksheet is broken into five rating categories. This example is indicative of issues with the other indicators.

D. Village Gs reduced somewhat in men and defenses; can attack in platoon strength from within [village] or 1–2 hours travel to hamlet.

E. Village Gs combat effective though some identified or eliminated; VC village defenses largely intact.[72]

This is very similar to the IJC district assessment rating definition levels in Afghanistan, which also use in a pre-coded 1–5 scale and provide standards for each level.[73] For example, a notional IJC district assessment metric used to determine the progress of local Afghan government leadership might have rating definition levels that span from 1, "Leader is *effective* and *not corrupt*," to 5, "Leader *ineffective* and *corrupt*."[74] But it is possible for a leader to be effective and also corrupt (especially in Afghanistan); therefore, it is feasible for a leader's behavior to rate a 1 on the rating scale and simultaneously rate a 5. Would this equal a 3? There is already a distinct definition for a 3 (in this example, "Leader is moderately effective and moderately corrupt"), so it could not. Therefore, in this notional Afghanistan case, the district assessment does not include an option that permits accuracy. The section of the HES on VC activity (which is representative of the other sections of the HES) was similarly restrictive. It might have been possible to have no threat of attack or harassment from adjacent villages (an "A" rating, equivalent to a 1) but also face intact VC defenses in the village in question (an "E" rating, or 5). This situation cannot equal "C" because that rating has a separate definition. In these cases, respondents probably would have chosen to average the two options together, despite the fact that averaging is illogical in that the resulting report would not reflect reality.[75] The process of forcing respondents—each of whom is facing very different circumstances—into narrow metrics and definitions results in inaccurate data.

It also appears that the authors of the definitions in the HES were attempting to elicit positive input from the respondents. In the best-case scenario, or "A" rating, the village (an area larger than a hamlet) and all surrounding villages are not only free of guerrillas but also free of harassment. This is, in essence, a peacetime situation and a laudable goal for counterinsurgents. But in the HES example presented here, the first two rating options both depict optimistic situations, and even the third (C) option shows a rather impressive degree of success: The guerrillas cannot prevent GVN activities in the hamlet. The contemporary equivalent might be the ability of Afghan government officials to operate freely and without harassment in an Afghan village; this would be a dramatic success in Afghanistan and probably even in Vietnam. *In fact,*

---

[72] Thayer, 1975b, pp. 49–50.

[73] This reflects the process as of late 2010.

[74] This is similar but not exactly representative of the actual rating scale used in IJC documents.

[75] Interviews with tactical commanders and assessment staffs in Afghanistan support this assumption. Quantitative inputs are routinely averaged for aggregation, regardless of context.

in Vietnam, a "C" rating showed the hamlet as being "secure" in aggregated reports and on maps.[76] The last two ratings (D and E) both describe the guerrillas in terms of how they have been degraded, and in the "E" rating—the worst possible rating for a GVN-controlled hamlet—some of the guerrillas have been identified and killed ("eliminated," in the parlance of the HES). Therefore, even in the worst-case situation, there are signs of success.[77]

Might a retooling of the definitions and rating levels make this system, or any similar system (e.g., the District Assessment Model used in Afghanistan) more realistic? The right definition would give sufficient leeway to account for any and all situations faced by province advisors. But in practice, any definition is channelizing and will create incongruities among the respondents; these are core metrics. The only way to allow for any and all types of input and to eliminate bias from the formula is to clearly define the data requirement (if this is possible) and then use a rating scale to "rate the hamlet in this category of assessment from 1 to 5 based on your understanding of the situation." This kind of dissociated quantitative response would make the aggregated data all but meaningless because there would be no standard associated with the numbers. Subjective bias would guide the selection process absolutely, and the consumers of the data would have no way to determine whether a 3 from one rater was equivalent to a 3 (potentially an average of 1 and 5) from another rater. The kinds of quantitative and graphic reports produced from the data would be meaningless. Once this was realized, the process of defining rating levels would begin anew and would almost certainly result in the same muddled and unsatisfactory definitions already found in the HES and the District Assessment Model. There was so much disagreement over the HES definitions that as late as 1973—six years after the implementation of the program—external auditors were still recommending revisions to the pre-coded ratings.[78]

Because each hamlet in the HES was rated on an A–E or 1–5 scorecard on several indicators (rather than just one), it was necessary to average all the scores to come

---

[76] This was true as of 1967, according to Komer, and late 1968, according to Tunney. This may have changed with later modifications to the program. See Komer, 1967, p. 4, and Tunney, 1968, p. 5.

[77] Komer states that e-rated hamlets did not include those hamlets controlled by insurgents or VC and that there should have been an "f" rating to show VC-controlled hamlets. There was no such rating in the HES at that time. Brigham (who attended the Komer news conference) stated that there should also be another category for "not rated," but that category did not exist either. These categories were later added (at least by 1969) as "VC" for VC-controlled, or "other" (see Komer, 1967, p. 14; Brigham, 1968, p. 12). For another official interpretation of the rating system and its relation to pacification objectives (i.e., the MACV Strategic Objectives Plan), see Sorley, 2004, p. 203.

[78] See, for example, Prince and Adkins, 1973.

up with a single overall rating.[79] Colonel Erwin R. Brigham, then an HES manager, explained how this worked: "A hamlet rated in security with 3 As, 4 Bs, and 2 Cs; and in development with 2 Cs, 4 Ds and 3 Es would be given an overall [unweighted] rating of C. Similarly, A, B, D, and E, hamlets may have individual ratings of A, B, C, D, and E."[80] There are at least two problems with this system. It can be misleading because it is impossible to know what is most important in this particular area at any one point in time. For example, is development inconsequential because security is insufficient? What do the people of the village actually think about their own security? The system begs for weighting, but weighting would be subjective and reduce transparency. This kind of system would make the data murky at best and opaque at worst. Ambassador Robert W. Komer helped develop the HES and established and ran the CORDS program in the late 1960s. He told reporters at a 1967 news conference that the HES was so complicated that it would take much longer than a 20-minute briefing and a 30- or 40-minute question-and-answer session to explain how it worked and how it compensated for the complex realities of South Vietnam.[81] This is not a transparent process.

Once the CORDS staff collected the data and averaged the responses to produce a rating for each hamlet, they would then average all of the hamlets together to develop a progress number that would represent overall pacification in the country.[82] According to Komer,

> Now it's pretty hard to measure pacification results. . . . [Y]ou've got 12,700 hamlets in this country that we've identified so far, and no doubt there are some others too. So you've got to look at the averages.[83]

It was this averaged, overall pacification number that was typically delivered to senior decisionmakers. Therefore, the overall score used to shape policy was composed of averages within an average. Averaging the total score further degraded the transparency and accuracy of the HES data to the point that the overall number was relatively meaningless.[84] William E. Colby, who was the CORDS director from 1968 to 1971

---

[79] Various descriptions of the HES also include some mention of weighting the numbers, but these descriptions are not consistent across all sources. It is not entirely clear how averaging and weighting figured into the final product from version to version.

[80] Brigham, 1968, pp. 9–10.

[81] Komer, 1967, p. 16.

[82] Thayer, 1975b, pp. 53–58. Note that there are sometimes two mismatched page numbers on some of the pages in this report. These pages are the ones on the lower right of the page.

[83] Komer, 1968a, p. 4.

[84] Further complicating this calculation was the inclusion of a "confidence rating" on the HES worksheet (at least in the 1969 version). The advisor would use a 1–5 scale to rate confidence in the information provided on the worksheet. If the reports were not aggregated, this rating might have been a valuable tool to help analysts assess

and responsible for the HES during that period, said that "a mathematical complication of the responses from all the hamlets suggested a false average instead of the wide individual variations that lay beneath it."[85]

It would be possible for the single, top-level average to hide a major security imbalance. For example, the heavily populated areas around Saigon could be improving, while the less densely populated areas in contested spaces along the demilitarized zone in the north or in the Mekong Delta in the south—the populations that probably mattered most to the COIN effort—might be declining rapidly; but in this case, the average would show only improvement.[86] OSA analysts could address this problem by delving into deeper layers of data, but by producing a single, quantitative finding, they made their own analysis irrelevant to busy policymakers or interested members of the public who did not have more than an hour (by Komer's reckoning) to figure out the HES system.

Every month, the HES produced approximately 90,000 pages of data and reports.[87] This means that over the course of just four of the years in which the system was fully functional, it produced more than 4.3 million pages of information, and each page may have contained ten, 20, or more discrete elements of data—perhaps 40 million pieces of data, as a round estimate.[88] This is a remarkably dense and complex reporting process. Such a vast quantity of information presented extraordinary logistical and analytical challenges for Komer's, Colby's, and Thayer's analysts in the 1960s and early 1970s. Computer-aided analysis has improved exponentially since 1971, of course, but even now it would be difficult to account for the possible reading, coding, aggregating, or transmission errors that could creep into these pages as they are written by advisors in the field, passed up to the province level for collation and retransmission, passed to the division level for another round of collation and retransmission, and then provided to MACV for a final round of collation and retransmission to OSD.

Despite obvious concerns about data accuracy as voiced by the program director, HES reports as delivered were said to be accurate to within a tenth of a percent-

---

the validity of the reporting for an individual hamlet. However, it is not clear that confidence ratings were incorporated into the total HES number or, if they were, how the variations in confidence ratings could be accounted for in an aggregated total. An averaged confidence rating could be misleading in the absence of disaggregated context. See MACV, 1969d.

[85] Colby and Forbath, 1978, p. 259. Colby, who would later become director of the CIA, believed that the HES was most valuable as an organizing tool at the local level; he thought that it helped motivate GVN officials to improve hamlet security and development.

[86] This is a notional example. In reality, the 1967 version of the HES did not even address the population in Saigon. Therefore, the 1967 data did not include approximately 3.6 million people out of a population that may have been about 17 million (Thayer, 1975b, p. 12).

[87] Brigham, 1970, p. 48.

[88] Some of these pages may have included reports that repeated or re-reported raw data from the field, but for the purposes of information management and analysis, the overall quantity is what is relevant.

age point. A January 11, 1969, HES report to military and civilian leaders in Saigon reported a "record gain" of 3.5 percent in secured hamlets and a decrease of 1.1 percent in the VC-controlled population. These figures were presented without caveat for accuracy or a margin of error. The way in which the figures were presented implied a degree of accuracy that would not be achievable even under ideal conditions. None of the assembled senior leaders, including the commanding general of MACV and the U.S. ambassador, questioned the report or the accuracy of the data.[89] This unquestioning acceptance of HES data may have been due to the pressure on senior leaders and HES program officers to meet quantitative pacification thresholds. For example, in a July 17, 1969, briefing, CORDS director Colby stated that the national objective was to achieve at least a C-rating for 90 percent of the total number of hamlets by the end of that year, and at least B-ratings (if not A ratings) for 50 percent of the total number of hamlets. According to Colby, "The President very clearly says that the purpose of that is to get it so the government can absolutely count on at least 50 percent of the population to vote for them in a challenge with the Communists."[90] Since the HES did not reflect anywhere near 100 percent of the South Vietnamese population (even when taken at face value), there is no way that the President's 50-percent objective for A- and B-rated hamlets could have guaranteed that 50 percent of the population would do anything.[91] This statement shows how policymakers can be misled by or misunderstand complex quantitative COIN data.

Both Thayer and Brigham cite two independent studies of the HES conducted in the field in Vietnam. The purpose of both studies was to substantiate the reliability of the HES model and to improve the system.[92] One of these studies was managed by the Simulmatics Corporation but conducted primarily by U.S. military and Vietnamese field collectors and analysts. It examined only the reliability of the HES system and not (according to the report) the validity of the data. It describes the system as generally reliable and the information produced by the HES as reasonably accurate.[93] This study is important because it ostensibly validates the HES methodology for campaign assessment.

The security conditions and lack of access undermine the study's reliability, however. According to the report, it covered 18 percent of the districts (40 out of 244 total districts) in which advisors were "present," and it is not clear how many hamlets were

---

[89] Sorley, 2004, p. 96, transcript of audio recording of a commander's conference in Saigon, South Vietnam.

[90] Sorley, 2004, p. 220. It is not clear whether he is referring to a benchmark established by Lyndon Johnson prior to the end of his term or by Richard M. Nixon during the first part of his first term.

[91] Colby stated that, as of 1970, 40 percent of the South Vietnamese population lived in urban centers, and the majority were not covered by HES reporting (U.S. Senate Foreign Relations Committee, 1970, p. 420). The propriety of the President's effort to fix the South Vietnamese election is immaterial to this discussion.

[92] The Simulmatics Corporation and Pacific Technical Analysis, Inc., conducted these studies in 1967 and 1968.

[93] Pool et al., 1968, p. 1.

located in these districts.[94] It is unclear what is meant by "present," but the sample could not then be extrapolated to cover the hundreds or perhaps thousands of hamlets that were rated by advisors but not visited. The samples primarily included heavily populated areas, areas that tended to receive more attention from MACV than the more contested rural districts and that would be safer and easier to rate. U.S. military officers who outranked the district and province advisors managed and coordinated the study, and two special forces officers of unknown rank (but who may have outranked the district advisors) helped conduct the field interviews. There is at least some impression that the military was "grading its own work," and there may have been some pressure on the relatively junior advisors to respond positively to interview questions.

Regardless of the reliability of the Simulmatics sampling process, at the end of the day, the study relied on district and province advisors to tell the interviewers whether or not their own reporting was accurate.[95] While this reliability report provided some excellent recommendations for program improvement, it is insufficient to prove the reliability of the HES; ultimately, the reliability of the overall system was never clearly established.[96] The next section addresses the validity of the HES data and the ways in which those data were used for pacification and strategic analysis.

Hamlet Evaluation System Data and Analysis

[T]he greatest possible restraint should be exercised by senior officials and policy-makers in using highly aggregated reporting results in which vital qualifications have been "summarized out."

—*Report on Vietnam hamlet pacification assessment, 1972*[97]

---

[94] Pool et al., 1968, p. 14; Brigham, 1968, p. 3. Brigham reported that there were 244 districts in Vietnam. The Simulmatics Corporation stated that it covered "approximately" 40 districts.

[95] The study did include some observations by the research staff, but the staff was small and consisted of either officers from out of country or Vietnamese officials of unknown affiliation; the visits were structured and of insufficient length to determine actual survey reliability or data validity. The study also included interviews with hamlet chiefs in 22 of the 40 districts covered. While there was coincidence between the responses of the hamlet chiefs and the ratings for the hamlets, this is a nonscientific sample. It also does not show whether district advisors relied heavily on the hamlet chiefs to provide them with their original ratings; this was a common tactic. In these cases, there would be natural coincidence between the original ratings and the field observations that would not necessarily speak to the reliability of the HES.

[96] RAND conducted a statistical item analysis of HES in 1968 that showed the system to be "meaningful and statistically stable and can be used as the basis for developing extensions to the construct of pacification" (Sweetland, 1968, p. ii). That study did not address validity. The Simulmatics study found problems with the development scores, and a survey of all information relating to the HES available in 2011 showed that the item analysis may have implied meaningful relations and statistical stability based on a steady stream of precise but inaccurate reporting. That said, any analysis of the overall reliability of the HES should include Sweetland's findings.

[97] Cooper et al., 1972b, p. 234. The source is a comprehensive report on the Vietnam civil pacification programs published by the Institute of Defense Analyses for the Advanced Research Project Agency.

[HES will become] the body count of pacification.

—Attributed to Vincent Davis, friend of John Paul Vann[98]

Some district and province advisors made every effort to fill out the HES worksheets accurately, and many HES reports probably contained some data that were very accurate. One district advisor described the HES as "a valid system which is correct in my district."[99] But the preponderance of available evidence seems to show that much of the information in the HES reports—it is impossible to know how much—was incomplete and inaccurate. Komer (the head of the program from 1967 to 1968) stated that HES data were "not glamorous."[100] Anecdotal evidence that a sizable amount of HES data were inaccurate or simply fabricated is similar to anecdotal evidence showing officers fabricated data for body counts in Vietnam. Environmental and bureaucratic complexities also eroded the quality of HES information. Instructions for filling out the four worksheets were long and complicated: The simplest instructions were about 25 pages in length.[101] The collection of HES data was necessarily ad hoc and differed from location to location. Thayer stated that it was infeasible to define HES data as completely accurate at any one point in time, and he described the factors that "combine to make HES reporting difficult":

1. Only a few of the 18 indicators can be rated on the basis of direct observation of a clear-cut condition.

2. Much of the HES information can be obtained only from Vietnamese [sources]. Surveys to date [late 1967] indicate that advisors rely on their Vietnamese counterparts for at least half of the raw data they use in answering the HES questions.[102]

---

[98] Sheehan, 1988, p. 697.

[99] Arthur, 1970, p. 193.

[100] Komer, 1967, p. 21.

[101] Mission Coordinator, U.S. Embassy, Saigon, South Vietnam, 1969, Annex C and appendixes. See also the *Hamlet Evaluation System Advisors Handbook* (MACV, 1971c), at 40 pages, and the *Hamlet Evaluation System District Advisors' Handbook* (MACV, 1971a), at 50 pages, including a helpful Vietnamese translation. Komer and others state that advisors were trained to fill out the HES worksheets prior to deployment, and mobile training teams that provided field instruction visited them occasionally. There was also an HES "newsletter" written by an officer at MACV CORDS to address ongoing issues with worksheets. A sampling of these handbooks, worksheets, and newsletters can be found in the Texas Tech University Vietnam Center and Archive.

[102] Despite this, the system prominently reported findings in statistical analyses from data that were gathered from single-source counterpart reports . For example, the HES showed that, in 1970, hamlet chiefs governing 80 percent of the population were regularly present both day and night in the hamlets. Since the vast majority of hamlet visits were often brief and infrequent (a few hours per month or once every few months) and did not occur at night (and district advisors could live in only one hamlet at a time), this information had to have come primarily from Vietnamese sources. Yet, the data are presented without caveat. Further, Thayer compares these data to perception data from PAAS showing that only 59 percent of the population *believed* (and could not confirm) that

3. Most advisors cannot visit all of their hamlets during one month. In [one area], advisors reported that US personnel visited about two-thirds of the accessible hamlets in their district in any given month, but managed to cover all but 6% over a four-month interval.[103]

One district advisor claimed to have visited two different hamlets per day, approximately every day of every month (there were 60 in his district) for the duration of his tour. This is an incredible pace that, if true, probably made him one of the top HES reporters in Vietnam. Still, sometimes these visits lasted 15 minutes and sometimes they lasted "a couple of hours," and they rarely took place at night when the VC were most active.[104] This means that if an advisor were able to visit every hamlet in a district every month (an anomaly, according to Thayer), the observation period would consist of (at best, according to Arthur) approximately three hours out of a potential ~720 hours in a monthly reporting cycle, or 0.41 percent of the month for a given hamlet.[105] It would be unreasonable to claim that such a limited observation period could prepare an advisor to write a comprehensive assessment of a hamlet, or that this assessment could be deemed either precise or accurate. In a monograph on pacification published by the U.S. Army, South Vietnamese Brigadier General Tran Dinh Tho questioned the validity and accuracy of HES data and addressed the ability of advisors to consistently visit and rate all the hamlets in their respective districts:

> The data recorded by district senior advisers were obtained partly from village and hamlet officials' reports, partly from information provided by friendly forces or through the advisory channel, and partly from actual visits to villages and hamlets by the senior adviser and the district chief. A question arose, however, as to the validity and reliability of the reports thus obtained; doubts about accuracy and timeliness. The most reliable way to have accurate data was to make visits to the villages and hamlets and see for oneself. But the truth was that even if all the time available were devoted to visits, and even if road communication and transportation facilities and helicopters were available all the time, no one could possibly cover all the villages and hamlets of a district in a single month.[106]

---

their local officials remained in or near the hamlets overnight (Thayer, 1975b, p. 168). Popular polling often failed to substantiate HES findings.

[103] Thayer, 1975b, p. 61.

[104] See, e.g., Arthur, 1970, p. 3. The transcript of Arthur's testimony showed that he was reluctant to state that he did not visit the hamlets at night. He eventually admitted that he received reports on nighttime activity from South Vietnamese militia units working for the GVN. Some advisors slept overnight in specific hamlets from time to time or lived in one specific hamlet, but the HES study was concerned with the typical coverage of all the hamlets in a district during any given month.

[105] This calculation is meant only to describe an average overview of HES data collection, not the overall effort of the advisors or other officials working at the hamlet level. In most cases, advisors lived in hamlets or spent the majority of their time working in hamlets.

[106] Tho, 1980, pp. 107–108.

Observation visits may have been structured and sometimes announced in advance, so it is not clear whether the advisors would get an accurate picture of life in the hamlet. (It would be easy to put on a show for a visiting advisor.) Other advisors flew over hamlets in helicopters. Sometimes, they counted huts from above and then multiplied the number of huts by the average number of residents per hut (eight) to build population estimates. This process might work in peaceful areas but is suspect in war zones with high rates of unpredictable civilian movement and millions of displaced persons. With few exceptions, none of the advisors spoke fluent Vietnamese. They could communicate with the population only through an interpreter, each of whom would vary in terms of capability, motivation, and availability.[107]

A psychological operations advisor in Phong Dinh province was tasked with filling out the HES reports for the province with no input from the district level.[108] His HES training consisted of a few minutes of overview during a 15-minute turnover from his predecessor.[109] There were between 250 and 300 hamlets in his area of operations, and he had traveled to only a handful of them. In the absence of accurate input, he sat down with his ARVN counterpart in front of a map of the province and, pointing to each hamlet in turn, asked, "Can you go here?" If the answer was yes, he coded it as friendly; "maybe" rated a neutral code; "no" rated an enemy code. He had no way of verifying this officer's subjective opinions and had no other input to work with. He sent the data to the division headquarters and was castigated for altering the report and changing the reporting pattern set by his predecessor; an officer on the staff told him to change the results back to avoid any controversy that might arise from a significant shift.[110] If he rated approximately 250 hamlets (at the low end), and fewer than

---

[107] Komer (1967, p. 12) stated that, in 1967, somewhere between 20 and 50 percent of the advisors spoke some Vietnamese with some (undefined) level of proficiency. Gibson (2000, p. 308) cites an informal study showing that most Vietnamese interpreters spoke fewer than 500 English words. Official statistics recorded during a March 1970 series of congressional hearings on CORDS (the "Fulbright Hearings") showed that 36.4 percent of district senior advisors had achieved level 1 on the Interagency Language Roundtable for reading and speaking, that 6.6 percent had achieved level 2, and that none had achieved levels 3, 4, or 5 (with 5 being the highest level of proficiency). Only 18.7 percent of assistant district senior advisors achieved level 1, 5.1 percent achieved level 2, and none achieved higher. A level 1 speaker has only elementary proficiency and cannot hold an extended conversation on a complex subject. Level 2 speakers are more capable of holding complex conversations on limited subjects but may still miss critical nuance and vocabulary. By these standards, no advisors were fluent in Vietnamese, and only about 6 percent of all advisors could be considered to have linguistic abilities that would allow them to function with only minimal assistance from an interpreter. On the other hand, 28 percent of civilian advisors could communicate at level 3 or better (U.S. Senate Foreign Relations Committee, 1970, p. 418). See also Interagency Language Roundtable, undated. The U.S. government implemented the program in stages from the late 1950s through the 1980s, and the system remains in use as of 2011.

[108] The fact that he received no input from the district level may have been indicative of structural problems in the reporting system only in this province and only during this period.

[109] Training improved over time as MACV and CORDS implemented training courses for advisors.

[110] Former U.S. Army psychological operations advisor to Phong Dinh Province, Vietnam, interview with the author, Washington, D.C., April 19, 2010.

9,000 of the known 12,000 or so hamlets were evaluated through the HES in 1967,[111] *this single anecdote* shows that nearly 3 percent of the overall HES data during this period were falsified (because his changes were not accepted), inaccurate, or based on extraordinarily poor collection methods.[112]

The story of another advisor reinforces the idea that it was very hard to downgrade an HES rating. In a conversation with Congressman John V. Tunney in 1968, this advisor stated that he downgraded four hamlets after the Tet Offensive in February of that year; these downgraded reports would have generated an automated PACES message to the MACV headquarters. He was "immediately hit with a barrage of cables from Saigon demanding a full explanation for downgrading them." He claims that he spent the next couple of weeks justifying his report, a time during which he says he neglected some of the other pressing demands of his job. The cables from his higher headquarters may have been sent in an earnest effort to get more context from the advisor, but they had a chilling effect: "I believe I am an honest man, and although I hate to admit it, it may be a long time in hell before I downgrade another hamlet."[113] Tunney presented this anecdote in support of his argument that, no matter how the HES is programmed, it will tend to elicit biased responses.

David Donovan's experience with HES was similar to the advisor's in many ways. Donovan was a district-level military advisory team leader in Vietnam in the late 1960s.[114] In his memoir, he provides his observations of the HES system. While Donovan thought the intent of the reports was good, "like so many good bureaucratic intentions, the idea was weakest at the point of practical application." He stated that he saw district senior advisors delegate the HES worksheet to "less informed and less experienced subordinates," sometimes with instructions to "just fill in the blanks with anything that seemed reasonable." Donovan identified timeliness of submission as having precedence over reporting accuracy, and many reports on hamlets were filled in "when the hamlet had never been seen by the [district senior advisor] or any of his team members." In many cases, overworked advisors would simply ask their Vietnamese counterparts for their opinion so they could fill in the HES (like the Phong Dinh Province advisor did), though the Vietnamese had good reason to be biased.[115]

Lewy cites one case of outright cover-up by Vietnamese officials who were feeding data into the HES 70 report, a case that reinforces Donovan's suspicions regarding host-nation inputs to the HES. The senior advisor for Phu Yen Province discovered that local officials in the district of Hieu Xuong were hiding incidents and abductions to

---

[111] Tunney, 1968, p. 1.

[112] He reported that he made only incremental changes after this point, but the accuracy of the report never changed much during his tour in late 1967.

[113] Tunney, 1968, p. 8.

[114] Donovan, 1985, pp. 157–158. David Donovan is the pen name of Terry T. Turner.

[115] Donovan, 1985, p. 158.

retain the district's "pacified" rating. Lewy also points out that it was still possible for even U.S. advisors to "game" the HES 70 and 71 reports. He states that while advisors no longer assigned their own ratings and that scores were added to the reports in Saigon, "it probably was not very difficult to figure out what kinds of conditions in the hamlets would result in a high rating, and advisors, in effect, were still rating their own success in achieving pacification."[116]

Douglas Kinnard relates a conversation with an official from the U.S. Embassy in Vietnam, who told him that "the gaps between the HES numbers and actual conditions in the provinces had reached ludicrous proportions." Kinnard and the official determined that, out of the five A-ranked hamlets in their area of operations, they could safely drive through only two. U.S. Marine Corps Col. William A. Corson served as a pacification advisor in Vietnam and as a systems analyst in OSD's Southeast Asia Programs Division in 1968. He states that the C-rated hamlets did not pass the pacification advisors' "sleep test"—they were not safe to sleep in—despite the fact that they were labeled secure.[117] He also stated that the GVN gamed the HES to protect its own people by putting its development cadre in A- or B-ranked hamlets, undermining one of the key purposes of the HES survey program; in theory, the HES should have been used to shift support to less secure or less developed hamlets an effort to bring them up to the A or B level.[118]

John Paul Vann, perhaps the most successful and well-known pacification officer in CORDS and the subject of Neil Sheehan's *Bright Shining Lie: John Paul Vann and America in Vietnam*, discovered that the HES could also overstate the degree to which the VC controlled certain hamlets. Sheehan describes Vann's discovery upon arriving at a new posting in Can Tho, the capital of Phong Dinh Province in the MACV IV Corps military region:

> He discovered that many of the 2,100 hamlets the HES listed as under Viet Cong ownership in IV Corps in February 1969 (another 2,000 hamlets were listed in varying degrees of Saigon control) were actually held by half a dozen guerrillas.[119]

Jeffrey Race was a civilian province advisor in Long An Province in 1968 and 1969. In his classic *War Comes to Long An*, Race describes the HES as useful in that it showed changes in some indicators over time. However, he also called the system's data a "misleading indicator of progress." He points out that, in many of the C-rated hamlets in his province—those considered pacified by HES standards—there was actu-

---

[116] Lewy, 1978, p. 192.

[117] Corson, 1968, p. 236.

[118] Corson, 1968, pp. 236–237.

[119] Sheehan, 1988, p. 732.

ally more revolutionary activity than government activity.[120] He describes in detail the efforts to pacify a hamlet called Hoa Thuan II. He states that local GVN officials could not travel in and around the hamlet, but clandestine VCI party officials had the freedom to move about off the main roads. Despite these conditions, the U.S. advisory team "upgraded the status of Hoa Thuan II to the C category because of the 'progress' it had shown."[121] Race considered the Hoa Thuan II assessment a cautionary tale.

Because the HES was rigidly structured (even with the HES 70 and 71 modifications), it did not allow the *systematic* reporting and analysis of hidden problems like those described by Race.[122] The lack of obvious VC activity could (and sometimes did) hide a very sizable insurgent presence in the hamlets; in some cases, the hamlet chiefs and GVN officials simply came to an arrangement with the VC, allowing them to operate with impunity. For example, in Vinh Binh Province (in 1971, so probably using HES 71 or at least the HES 70), there was a B-rated hamlet that also contained the 312 Main Force VC Battalion. The VC held "veto power over life and death issues" in the hamlet, but since there had been no overt incidents and the people had responded positively to the GVN in public, "the HES marks this a B hamlet."[123] Only detailed and contextual all-source intelligence could have identified and reported this situation in a format that was meaningful to operational decisionmakers.

Did any of this ultimately matter? Did policymakers take the HES data seriously? The evidence shows that the HES was not only a central pillar of the Vietnam-era assessment process, along with body counts, but that it also shaped both strategy and combat operations. A review of official message traffic from the U.S. Embassy in Saigon to the President and DoS showed that senior officials like Komer and Ambassador Ellsworth Bunker based their strategic recommendations at least in part on hamlet security data.[124] Brigham describes the HES as a "management tool" and explains how HES reports were used to direct combat operations and, alarmingly, also used by artillery units to determine population density in what may have been an effort to avoid civilian casualties.[125] At least some policymakers took the data seriously, relying on them to show progress to Congress and the public. Secretary of Defense Robert S. McNamara reported pacification data to the press and the public in news conferences, highlighting them as a major indicator of progress.[126] George Allen had 17 years of

---

[120]Race, 1972, p. 214.

[121]Race, 1972, p. 260.

[122]Later versions of HES allowed for subjective input, but it is not clear that this input was seen or analyzed at the MACV or OSD level due to data aggregation.

[123]Lewy, 1978, p. 194.

[124]See BACM Research, Disk 2, as well as Bunker, 1968a and 1968b, and Komer, 1968b.

[125]Brigham, 1968, p. 21; Brigham, 1970, p. 53.

[126]See, for example, McNamara, 1967.

experience studying Vietnam for various intelligence agencies. His account provides some insight into the way the system was created and how HES results were misunderstood and misused. Here, he describes pressure from the executive level to show progress through using the aggregated and twice-averaged HES total score:

> [W]e got pressure from the White House, the way they used the Hamlet Evaluation System. They were the ones that decided this A-B-C-D-E, or 5-4-3-2-1 scale, you could come up with an aggregate average score for all the hamlets in Vietnam, monthly, and it was 2.163 maybe this month. And they would sit there with [bated] breath waiting to see what next month's score would be. And if it were 2.165, two one-thousandths higher than last month, "Great, we're progressing, we're progressing!"[127]

This anecdotal insight into the strategic use of HES data by senior policymakers is reinforced by other sources. For example, Vietnam-era CIA officer and hamlet pacification expert Rufus Phillips stated that, in 1965 (pre-HES), there was a "demand from Washington for instant or nearly instant pacification results."[128] John Tunney was both a critic of the HES and, as a sitting U.S. congressman, a consumer of HES data. He conducted a special study mission to Vietnam to investigate the HES in May of 1968. Reinforcing Allen, Tunney states that on numerous occasions, "Presidential advisors have briefed Members of the Congress and told them that a certain percentage, usually carried to one or two decimal places, of the South Vietnamese people were living in relatively secure areas under the control of the Saigon government." Tunney went on to state that "this information was being used at the highest levels of the U.S. government as a guide to operational planning and as a means of informing the American people of the progress of the war."[129] That Sharp and Westmoreland highlight a full-page HES chart showing near-steady improvement in HES reporting from 1965 through 1968 in the conclusion of their *Report on the War in Vietnam* reinforces this supposition.[130]

This critique of the HES is not intended as an indictment of CORDS or the overall pacification strategy; both programs were probably successful to some degree, and the theory behind bottom-up COIN has generally proved to be sound. Nor is this section intended to criticize the personal efforts of any of the key players in the HES assessment chain.[131] *It is quite clear that the HES was a best try in the absence of*

---

[127] Rehm, 1985, pp. 6-14–6-15.

[128] Phillips, 2008, p. 267.

[129] Tunney, 1968, p. 1.

[130] Sharp and Westmoreland, 1968, p. 199.

[131] To what degree pacification succeeded will never be known because the hard work by CORDS teams literally went up in smoke in 1975. Colby claims that pacification was a clear success, citing the lack of VC activity during the 1972 and 1975 NVA invasions (Colby and Forbath, 1978, pp. 285–286). However, this shows only that the VC may have been largely defeated by 1972. This is coincident with half of the objectives of pacification accord-

better options. No similar or better options had been articulated by 1967.[132] The MACV CORDS staff seems to have made every effort to ensure the accuracy of the HES data and to improve the program over time. Komer, the man best positioned to comment on the system from the inside, stated, "All that we think we can say is that it seems to be much better than anything we have had before. It's more objective, it's more accurate, it's more systematic [than previous efforts]."[133] Komer points out that there was insufficient analysis of the war (and he had only ten analysts in the CORDS Operations and Analysis Division to pore over the HES data),[134] but it is not clear that more analysis of aggregated data like that from the HES would have better served policymakers than some kind of contextual analysis.[135] Kinnard's survey showed that 75 percent of general officers believed that the HES "had weaknesses but was about as good as could have been devised." This far-from-ringing endorsement seems to support the "good-enough" proposition for COIN assessment. However, only 2 percent of these senior leaders found the HES to be "a good way to measure progress in pacification."

Because the HES required narrow and centralized quantitative inputs and aggregation, it reflected the most misleading elements of contemporary centralized assessment and EBA. In fact, it was an EBA system in all but name: It used a set of core metrics to gather and aggregate standardized data to measure the effects that U.S. and GVN pacification programs were having on the population of South Vietnam, down to two or three decimal places.[136] EBA promises and depends on accuracy, though neither Thayer nor Komer thought that HES data were ever truly accurate. But the structured collection of HES data, the statistical analysis, and the defense of the aggregated HES findings are collectively equivalent to the process of producing an EBA report. As Allen and others showed, policymakers took the findings quite literally, even when they were warned not to. Thayer, Komer, and Colby cannot be wholly to blame for this overreliance on the data; the policymakers who seized on the reports as a quantification of the war are also somewhat to blame. But no one involved fully appreciated that the act of aggregating and publishing *statistical* HES data was enough to give those numbers a life of their own. Caveats should have extended from data quality to the validity of the overall process and findings.

---

ing to MACV orders and the HES; it does not prove the success of pacification programs. Both the Simulmatics Corporation (Pool et al., 1968) and Bendix Aerospace (Prince and Adkins, 1973) reports hint at a split between the relative validity of HES security statistics (more valid) and development statistics (less valid).

[132] Ahern (2010, p. 252) describes the Saigon manager's statement that the HES was "far and away superior" to any other method, a "modest claim given that there was no other method."

[133] Komer, 1967, p. 9.

[134] MACV, 1968b, p. 16.

[135] Komer, 1970, p. 70.

[136] Even a quick survey of official publications and transcripts of official statements on the HES from 1967 to 1972 will show that the program was intended as a quantitative tool to measure effects.

By the end of the war, it was clear that the HES had not successfully informed policy (or, at the very least, had not informed a successful policy). Because they were presented as scientifically accurate, the quantitative results, with their false precision, misled the executive branch, Congress, and the American public as to how the United States was actually performing in Vietnam. The lessons from the HES alone cannot be empirically generalized to all similar efforts, but they are similar to those generated by an examination of body counts, operational days of contact, hostile incidents, and other categories of assessment reporting. Weighted by these other examples, the HES appears to show that the most extensive and structured effort to obtain ground truth in a centralized quantitative system likely produced fabricated or irrelevant data that had no real utility at the tactical, operational, or strategic levels of operation and decision-making. It also shows that these efforts potentially distracted tactical-level counterinsurgents from focusing on what should have been their overarching purpose: mission accomplishment.

Pattern and Trend Analysis of Vietnam War Data

> Perhaps only a single thread separates us from the truth, or perhaps an entire ream, but we will know for certain only when we look at the whole weave.
>
> —*Hassan, in* The Storyteller of Marrakesh[137]

> If you want [the statistics] to go up, they will go up. If you want [the statistics] to go down, they will go down.
>
> —*A South Vietnamese general officer*[138]

This section considers the value of aggregated Vietnam COIN data to campaign assessment and the possibility that useful patterns and trends could be found in what tended to be inaccurate, incomplete, and aggregated quantitative data. Efforts to apply pattern and trend analysis by MACV and OSA provide good insight into the ways in which assessment staffs use scientific (in this case, statistical) analysis of data collected and reported through nonscientific means to measure shifting objectives.

Throughout *War Without Fronts* and in his 1985 presentation to analysts on assessing the Soviet war in Afghanistan (included in Rehm, 1985), Thayer asserts that, despite obvious flaws, Vietnam-era COIN data were reasonably accurate and sufficient to find useful patterns and trends. Specifically, "the patterns we found were very

---

[137] Roy-Bhattacharya, 2011, p. 222. In the novel, the storyteller Hassan retells the story of the disappearance of a pair of tourists, each time embellishing the story with additional details and contradicting the memories of his listeners, who contribute their own (subjective) narratives, further muddying the waters.

[138] Gibson, 2000, p. 314.

definite."[139] He mentions *patterns* (consistent arrangements or behaviors, not necessarily over time) far more frequently than *trends* (prevailing tendencies over time) and, at times, he conflates these two concepts.[140] Were the aggregated quantitative data from the Vietnam War reasonably accurate and therefore useful for assessment? Was it possible to find definite and useful patterns and trends in these data? Which elements of Thayer's analyses might show actual patterns or trends, and how might they show progress in a COIN campaign? Which elements probably do not show actual trends, and why not? What are the dangers of reporting trends that probably do not exist?

While "reasonable" and "accurate" are loosely defined terms, it makes sense to define them as distinct assessment objectives. Absolute accuracy is, of course, more difficult to achieve than reasonable accuracy. *Accurate* pattern and trend recognition requires accurate data. But Thayer uses terms like "unofficial guess," "notoriously unreliable," and "tenuous estimate" to rate the quality of key categories of Vietnam War data. He describes the PAAS polling as "much less precise" than polls taken in the United States.[141] The evidence clearly shows that the COIN environment played havoc with data collection and reporting in Vietnam. Both aggregated body count and HES data were significantly inaccurate and incomplete, but all the data sets contained substantial gaps and inaccuracies. Except in the most exacting and controlled of circumstances, no aggregated data set is ever 100-percent accurate or complete; as McKnight et al. show, even most scientific studies may have data gaps that exceed 30 percent.[142] But the degree of inaccuracy in Vietnam-era assessment data—as described by Thayer, who supervised the analysis of those data for six years—strongly suggests that no policymaker should have placed strategic value on a single chart or graph.

Does reasonably accurate pattern or trend recognition for COIN require accurate or only reasonably accurate data? And if it is possible to produce patterns from COIN data, are they useful to COIN assessment let alone operational planning? Can behavioral patterns drawn from reasonably accurate data—or any data for that matter—be equated with progress or lack of progress in a COIN campaign? Chapter Three showed that patterns and trends can be found in data that are somewhat inaccurate but consistently so. All Thayer was looking for was consistency in reporting, with the idea that consistency shows through inaccuracy. A consistent pattern might or might not be revealing by itself, but it might be useful when compared to other data, or it might serve to highlight an area of concern for further analysis. For example, Thayer shows

---

[139] Rehm, 1985, p. 4-4.

[140] A pattern can also be a trend if it is a pattern over time, but a pattern does not necessarily need to be identified in time series. For example, the drop in violence in Afghanistan over the winter, year after year for many years, is a pattern over time; the coincidence of decreased civilian activity in the presence of an impending insurgent ambush is a general pattern; the steady increase in oil revenue over time is a trend but not necessarily a pattern.

[141] PAAS data and related analysis can be found in Thayer, 1975b.

[142] McKnight et al., 2007, p. 3.

that friendly casualty levels always went down in the rainy season and always rose in the dry season—a pattern.[143] The one aggregated metric that is almost always highly accurate in U.S. COIN assessment is U.S. casualties.[144] Historical weather reports from Vietnam were reliable (it is always possible to know that it rained after it has rained). If Thayer's analysis uncovered a genuine pattern, was this identified pattern useful to policymakers trying to determine success or failure in the Vietnam War? By itself, could it show the American public that the United States was achieving its objectives? It may have been useful to operational planners in the same way that the pattern of lower violence during Afghan winters is helpful to ISAF planners. But this finding is of questionable strategic assessment value, since, *by itself*, it reveals nothing about the progress of the campaign. This pattern is not reliably predictive because historical analysis of human behavior in warfare (even when matched to weather patterns) is not predictive. In an example that parallels the weather-casualty pattern, a July 26, 1970, MACV intelligence report stated,

> In the Laotian panhandle, the rainy season runs from the middle of May through the middle of October. During that period the enemy logistical activity along the [Ho Chi Minh] trail is almost at a stop. *This* year he is making a very deliberate effort to continue his logistical effort into the rainy season.[145]

Recognizing that there are traditionally fewer casualties and less fighting during heavy rains might help a policymaker understand historical aggregated casualty data by showing why casualties may have risen or fallen during specific periods. But this kind of historical analysis, while it can have clear statistical correlations, cannot show clear cause-and-effect relationships: While it may be true that casualties rise and fall with the rains, casualty levels over any one specific period will rise or fall for a complex web of reasons. At best, it would be possible to say the following about Thayer's weather pattern analysis:

- There *typically* seems to be less fighting during heavy rain than during dry periods.
- Historically, friendly casualties have dropped during periods of heavy rain.
- It seems likely that this drop in casualties is due to reduced fighting.
- However, we cannot show clear cause and effect in all cases.
- This analysis is not reliably predictive.

---

[143] Rehm, 1985, p. 4-5.

[144] Although Thayer shows that even these data were often lagging and inaccurate at any one point in time.

[145] Sorley, 2004, p. 445 (emphasis in original). In this case, the pattern could be useful in showing how the North Vietnamese were applying additional effort to logistics. This information could be used to support an all-source intelligence analytic finding that could be incorporated into campaign assessment, but it would be one of many inputs in the assessment.

Even if this pattern could be predictive, predicting insurgent behavior based on patterns (or trends) might be a recipe for disaster if insurgents are savvy: Insurgents are also capable of identifying the same pattern and can take advantage of pattern analysis. The North Vietnamese achieved strategic surprise against the United States and GVN by launching a major countrywide offensive during the traditional Tet holiday in 1968. In the years prior to this offensive, North and South had observed a cease-fire during Tet. The historical pattern showed that there was much less fighting during Tet than immediately before and after the holiday. This pattern analysis was so convincing that it likely precluded the United States and GVN from capitalizing on intelligence reporting that indicated that an attack was likely.[146] It would not be much of a stretch for the North Vietnamese to take advantage of predictable U.S. behavior (e.g., a maintenance stand-down when VC units are assumed to be less active) during the rainy season.[147]

Could the inclement weather and casualty pattern have been useful as a coincidence indicator to support a broader assessment finding? Perhaps, but because it was a nonspecific pattern and not a time-series trend based on information accumulated over a specific period, it could not be safely compared to the kind of time-series trends that are used for centralized assessment (e.g., attack data or casualty data over time, updated to the date of the assessment). For example, one could compare the weather and casualty pattern to current time-series casualty data to show that the rainy season might have led to a reduction in casualties. But this analysis would not stand up to scrutiny because the nonspecific pattern could not show clear causation. From an analytic standpoint, it would be dangerous and potentially misleading to claim that casualties were dropping because "they always drop this time of year due to the rain." In this same case, then, could the pattern be used as a "flashlight" to point analysts in the direction of more explicit information? It could be used as a tool to help discern patterns from trends, but this kind of examination is better suited to intelligence analysis. In practice, Thayer presented these patterns in his *Systems Analysis* series of reports as

---

[146] Sharp and Westmoreland, 1968, p. 41. See also Wirtz, 1994, p. 66. Former CIA analyst John T. Moore testified before Congress in the *Westmoreland v. CBS* case that the Defense Intelligence Agency had suppressed pre-Tet warnings based on a disagreement over troop strength estimates between the two agencies and MACV (*Westmoreland v. CBS*, 1984; see, specifically, Westmoreland's memorandum in opposition to CBS's motion to dismiss, October 19, 1982).

[147] See Gayvert, 2010, p. 7. In a thought-provoking article on the HES, Gayvert points out that HES data showed a decline in GVN control across the countryside in the six months prior to the Tet Offensive and that the HES might have clued analysts in to impending trouble. The HES data may have shown an increase in insurgent activity, but the gross HES graphics presented to the public showed an almost steady increase in GVN control across the population and only flat levels of VC control from January 1967 to January 1968 (Sharp and Westmoreland, 1968, p. 199). OSA analysts described the pre-Tet decline in support as a "slight net downgrading" and not a clear indicator of trouble (Thayer, 1975b, p. 78). HES data also showed a very brief decline in GVN control after Tet. This decline may have reflected the effects of the Tet Offensive, or it may have been an indication that the pre-HES data were inaccurate. There is no way to tell whether the drop in control as reported by the HES in the six months prior to Tet was in fact accurate (and therefore indicative of problems) or whether it reflected a shift in reporting criteria, a change in personnel, or any number of other factors.

key components of his assessments, and in his own words, he based his entire assessment process on the ability to find useful patterns in reasonably accurate COIN data.

What about *trends* identified by OSA analysts? One of the standard methods of analysis was to compare U.S. operations to either enemy operations or environmental changes. This kind of analysis lines up well with contemporary trend analysis. Yet, two of the most confounding and inaccurate aggregated metrics used to measure progress in the Vietnam War were "battalion days of operation" and "operational days of contact." The battalion days of operation metric purported to show how frequently U.S. and GVN military units were operating and at what scale (large or small). Operational days of contact purported to show only those days that battalions actually saw combat.[148] In *War Without Fronts*, Thayer finds that "the figures are *not useful for even simple analysis of trends and comparisons*."[149] He also calls any attempt to use these data for statistical analysis futile.[150] However, he uses them for statistical trend analysis on the following page to produce what he calls tentative conclusions. Later in the book, he directly correlates statistical data on operational days of contact with casualty data in a table titled "Correlation Analysis: Friendly Battalion Operations with Contact Against Combat Deaths."[151] This statistical correlation analysis is presented by Thayer without caveat. Anyone reading this table without the context of Thayer's earlier comments (pp. 55 and 56; the table appears on p. 93) would have every reason to believe that these data were accurate and that the correlation analysis was derived through a sound statistical method using valid data. The method may or may not have been sound, but the fact that OSA would use data described as unsuitable for comparisons to produce a correlation analysis shows how compelling aggregated quantitative data can be to both analysts and policymakers. The trends shown in the table are not useful for *any* analysis by OSA standards, so they should not be used for campaign assessment. The table on casualty comparison in *War Without Fronts* also compares bad operational data not just with accurate U.S. casualty data but also with bad enemy body count data.[152] When data are used in correlation analyses (like the comparison of friendly operations to combat deaths), the most inaccurate and gapped data shape the overall analysis. For example, if it were possible to know the absolute accuracy and consistency

---

[148] Thayer points out that a battalion "day" was actually an aggregation of company days and that one battalion could have 1.33 days of potential activity. The entire system was nonsensical (Thayer, 1985, p. 56). Further, Sheehan (1988, p. 636) describes how Marine Corps units were forbidden from counting pacification missions as battalion days of operation, despite the fact that pacification was part of their mission.

[149] Thayer, 1985, p. 55 (emphasis added).

[150] Thayer, 1985, p. 56.

[151] Thayer, 1985, p. 93.

[152] It also calculates findings using assumed troop strength for a Republic of Vietnam Armed Forces infantry battalion and a U.S. infantry battalion. These numbers do not reflect reality; very few Vietnamese or U.S. infantry battalions were ever close to full combat strength.

of a data set, it might be possible to determine the overall inaccuracy of analysis based on the correlations among the data elements in two or more sets. If the operational days of contact data set was 40-percent accurate and the U.S. casualty data set was 95-percent accurate, the correlation analysis based on the two data sets would still be, at best, only about 40-percent accurate. This means that any effort to analyze inaccurate aggregated data by correlating them with other data—the most common way to place aggregated data in context or to identify cause-and-effect linkage—would be counterproductive. Since there is no way to determine absolute data accuracy in COIN, there is no way to determine the accuracy of any correlation-based analysis for COIN assessment.

It is also possible to show patterns and trends in the quantitative analysis of Vietnam War data that are precise and consistent but misleading. The HES may have produced some consistent data, but those consistencies probably reflected a good deal of consistent inaccuracy, or what Downes-Martin refers to as "precision without accuracy."[153] For example, the advisor in Phong Dinh Province consistently reported most of the same erroneous data for almost his entire tour. Thayer contends that HES data correlated well with non-HES data to show definite patterns. For example, he shows that anti-aircraft incidents "tend to occur over VC controlled hamlets, and incidents around A or B hamlets tend to feature terror."[154] This may have been true, but those A- and B-rated hamlets may also have been hiding large VC units or VCI elements that were savvy enough not to draw attention to their presence by conducting anti-aircraft or large ground attacks nearby (e.g., Vinh Binh Province in 1971). The VC could easily figure out that advisors rarely visited hamlets and then typically did so in daylight; it would not be difficult to deceive them by simply terrifying the population into putting on a good face and hiding for an hour or two once a month. It is very possible that many A- and B-rated hamlets were, in fact, sanctuaries for insurgents.[155]

The steady rise in population security recorded in the HES data could have been due to an actual improvement in pacification, but it could also have been the result of consistent but inaccurate reporting from advisors who were unofficially deterred from downgrading reports and simultaneously rewarded for showing at least incremental improvement. This kind of pressure could lead to a slow but steady improvement in the percentages, and this trend could have been sustained as U.S. advisors transferred reporting responsibility to GVN partners who may have been more likely to report

---

[153] Downes-Martin, 2010b, p. 7.

[154] Thayer, 1975b, p. 61.

[155] Both Corson (1968) and Ahern (2010) substantiate this assumption. Ahern states that the VCI may have been present when main-force VC units were not. He describes an operation in a hamlet that sat only a few hundred yards from a standing Marine Corps infantry battalion's headquarters. The Marines and accompanying GVN civil and military forces discovered "booby traps, concrete-lined bunkers, camouflaged pits concealing sharpened bamboo stakes . . . and caches of documents and weapons" (Ahern, 2010, pp. 249–250).

improvement as a matter of course.[156] Trends that emerge from consistently poor and misleading data are poor and misleading. The assertion that definite and *useful* patterns and trends for COIN campaign assessment could be gleaned from aggregated quantitative data of notably poor quality is not proven. Leslie H. Gelb and Richard K. Betts, authors of *The Irony of Vietnam: the System Worked*, believe that these analyses were often mistaken and misleading:

> [S]tatistics by their very nature could not go deeply enough. Much of the most important information about Vietnam was essentially unquantifiable, and even when it could be quantified the dangers of misinterpretation and overinterpretation were ever present. Comparison with years past was an illusory measure when it was not coupled with judgments about how far there still was to go and how likely it was that the goal could ever be reached. It was all too easy to confuse short-term breathing spells with long-term trends and to confuse "things getting better" with "winning."[157]

The last sentence in this quote is particularly relevant for pattern and trend analysis. Because pattern and trend analysis in Vietnam was necessarily imprecise, it was not possible to establish meaningful time-series thresholds for operational or strategic patterns and trends. Consequently, analysts could not plot an accurate, definable path to victory along time-series graphs because there was no reasonably defined quantitative objective. This means that all pattern and trend analyses required interpretation that was almost entirely subjective. The degree of subjectivity inherent in these analyses undermined their value in the absence of extensive narrative explanation, but narrative explanations (e.g., province advisor reports) became superfluous when quantitative data (e.g., HES data) were made available. Policy debate then boiled down to the subjective interpretation of inaccurate data by people far removed from the battlefield. In the few cases in which quantifiable objectives or thresholds were established during the Vietnam War, they proved to be both meaningless and misleading. Population security objectives established at various times by Komer and Westmoreland did not produce victory even when they were surpassed, and, as discussed later in this chapter, enemy casualty thresholds failed to deliver a quantifiable milestone for success.[158]

Thayer recognized the inherent subjectivity in pattern and trend analysis, and he always couched his analyses with narrative. He had experience in Vietnam and continued to draw on viewpoints from the field to support his analysis at OSA.[159] Even a

---

[156] Details of this transition to GVN control can be found in Cushman, 1972, p. 27.

[157] Gelb and Betts, 1979, p. 303.

[158] For example, Komer described a pacification threshold objective of 1,800 hamlets in 1966. This particular threshold was not met, but others were (Komer, 1968a, p. 2).

[159] See, e.g., Sheehan, 1988, p. 683. Sheehan describes Thayer's debriefing of Army LTC Hal Moore, commander of a U.S. Army infantry battalion that had recently been involved in heavy fighting in the Ia Drang Valley.

quick review of the *Southeast Asia Analysis Reports* shows that Thayer and his analysts sometimes had a much more in-depth understanding of the issues than the aggregated data alone could reflect.[160] They had obviously done extensive research on specific subjects and relied on multiple sources for some elements of their work, although these other sources were rarely cited. While some of the assessments in the *Southeast Asia Analysis Reports* reflected shallow quantitative interpretation, others were nearly (for lack of sourcing) on par with high-quality all-source intelligence analyses. For example, Thayer's analysis of the security situation along Route 4 between the Mekong Delta and Saigon in 1967 contained what appears to be useful insight that might have supported operational decisionmaking. The core of this analysis hinges on two captured enemy documents that, if accurate, showed clear intent to disrupt traffic on Route 4. The HES and security force data presented in charts and tables add almost nothing to the contextual narrative; they could have easily been left out of the report. The important quantitative data are those that show a twofold increase in VC attacks along Route 4 in two specific provinces. In this case, Thayer showed a useful trend— one that helped verify the validity of the captured documents.[161] The important distinction between this kind of analysis and his analysis of broader statistics in the *Southeast Asia Analysis Reports* and *War Without Fronts* is that (at least in this example) the quantitative data set was small, locally focused, and contextualized with other sources and narrative. Even this rather useful analysis would have benefitted significantly from citation and added subordinate context.

OSA pattern and trend analysis was least useful and potentially most misleading when it was highly aggregated and presented without sufficient context. Komer, who was working out of Saigon, took Thayer's analysts to task for failing to conduct a more detailed analysis of the pacification program. He questioned why OSA was not incorporating reams of other available data into the analysis published in the *Southeast Asia Analysis Reports* and asked the analysts not to "over-read" the HES data. From his vantage point in Saigon, Komer recognized that assessment requires context. Yet, the kind of contextual analysis he was demanding from OSA was difficult to develop for a small, centralized staff. The OSA analysts replied that they did not have access to detailed data streams and then fell back on their quantitative processes to defend their findings.[162]

---

[160] The reports were later republished in the 12-volume series *A Systems Analysis View of the Vietnam War 1965–1972*.

[161] Thayer, 1975c, pp. 9–22. This analysis would have been much more credible if it contained source notes. In the absence of sourcing, it is unclear how much of this analysis was Thayer's opinion and how much was grounded in specific reporting. It is also unclear to what degree the reporting he did use was accurate, although he does caveat some of his information.

[162] Thayer, 1975b, pp. 75–83. This analytic exchange offers fascinating insight into the inner workings of both CORDS and OSA. It reveals Komer's deep dissatisfaction with elements of OSA's analysis. (At one point, he calls the analysts' assumptions "silly.") He directly and angrily confronts the analysts, identifying them as "desk types

OSA analysis was most useful and least likely to be misleading when minimally aggregated and presented in context; the Route 4 study is a case in point. But these detailed narrative assessments crossed the line from strategic campaign assessment into the realm of operational intelligence analysis.[163] The study of VC activity along Route 4 could have been useful for operational planning, and it certainly *informed* aspects of COIN assessment. But the report was no more a holistic assessment than any other intelligence analysis report. By itself, it did not show strategic progress, and it did not present a sufficient stand-alone case for increasing or decreasing troop deployments. The report, and others like it, would have been more appropriately written by an operational intelligence staff and then incorporated into a layered contextual analysis of the war. There is clear indication in Thayer's writing and postwar statements that the more he became immersed in the data, the more he and his analysts slid away from comprehensive assessment and into the realm of intelligence analysis. The desire for meaning and context in the aggregated data drove them to seek more and more contextual detail even as they continued to try to find broad meaning in gross quantitative patterns. Thayer analogized the process to cooking.[164]

Thayer and OSA made a valiant effort to pull together data and analysis from a cold start. They had no COIN assessment doctrine to work with, and they had no historical case studies to rely on (at least not one that compared to the complexity of the COIN campaign in Vietnam). They were not always fully assisted in their efforts by MACV and other military organizations, and they often had to hunt down their own data: In April 1968, they stated that they were "hampered by an almost total

---

in Washington," and challenges them to visit Vietnam to check their work. The analysts' reply was detailed. This back-and-forth debate has been replicated in both form and (with necessary changes in terminology) context in both Iraq and Afghanistan. Some might argue that debate of this sort is healthy, and from a strictly analytic standpoint they would be right: It is healthy for analysts to have iterative engagement with people in the field to gain ground-truth perspective on their reporting. However, as the head of the pacification program, Komer was necessarily an interested party in the results of the analysis. This two-way debate would have been well served by other input.

[163] This may have technically fallen under the purview of OSA's Southeast Asia Intelligence Analysis and Force Effectiveness Division, but it is not clear from the literature whether the office was specifically conducting intelligence or assessment. However, Thayer's interpretation of his division's work seems to make it clear that it was focused on assessment and systems analysis for strategic decisionmaking.

[164] Rehm, 1985, p. 9-11. In another example of the shift to intelligence analysis, Thayer stated,

> If we had known before the Tet Offensive . . . that Westmoreland thought there were 25,000 infiltrators a month coming down we would have flipped to McNamara or whoever was there at the time. We would have said, "there is something coming. We don't know what it is or where or when, but we suggest that you want to ask the Command or the Joint Chiefs of Staff the following questions." (Rehm, 1985, p. 9-13)

This is clearly an intelligence indications and warning analysis by all definitions. Thayer may not have been precluded from making these analyses (his division had the word "intelligence" in its title), but if this was the focus of OSA assessment then the office should have been resituated within an intelligence analysis organization, such as the CIA or Defense Intelligence Agency, or intelligence analyses should have somehow been segregated from assessments.

absence of meaningful data."[165] The data they did receive were significantly inaccurate and incomplete. Strategic objectives were often unclear. The systems analysts and operations researchers at OSA reverted to centralized quantitative analysis by default. As a result, OSA analyses were presented in slices that did not form a comprehensive picture. The patterns and trends that were identified could not be defended as accurate and did not build an overarching understanding of progress in comparison to strategic objectives. Policymakers and the public were left to contend with vague patterns and misleadingly precise trends drawn from quantitative systems analyses that were so complex as to be opaque to the average American.

This was a best effort in the absence of other clear options, but it was not transparent, credible, relevant, balanced, or parsimonious. One could argue that the data were analyzed, but analysis was conducted without an overarching method designed to produce a holistic impression of the campaign. Even during this period, ample COIN doctrine and literature were available to OSA.[166] It would have been possible to create an analytic process that was congruent with a more realistic appreciation for the COIN environment, but probably not from a top-level staff.

Casualty Data and Impressionistic Assessment of the Vietnam War

> Every quantitative measurement we have shows that we are winning the war.
>
> —*Secretary of Defense Robert S. McNamara, 1962*[167]

Building on the discussions of Vietnam-era assessment presented in this chapter, this section develops a case example showing how senior leaders used aggregated data (specifically casualty data) to shape strategy and support decisionmaking during the Vietnam War. The example revolves around the strategic assessment of aggregated body count data by both military commanders (e.g., GEN William C. Westmoreland) and civilian leaders (e.g., Secretary of Defense Robert S. McNamara and National Security Advisor McGeorge Bundy). It refers to General Westmoreland's analysis of a casualty data chart in his *Report on the War in Vietnam as of 30 June 1968*, which he (and his staff) coauthored with ADM U. S. G. Sharp, commander of U.S. Pacific Command.[168]

---

[165] Thayer, 1975b, p. 77.

[166] This includes the U.S. Marine Corps' *Small Wars Manual*, the Army's FM 31-series publications dating back to the 1950s, Mao Tse-Tung's *On Guerrilla Warfare*, and Che Guevara's *Guerrilla Warfare*.

[167] Sheehan, 1988, p. 290.

[168] As commander of the U.S. Pacific Command, Sharp was effectively the next level of command above Westmoreland. The specific analysis of casualty data was probably authored by Westmoreland and MACV staff, not by Sharp and his staff. The layout of the book seems to show that Sharp addressed "area of responsibility" issues outside of Vietnam, while Westmoreland authored or edited the portions on the war within Vietnam. The discus-

At more than 300 pages, the report is replete with statistics, charts, and photographs. It represents the consolidated military data on the war up to mid-1968. It also contains contextual narrative of the war effort and what amounts to the holistic assessment of the U.S. COIN effort in Vietnam. Westmoreland conveys most of his key assessment points through annotated charts.

Westmoreland's assessment is predicated on a theory of attrition warfare and not population-centric warfare as described in the 1940 Marine Corps *Small Wars Manual* or in the 2006 COIN manual, FM 3-24.[169] After Harkins, he was the leading proponent of the attrition warfare strategy and believed that when the enemy "loses one man it's equivalent to our loss of more than ten."[170] The entire attrition strategy was based on unsustainable quantitative assumptions like this. The purpose of this section, however, is not to debate the relative merits of attrition versus population-centric COIN strategies. Instead, it is intended to explain how centralized quantitative assessment can present and sustain a misleading understanding of a COIN campaign. It shows how time-series analysis, trend analysis derived from time-series data, and quantitative thresholds can both confuse policymakers and mask gaps in knowledge and understanding.

Figure 6.3 presents a comparative casualty ratio. It shows the ratio of enemy casualties (body count statistics) compared to allied casualties (U.S. casualty statistics) between January 1965 and December 1968. Throughout their report, Sharp and Westmoreland use these types of charts to support their narrative comments, and the narrative tends to rely heavily on the charts.[171]

Westmoreland annotated this graph with an assessment on the opposite page (p. 190). He does not describe how he reached his conclusions, what other inputs he might have used to analyze these data, or how he drew a connection between cause and effect:

> The chart on the facing page shows the general upward trend in the ratio of friendly to enemy casualties. . . . The trend is clearly up, and at the present time it has reached a ratio in which six of the enemy are killed for each allied soldier lost. From a purely military standpoint this trend shows the impact of the introduction

---

sion here does not present a comprehensive analysis of the *Report on the War in Vietnam*; such an examination was beyond the scope of this study.

[169] See Headquarters, U.S. Marine Corps, 1940, and HQDA, 2006c, respectively.

[170] Westmoreland, 1967b, p. 5. On page 31, he explains that this ratio was derived from the difference between North Vietnamese and U.S. population figures. He should have stated in his earlier remarks that his ratio was purely mathematical. In this press briefing, Westmoreland clearly defines and explains the attrition strategy and his theories of counterrevolutionary warfare. Westmoreland reads from an operational update that he received that morning. It described a captured NVA officer who had official military orders on his person to fight to the last man. Westmoreland also states on page 37 that the Battle of Dak To was "the beginning of a great defeat of the enemy."

[171] Sharp and Westmoreland, 1968, p. 191.

of U.S. troops, the steady improvement in performance by all allied forces, and the steady decline in battlefield performance by the enemy.[172]

When parsed, this paragraph reveals some of the limitations of operational time-series assessment as a holistic assessment tool. Because one of the data sets (enemy body count) was inaccurate to a significant degree, the entire chart is inaccurate to a significant degree. If the reporting had been accurate it *might* have helped Westmoreland support the supposition in the final sentence: Allied forces were steadily improving while the enemy was steadily declining in performance. However, even a reasonably accurate trend could be the result of other factors. A higher VC and NVA casualty rate might show increasing boldness among the enemy or an increase in active manpower (more troops in the field leads to more casualties). This would be a negative and not a

Figure 6.3
Ratio of Enemy to Allied Casualties, Vietnam, 1965–1968

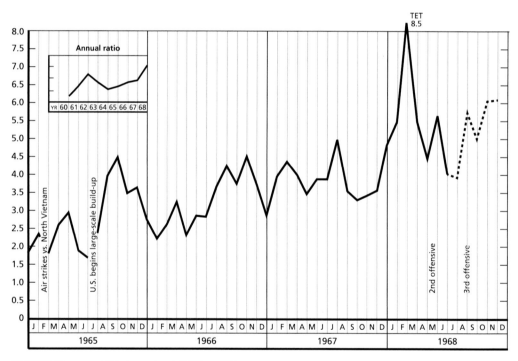

SOURCE: Sharp and Westmoreland, 1968, p. 191.
RAND *MG1086-6.3*

---

[172] Sharp and Westmoreland, 1968, p. 190. It is concerning that Westmoreland misidentifies his own chart as friendly-to-enemy casualties rather than enemy-to-friendly in this assessment. This shows either a simple editing error or a deeper failure to understand the data.

positive trend for the United States and GVN. Myriad other factors that play into a ratio like this are not—and could not be—depicted in such a graph.

Westmoreland caveats his finding by adding the phrase, "from a purely military standpoint." He seems to suggest that one should not draw any holistic (i.e., national strategic) conclusions about the COIN campaign from the graph.[173] However, despite his caveat, Westmoreland clearly implies that this improved ratio is a key indicator of success over time. His next time-series graph shows the ratio of enemy to allied weapon losses over the same period. In the text accompanying this graph, he states, "Surely the combination of these first two charts reflects a situation which must be of major concern to the enemy."

This supposition, offered without any clear analysis or corroborating evidence, was mistaken.[174] Neither President Ho Chi Minh nor the commander of the NVA, General Vo Nguyen Giap, was overly concerned with casualties or weapon losses.[175] They made this clear in statements that would sound boastful coming from anyone other than 1960s-era North Vietnamese communists. In a seemingly direct refutation of Westmoreland's casualty ratio assumption, Ho famously stated, "You can kill ten of my men for every one I kill of yours, but even at those odds, you will lose and I will win," while Giap stated, "Every minute, hundreds of thousands of people die on this earth. The life or death of a hundred, a thousand, tens of thousands of human beings, even our compatriots, means little. . . . Westmoreland was wrong to count on his superior firepower to grind us down."[176] While the VC and NVA may have been losing more weapons in 1968 than in 1965, China was pouring materiel into the North, replenishing supplies perhaps as fast as they were consumed.[177] By 1965, the communists had settled in for a long war, and Ho Chi Minh was prepared to fight for "five, ten, twenty-five years, or even longer."[178] Westmoreland looked at the graph, saw what appeared to be a positive trend, and then wrongly extrapolated what should

---

[173] Even taking into account the potential contextualization provided by the rest of the nearly 300-page report would likely leave the consumer with little more than a vague impression of progress and no comprehensive understanding as to how this progress fit into a strategic framework.

[174] To be fair to Westmoreland, as late as 1970, the intelligence community continued to assert that the leadership in Hanoi could not sustain manpower losses indefinitely ("The Outlook from Hanoi: Factors Affecting North Vietnam's Policy on the War in Vietnam," Special National Intelligence Estimate 14.3-70, February 5, 1970, in National Intelligence Council, 2005, pp. 506–526). This reporting may not have been accurate, but it probably influenced Westmoreland's thinking.

[175] It is difficult to analyze Westmoreland's comment because the phrase "of major concern" is a vague and meaningless qualifier to a policymaker attempting to determine courses of action. There may have been times that both leaders were concerned about the logistics of manpower and materiel, but these concerns did not deter them from believing in eventual victory against the South and the United States.

[176] Karnow, 1984, p. 20.

[177] Zhai, 2000, pp. 179–180.

[178] CIA, 1966, p. VIII-9.

have been a very narrow set of data—or a trend—into a broad conclusion about the overall success of his COIN efforts. He had reams of other analyses, raw reports, and advice at his fingertips, but his strategic bottom line was derived immediately and incorrectly from centralized quantitative charts. This episode shows the degree of influence that aggregated and decontextualized charts can have on even seasoned and well-informed decisionmakers.

McNamara also used body count statistics to assess the war and to inform policy-makers and the public. As Secretary of Defense during the period that Westmoreland was in command in Vietnam, he is ultimately responsible for the *Report on the War in Vietnam* and for the analysis coming out of OSA. He defended his role in count-ing VC and NVA casualties, stating, "We tried to use body counts as a measurement to help us figure out what we should be doing in Vietnam to win the war while put-ting our troops at the least risk."[179] He generally attributed the details of the attrition-based assessments to Westmoreland. He described Westmoreland's conviction that the North would hit a "crossover point" at which it could no longer sustain the level of casualties inflicted by U.S. and South Vietnamese forces. Westmoreland thought that the point was reached in the spring of 1967, a year before the strategically disastrous Tet Offensive.[180] It is obvious in retrospect that they were wrong, but how were West-moreland and McNamara determining that this crossover point existed or had been reached? McNamara stated that "we needed to have some idea what they could sustain and what their losses were," but he failed to explain how they determined the first part of the equation.[181]

This crossover point is simply a time-series threshold or milestone.[182] In the case of Westmoreland's casualty ratio graph, the determined threshold would appear as a number along the y-axis of the graph. At that specific quantitative point, the North Vietnamese would lose the will to fight or would be physically defeated. Identifying that point would require knowing that (1) this method could have the desired effect (killing wins the war); (2) the North Vietnamese leadership had a specific casualty limit in mind, and when that number was reached they would quit; or (3) the United States knew more about the willpower and intent of the North Vietnamese leaders than those leaders knew about themselves, down to the specific ratio of casualties they were willing to accept. Any shift in the Vietnam casualty ratio would result in a change to the assessment of both the effect and the overall strategic outlook. This change would

---

[179] McNamara, 1995, p. 238. This statement is sadly ironic, considering that the body count metric put troops at risk.

[180] McNamara, 1995, p. 238.

[181] McNamara, 1995, p. 238.

[182] If Westmoreland had predicted that the threshold would have been achieved at a certain point in time, it would also have been a milestone. His "tipping point" became a rather dubious threshold and milestone only in retrospect.

have to be justified in some detail to retain transparency and credibility, assuming that it was possible to justify the quantitative threshold in the first place.

National Security Advisor McGeorge Bundy played a key role in shaping the Vietnam War strategy and particularly in estimating enemy strength and resiliency. He is generally recognized as one of the most influential advisors in the White House in the early to mid-1960s. In a self-critical statement, he admitted that there were insufficient analyses at all levels of the national security apparatus and that all analytic staffs failed to grasp "how the enemy would take it and come back for *more*."[183] According to Bundy biographer and confidant Gordon M. Goldstein, "There was no analysis or evidence to validate Bundy's expectation that Ho Chi Minh and his fervent followers would capitulate."[184] Not only did Bundy fail to adequately examine the policy of attrition, but he also failed to analyze or shape the assessment of the attrition strategy. While it is not clear what role Bundy had in shaping or approving Westmoreland's ill-conceived body count threshold, Goldstein takes him to task for failing to follow through on the mechanics of the attrition assessment process:

> How many casualties would be required to compel them to quit? How many years would it take? Where, exactly, was the tipping point—the threshold of pain and loss that would extract a fundamental reversal in Vietnamese nationalist ambitions?[185]

Goldstein's questions are rhetorical, even sardonic. But to Westmoreland, McNamara, and Bundy, each struggling in his own way to win the war, these questions might have seemed quite reasonable early on. Vietnam arguably was the most complex irregular war in which the United States had engaged to that point: The military was battling against both a conventional army supported by powerful external sponsors and a full-strength insurgency consisting of tens (if not hundreds) of thousands of experienced and aggressive combatants. The U.S. military was simultaneously trying to hold together a faltering ally while its own morale ebbed. There was extraordinary pressure on the top advisors and decisionmakers in the military, the Pentagon, and the White House to bring the war to a successful conclusion. Each of these three men, in his own way, bought into the idea that a finite solution might exist, that this solution could be achieved by applying graduated pressure and specific amounts of resources against the threat, and that the solution could be measured through systems analysis

---

[183] Goldstein, 2008, pp. 178–179 (emphasis in original).

[184] Goldstein, 2008, p. 179.

[185] Goldstein, 2008, p. 179.

and displayed on a time-series graph. In the end, they were all proven wrong, and both McNamara and Bundy agreed in retrospect that theirs was a failed approach.[186]

This example shows the dangers of relying on a narrow set of aggregated and decontextualized data to inform strategic decisionmaking in COIN. It is not a precise allegory for either Iraq or Afghanistan because neither of these wars has been prosecuted according to attrition warfare theory and neither reflects the kind of hybrid conventional/COIN war fought in Vietnam. However, it does show how even "the best and the brightest" can fail to understand the complexities of COIN and COIN assessment. It also shows that despite the availability of massive amounts of aggregated quantitative data and finished quantitative analyses from MACV and OSA, both senior military leaders and policymakers relied on supposition and overly optimistic readings of quantitative charts to reach at least some of their major decisions. Some of the criticisms aimed at these senior leaders may be valid, but it is also apparent that there was something missing from the assessment process.

Is Analysis of Vietnam Assessment Germane to Iraq and Afghanistan Assessment?

Comparisons between Vietnam and contemporary COIN campaigns suggest that it is reasonable to equate the Vietnam-era assessment processes and those used in Iraq and Afghanistan for the purposes of research:

- There were significant advances in technological capability between 1972 and 2001–2011, particularly in the fields of communication and surveillance. While technology can certainly account for some improvement in data reporting, particularly in the speed of reporting, units in both Iraq and Afghanistan continue to pass reports in written form (even if it is electronic) and via tactical radio.
- COIN is still conducted by infantry units patrolling on foot through underdeveloped urban and rural terrain, and these units face the same challenges that their counterparts faced in Vietnam. Their rifles, radios, rockets, hand grenades, boots, and helmets typically are improved versions of Vietnam-era equipment. There has been no "revolution in COIN affairs."
- Centralized core metrics have not changed significantly, and in most cases, Iraq and Afghanistan metrics appear very similar or identical to each other and to Vietnam-era metrics.[187] Units in contemporary campaigns are generally trying

---

[186] McNamara tended to equivocate on analytic methods and in several statements defended the application of systems analysis to war assessment. However, he also stated that the attrition and graduated pressure strategies were poorly conceived.

[187] For example, many of the questions in the HES are similar to both the security and development metrics in use in Afghanistan today.

to count the same things that their counterparts were attempting to count in Vietnam—including bodies in some cases.

- Polls are still taken using face-to-face engagement and human intelligence is still captured in face-to-face discussion; all collection and reporting continues to depend on people for precision and accuracy.
- Assessments in both Iraq and Afghanistan rely, in part, on host-nation reporting, which is often unreliable. Most host-nation reports are delivered in writing (typically on paper and not in English), in face-to-face meetings, or via radio.[188]
- Time has not changed the geometry of war. Accurate population-centric data cannot be collected in insurgent-held territory or in places where there is no friendly presence.[189] This was true in Vietnam and it is true in Iraq and Afghanistan.
- Vietnam-era assessment tools like the HES were effects-based despite the fact that the language of effects-based theory did not permeate U.S. doctrine until well after the end of the war.

It would not be far-fetched to state that the current process is immediately derived from, and closely mirrors, the Vietnam-era process. Similarities between some of these systems are general, while others are specific. This section concludes with a striking comparison of the 1964 pacification assessment process in Vietnam and the 2010 district assessment process in Afghanistan:

- In mid-1964, MACV implemented the Pacification Reporting System. The system purported to show the degree of control that the GVN had over key areas of the countryside. There were five rating codes in this system: secured, undergoing securing, undergoing clearing, uncontested, and VC-controlled. These five codes were represented by colors (respectively, dark blue, light blue, green, white, and red) on a map that was maintained by the Revolutionary Development and Support Directorate in Saigon.[190]
- In early 2010, IJC implemented the District Assessment Model (DAM), a system that purported to show the degree of control that the government of Afghani-

---

[188] Iraqi reporting might be more accurate than Afghan reporting because the Iraqis have a tradition of structured and centralized administration from the provinces to the center and the Afghans do not; the South Vietnamese had a highly structured bureaucracy designed by French colonial administrators. While host-nation reports during war should be considered suspect for a variety of reasons, there is no clear reason to value Iraqi or Afghan reporting over Vietnamese reporting. Cooper et al. (1972b, p. 235) advise, "Never again should the United States, while lending advice and assistance to a host nation fighting an insurgency, allow itself to become . . . dependent, for the vital information it requires for policymaking, on distorted and otherwise unsatisfactory reporting and evaluation by host nation agencies."

[189] It is always possible to collect intelligence from anywhere on earth, but the quality and completeness of the data drawn from intelligence collection is highly dependent on collection capabilities, access, and resource availability. Collection of accurate population data requires some kind of immersion in the population.

[190] Cooper et al., 1972b, p. 210.

stan had over key districts in the countryside. There were five rating codes in this system: population supporting the government, population sympathetic, population neutral or on the fence, population sympathizing with insurgents, and population supporting the insurgency. These five codes were represented by colors (respectively, green, blue, yellow, olive, and red, with white indicating "not assessed") on a map that was maintained by the IJC Information Dominance Center in Kabul, Afghanistan.

Chapter Summary

Assessment of the Vietnam War was haphazard before MACV became a theater combat command in the early 1960s. Between the mid-1960s and the early 1970s, MACV and OSD (with CIA assistance) created the single largest and most comprehensive military COIN assessment apparatus in the history of warfare. It involved the efforts of hundreds of thousands of military personnel, civilians, Vietnamese nationals, intelligence experts, and analysts over the course of several years. These contributors produced hundreds of millions of data items (in DoD parlance), tens of thousands of tactical and operational analytic reports, and hundreds of comprehensive assessments that addressed nearly every aspect of the war. While it is not possible to state that poor assessment led to the loss of the Vietnam War, the war was distinguished by internal confusion, poor decisionmaking, and, ultimately, strategic defeat. It is apparent from analyzing the way in which assessments were presented and used that they contributed to many of the poor decisions that led to this defeat. Why did operational COIN assessment of the Vietnam War fail? This chapter explored the following issues and reasons:[191]

- No adequate precedent or doctrine on assessment existed prior to the war.
- The assessment data were intrinsically flawed due to the complex and chaotic environment of Vietnam.
- Centralized pattern and trend analysis was often inaccurate or misleading.
- Top-down assessment requirements resulted in widespread data fabrication.
- Senior leaders and policymakers misunderstood the value and meaning of available quantitative data.
- Senior leaders, policymakers, and analysts overrelied on centralized, quantitative analyses of aggregated data that were devoid of useful context.

---

[191] This is not to imply that defeat resulted directly from poor campaign assessment. Policymakers, military officers, the North Vietnamese, the South Vietnamese, and many other players all had a role in shaping the end of the war.

Overreliance on centralized, quantitative analysis was commonplace. There seems to have been no *systematic* effort to integrate qualitative reporting (e.g., human intelligence reports, engagement reports) into the OSA assessment process. Similarly, the MACV assessments were heavily quantitative.[192] This is an unfortunate oversight. In contrast to both Iraq and Afghanistan, Vietnam was a holistic U.S. government campaign from the beginning of the war (and certainly by the mid- to late 1960s). Hundreds of civilian advisors worked at the district and province levels across South Vietnam.[193] These advisors wrote hundreds if not thousands of highly detailed narratives that placed both qualitative and quantitative reporting within (often) insightful and structured context. The MACV CORDS report on Phong Dinh Province for February 1968 is an excellent example of the kind of information that was available from the field.[194] Military advisors and intelligence officers also produced reams of detailed narrative analysis that contained useful field context. While analysts and policymakers certainly read some of these reports, neither the U.S. military nor the civilian establishment systematized the collation and analysis of these reports for campaign assessment.

Some of the narratives written by senior leaders were relatively useful in presenting a step-by-step history of the war (e.g., Sharp and Westmoreland's *Report on the War in Vietnam*) or in presenting senior policy positions. These senior-level reports did not deliver consistent, transparent, or credible assessments, however. Published reports were often suspect in the minds of consumers because they contained obvious biases or were not clearly tied to contextual field reporting.[195] This research shows that the data produced by MACV and OSA obviated the requirement for comprehensive qualitative insight or narrative context, at least in the minds of some policymakers.[196]

The section "Pattern and Trend Analysis of Vietnam War Data," earlier in this chapter, compared the assessment methodology employed by OSA to the standards for assessment framed in the introduction to this monograph. It showed that the kind of pattern and trend analyses applied to the Vietnam campaign did not stack up well against these standards. Table 6.1 shows how assessment of the Vietnam War compared to the standards.

Subsequent chapters show a clear connection between COIN assessment processes in Vietnam, Iraq, and Afghanistan. There are certainly some differences among the cases, and each campaign has unique characteristics: Vietnam is not Iraq is not

---

[192] See, for example, Sharp and Westmoreland, 1968. This refers to campaign assessment and not necessarily to MACV intelligence analysis.

[193] Some civilian advisors work closely with military units in remote areas of Afghanistan and some did so in Iraq. However, neither of these efforts compared well with Vietnam-era efforts.

[194] MACV, 1968a; see Appendix C of this monograph.

[195] See, again, Sharp and Westmoreland, 1968, p. 9.

[196] The notable exceptions to this finding are the comprehensive intelligence analyses written by the CIA and other intelligence agencies, as well as the Pentagon Papers.

Table 6.1
Performance of Vietnam-Era Assessment Against Assessment Standards

Standard	Vietnam-Era Assessment	Analysis
Transparent	No	Data aggregation and methods opaque
Credible	No	No clear method, data questionable, oversold to policymakers and public
Relevant	No	Did not effectively support decisionmaking
Balanced	No	Did not systematically incorporate qualitative and mixed data or subordinate reports
Analyzed	No	Analysis was not consistent or holistic
Congruent	No	Did not reflect understanding of COIN
Parsimonious	No	Data requirements were overbearing

Afghanistan. But the challenges of assessment in Iraq and Afghanistan are consistent with most of the challenges faced in Vietnam. While Vietnam provided a precedent for COIN assessment, the following points are consistent for Iraq and Afghanistan assessment:

- No adequate doctrine on assessment existed prior to the war.
- The assessment data were and are intrinsically flawed due to the complex adaptive environments of Iraq and Afghanistan.
- Centralized pattern and trend analysis was and is often inaccurate or misleading.
- Top-down assessment requirements have resulted in data fabrication.
- Senior leaders and policymakers misunderstand the quantitative data.
- Senior leaders, policymakers, and analysts overrely on centralized, quantitative analyses of aggregated data that are devoid of useful context.

The U.S. military and the policy community have not adequately incorporated the lessons of the Vietnam COIN assessment case into military doctrine or into the U.S. government's approach to warfighting. Or, quite possibly, they have incorporated the wrong lessons. As a result, military staffs and policymakers must continue to rely on assessment processes that are "better than nothing." It remains to be seen whether this prescient post–Vietnam War quote from Don Oberdorfer, *Washington Post* reporter and author of *Tet! The Turning Point in the Vietnam War*, will also apply to future U.S. COIN campaigns as it has to Iraq and Afghanistan:

Originally intended to place the conduct of the war on a "scientific" and thus manageable basis and to enhance public confidence, the practice [of aggregated quantification of war data] ultimately consumed vast amounts of time and energy, led to misconceptions and erroneous conclusions about the war and was a major factor in the erosion of public confidence. Nevertheless, the "numbers game" may be now so deeply imbedded in military, press and public thinking that it will persist in future military conflicts.[197]

---

[197] Oberdorfer, 2001, p. 262; also quoted in U.S. Defense Logistics Agency, 1980, pp. 15-26–15-27. Oberdorfer's account of the Tet Offensive was originally published in 1971.

CHAPTER SEVEN

Assessment in Afghanistan

Going forward [in Afghanistan], we will not blindly stay the course. Instead, we will set clear metrics to measure progress and hold ourselves accountable. We'll consistently assess our efforts to train Afghan security forces and our progress in combating insurgents. We will measure the growth of Afghanistan's economy, and its illicit narcotics production. And we will review whether we are using the right tools and tactics to make progress towards accomplishing our goals.

—President Barack Obama, in a speech on the U.S. strategy in Afghanistan, March 27, 2009

The Vietnam case study offers a comprehensive view of the COIN assessment process from data collection in the field to policy support. It is also a static historical case and is therefore relatively easy to dissect and explain compared to ongoing cases. The definitive scholarly work on assessment of the Iraq and Afghanistan campaigns has yet to be written. And due to classification and the ongoing nature of operations, neither of these contemporary cases can provide the kind of detail and perspective offered by the Vietnam case. Nevertheless, Iraq seems particularly suited to examination because the war there is nearing termination, with the U.S. withdrawal in progress as of this writing. Unfortunately, the tendency to classify Iraq assessment reporting precludes a useful examination of Iraq assessment here, even as an ongoing case study; this chapter and the next refer to some of the available unclassified Iraq publications. The Afghanistan case is far more accessible to researchers due to efforts by senior military officers to keep at least some assessment material unclassified. Because the war there is ongoing, examination of the Afghanistan case can contribute to improvements in current U.S. and NATO assessment efforts. Indeed, this was the original intent of this research effort. Thus, this chapter focuses on how ISAF and its subordinate staffs are conducting assessment in Afghanistan.

Assessment of the Afghanistan campaign did not begin in earnest until 2003. Still, there was no comprehensive effort by the military to gather information and assess the war until at least 2008, so this chapter does not address the assessment process prior to that period.[1] From late 2008 through 2011, commanders, policymakers, and assessment staffs looked at assessment processes and methods with greater interest. They had very little doctrine to fall back on, and they had only the Vietnam and Iraq cases to guide them through an "industrial-strength insurgency." By default, they tended to borrow from Iraq because many had experience in that conflict; these processes were, in turn, duplicative of many Vietnam-era efforts. But commanders and staff officers also did exactly what the Vietnam-era OSA did in the mid-1960s: They made due with whatever tools they had on hand. They borrowed from a range of disciplines and tried to come up with assessments that made sense through trial and error. They fell back on heavily quantitative processes at first and then struggled to figure out how to capture contextual narrative and popular sentiment. The number of assessments grew uncontrollably as various offices and commands within NATO, ISAF, the U.S. military, U.S. civilian agencies, and policy circles struggled to understand the war. At one point in 2010, all the interested parties were collectively producing nearly 50 assessments of all kinds, delivered at various intervals by staffs from the battalion to the NATO headquarters level.[2] The number of assessments produced reflects bureaucratic momentum to some degree, but it also reflects a general dissatisfaction with assessment reporting. As of early 2011, efforts to improve and refine the holistic campaign assessments of Afghanistan were ongoing.

This chapter first describes the assessment process as it developed between late 2008 and 2010 and then provides an original analysis of this process, supplemented by the perspectives of two independent researchers. The assessment process in Afghanistan is in a near-continuous state of flux: The descriptions in this chapter are current only through late 2010. By the time this monograph is published, an entirely different process may be in place. However, an examination of the late 2008–2010 assessment process is useful not only to support continuing improvements in Afghanistan assessment but also to contribute to future doctrine and practices.

The theater-level Afghanistan assessment process is derived from doctrinal EBA, but it does not strictly follow EBA guidelines that have proven difficult to implement. Instead, it borrows from EBA, pattern and trend analysis, and other scientific and statistical methods to respond to centralized requests for information. The formal assessment process does not, in and of itself, provide an overarching holistic understanding of the campaign; instead, these insights are contributed by senior commanders in narrative and subjective reports. The lack of a comprehensive methodology, the inability to account for inconsistencies in centralized COIN assessment, and the onerous demands

---

[1]   During this period, OSD continued to try to assess the war through various offices and programs. Most of the briefings from these offices are restricted and cannot be included in this discussion.

[2]   This number was identified through a collaborative effort by military officers and researchers working on a NATO project in support of ISAF assessments from early 2010 to early 2011.

of a coalition bureaucracy have resulted in an assessment process that arguably does not meet most of the standards for quality assessment as defined in this monograph.

Overview of the Assessment Process in Afghanistan as of Late 2010

Between early 2009 and early 2011, ISAF and IJC attempted to implement various iterations of centralized assessment. GEN Stanley McChrystal, ISAF commander in 2009 and early 2010, stated that assessments should be both transparent and credible.[3] The first effort to create a transparent and credible system was initiated in 2009 under the newly formed ISAF AAG. These assessment documents described an intricately detailed and highly quantitative effects-based process that would rely on the collection and analysis of approximately 100 discrete field indicators. One of the most senior and experienced operations research officers in the U.S. Army led this effort. The ISAF commander rejected the standard operating procedure and its requirements on the basis that it was overly complex and opaque.[4]

Subsequently, in late 2009 and early 2010, the IJC developed and implemented the DAM. The DAM, discussed later, is an effects-based process that incorporates the commander's qualitative input but relies heavily on a list of data-driven indicators to provide what FM 5-0 refers to as *quantitative objectivity*.[5] District assessments are translated into a color-coded map product similar to the one maintained by the Revolutionary Development and Support Directorate in Saigon, South Vietnam, as discussed in Chapter Six.[6] Both the AAG and the IJC Information Dominance Center have experimented with several versions of narrative assessment reports, as well as some

---

[3]   The director of the AAG stated, "Credibility and transparency are the watchwords for our process. COMISAF [commander ISAF] and COMIJC [commander IJC] often ask, 'Is this credible?'" (AAG director, interview with the author, Kabul, Afghanistan, May 2010).

[4]   Assessment analyst, interview with the author, Afghanistan, May 6, 2010. ISAF currently produces quarterly and monthly reports derived from IJC reports and reports from other functional commands in theater. ISAF reports include a monthly assessment report that is submitted to NATO Joint Forces Command Brunssum, as well as a quarterly report, a monthly report on Afghan National Security Forces (ANSF), and documentation to support semiannual 1230 and 1231 reports. (These reports are required to comply with sections 1230 and 1231, respectively, of the National Defense Authorization Act for each fiscal year.)

[5]   See HQDA, 2010. I spent a week with the IJC Assessment Section in early May 2010 and continue to review IJC assessment publications. Contrasting this finding, a senior U.S. military officer stated at a recent conference on assessment, "If the commander disagrees with the way the data is being interpreted at higher headquarters, his judgment is always given precedence" (statement by a senior U.S. military officer, Allied Information Sharing Strategy Support to ISAF Population Metrics and Data Conference, Brunssum, Netherlands, September 1, 2010).

[6]   Cooper et al., 1972b, p. 210; former U.S. Army psychological operations advisor to Phong Dinh Province, Vietnam, interview with the author, Washington, D.C., April 19, 2010; Sharp and Westmoreland, 1968; Donovan, 1985.

highly quantitative efforts to assess ANSF and transition.[7] An example of these efforts is the Transfer of Lead Security Responsibility Effect Scoring Model, which relies on a system of scores and weights to build a quantitative transition map.[8]

In addition to IJC and AAG, a number of other groups produce formal assessments of the campaign in Afghanistan, including the NATO Joint Forces Command Brunssum; the NATO Consultation, Command, and Control Agency (NC3A); U.S. Central Command Afghanistan-Pakistan Center of Excellence; several U.S. and foreign intelligence agencies; and various policy groups within DoD and the U.S. executive branch. As of September 2010, it was unclear to a group of 150 experts, including many NATO and ISAF officials, exactly how many or even what types of assessment existed.[9] Few of these reports are thoroughly coordinated outside their agencies, and their comparative value is minimized by broad inconsistencies in methodology, format, data content, and varying requirements for security access. For example, while the IJC DAM reflects some qualitative input from subordinate commanders, the NC3A assessment produced in 2010 was built from an almost purely quantitative model that used automated color-coding. The semiannual DoD reports on Afghanistan—the de facto U.S. government assessment of the COIN campaign—differ to some degree in terms of method and means of coordination from report to report.[10] Few of these reports reflect a methodical or comprehensive process that might tie requirements to objectives, or collection to requirements. There has been little to no effort to untangle the specified and implied data collection and production requirements that these reports place on subordinate staffs.

Aside from a few narrative products from AAG and some of the intelligence analysis reports, theater-level assessment reports on Afghanistan are effects-based. All refer in some way to effects and indicators, and most require subordinate commanders to provide a range of quantitative data to senior staffs. This approach reflects U.S. Army and joint Army–Marine Corps assessment doctrine as spelled out in FM 5-0 and FM 3-24.[11] Assessments conducted by both ISAF and IJC reflect the following dynamics:[12]

---

[7]   Commanders have significant input into the principal tool to rate ANSF effectiveness, the Commander's Unit Assessment Tool (CUAT) for the Afghan National Army and the Afghan National Police CUAT (P-CUAT).

[8]   ISAF AAG, undated.

[9]   To address this problem and improve coordination in the assessment process, ISAF and IJC established the Assessment Synchronization Board.

[10]   This discussion of the 1230-series reports is based on extensive conversations with assessment personnel in both Afghanistan and the United States who are familiar with the development of these reports. One person writes the 1230 and 1231 reports.

[11]   HQDA, 2010, and HQDA, 2006c, respectively.

[12]   These observations are based on ongoing interviews and interactions with ISAF and IJC assessment staff members, ten months of reviewing assessment products and methodologies, and a dissection of the doctrinal and professional literature used to develop current models.

- COIN assessment is effects-based and is used to gauge effects-based operations.
- Data flow up to feed a centralized assessment process.[13]
- Commanders' informed judgment is valued but remains almost wholly subjective.
- Quantitative data are inherently objective and useful in evaluating subjective judgment.
- Qualitative data (e.g., engagement reports) are subjective and not suitable for systematized, centralized analysis.

Regional command (RC)–level reports (those conducted just below the theater level but above the brigade or regimental level) are designed and written in accordance with the RC commander's preference, staff capabilities, and ISAF and IJC guidance. While all RCs must file centralized reports to IJC, as of late 2010, each RC used different methods, reported on different timelines for some products, and relied on different data sources. IJC had intentionally allowed each RC to develop its own methodology and reporting process to give commanders leeway in assessments, an approach recommended in FM 5-0. Table 7.1 compares how each RC (e.g., Capital, East, North) conducts its assessments. It shows the commands' reporting cycles, primary sources ("input elements"), and level at which they focus their assessments ("geographic level"). Each RC uses different sources, reports on different frequencies and in different formats, has varying focus within its respective region, and has varying assessment capabilities.

These RC reports are processed at the IJC Information Dominance Center and are used to feed various assessment reports, including the district assessment report. These reports are also captured by ISAF AAG, where they are compared to reams of quantitative data, including SIGACTs, economic data, reporting on the development of ANSF capabilities, and polling data. AAG develops a range of reports for ISAF, NATO, and U.S. customers at the military theater, combatant command, and alliance (NATO) levels and for policymakers at the National Security Council and in Congress. AAG analysts spend a great deal of time managing data, rectifying errors in databases, and producing time-series charts (sometimes annotated, sometimes not) to answer one-off requests for information from senior officers, congressional staffs, and policymakers. As of early 2011, AAG staff, including LTC Bret Van Poppel, Ryan McAlinden, Capt. Eric S. Gons, and Marcus Gaul, were attempting to tackle the challenges associated with centralized assessment, data aggregation, and the lack of holistic analysis methodology. These efforts may have significant impact on ISAF's approach to assessment in the coming year, but the work was not yet available as of this writing.

---

[13] ISAF and IJC intent, as published in the various briefings cited throughout this monograph, is to build assessments from the bottom up and not to conduct centralized analysis. This does not appear to be current practice.

Table 7.1
Regional Command Assessment Processes as of Late 2010

Regional Command	Frequency and Format	Input Elements	Remarks	Geographic Level
Capital	No assessment reported for RC-Capital. IJC receives threat assessment directly from the district. Also contributes to IJC CUAT.			
East	Quarterly brief	Perception surveys, line-of-operations ratings, commander comments	Robust assessment; recent assessment focused on perception of the population based on survey responses; also contributes to ICJ CUAT	District
North	Weekly slides	SIGACTs	Qualitative determination of the security situation; improving, stable, or deteriorating; also contributes to IJC CUAT	Province
South	Quarterly brief	Qualitative indicators, perception surveys	Robust assessment; assessment can be projected along effects or by geographical level; also contributes to IJC CUAT	District
West	Weekly slides	Staff and battlespace owner assessment	District-level assessment utilizes IJC definitions; also contributes to IJC CUAT	District
Southwest	Quarterly brief	Perception surveys, line-of-operations ratings, commander comments	Robust assessment; recent assessment focused on perception of the population based on survey responses; also contributes to IJC CUAT	District

SOURCE: Sok, 2010, p. 5.

Data Management Challenges: The Case of Afghanistan

> We're running the world's greatest OPSEC [operational security] campaign in the world—against ourselves.
>
> —*Senior ISAF information analyst, 2010*[14]

Efforts at the IJC Information Dominance Center and ISAF AAG to build a transparent, credible, and relevant assessment have been hampered by the lack of accurate and complete data. The spread of assessment requirements from the United States, NATO, ISAF, and IJC have created high demand for both raw and aggregated data from subordinate military units and civilian reconstruction units; there may be a requirement for

---

[14] Cited in a briefing delivered by Jonathan Schroden at the U.S. Naval War College on October 19, 2010c. The quote is derived from an interview that Schroden conducted in February 2010.

RCs to address as many as 2,000 metrics to feed these demands.[15] Data for these assessments are submitted through many different channels, and, because some NATO staffs (including some U.S. staffs) report directly to their national commands, data pipelines are often isolated.[16] Coalition civil-military operations also incorporate non-DoD civilians who report through nonmilitary channels. This stovepiped relationship prevents military commanders and assessment staffs from directly tasking some groups, such as coalition partner-led PRTs, with assessment information requirements. For some coalition members, these data management rules are cemented in diplomatic memorandums at the ministerial level. The PRTs sometimes control battlespace in Afghanistan, particularly in more stable areas. This means that there are areas under ISAF theater control that are essentially black holes for many types of assessment data.[17]

Until recently, there were 39 (known) separate databases in use just to store information on attack data, or SIGACTs.[18] As of April 2011, no officials interviewed for this study had a clear grasp of all existing data sources, let alone a method for metadata tagging, coding, and sharing of those data, but there has been a concerted effort to consolidate all information into a single system. Graphic data reports, typically in time-series format, are developed at multiple levels of command and are published and distributed by official email outside the theater, separately from formal assessments. Often, senior staff officers and policymakers use these time-series graphics as standalone campaign assessment devices.

Much of the data—quantitative and qualitative—required for comprehensive central analysis cannot be collected for a variety of reasons. In some cases, the data exist but never make it to the combined (U.S. and NATO) theater headquarters data networks. Problems with collection go well beyond any backlash against excessive data demands. Subordinate units may attempt to collect against centralized requirements but find data to be unavailable, intermittently available, or, when available, of poor quality. Collecting data in a chaotic environment shaped by high threat levels, broken lines of communication, and mountainous geography is necessarily difficult and often unfruitful. As a result, there are entire sets of data that cannot reasonably be collected, and commonly referenced data sets can be chronically incomplete and are probably not representative of reality.

---

[15] Midgrade U.S. military officer, email correspondence with the author, June 27, 2011.

[16] An August 2010 analysis of the data reporting processes in Afghanistan conducted by the National Defense University produced an unpublished data-sharing map that resembled the plumbing schematic on an aircraft carrier.

[17] Statement by a subject-matter expert on Afghanistan assessments and SIGACT reporting, Allied Information Sharing Strategy Support to ISAF Population Metrics and Data Conference, Brunssum, Netherlands, September 1, 2010.

[18] Information analyst, interview with the author, Afghanistan, May 6, 2010.

Despite leaps forward in information technology since the Vietnam War, there has been little progress toward creating an effective data collection, management, and sharing system for U.S. COIN operations. Afghanistan offers salient examples. Data collection and management problems in Afghanistan are persistent, and corrections to known capability gaps have been slow and are often marginally effective. Ongoing concerns were reflected in mid-2010 discussions of theater data management. For example, a briefing by the Center for Army Analysis concluded, "OEF [Operation Enduring Freedom] assessments . . . within the DoD and in theater have been hindered by a lack of necessary data: [There has been] no standardization over time for many data types [and there is] little to no readily usable historical data available."[19] The proceedings from a Military Operations Research Society workshop in April 2010 identified the following problems with data knowledge and management in Afghanistan:

• Mismatch between data required by analysts and data collected by operators

• Disconnect of data needs among operators—analysts—and commanders

• Lack of standard data architecture, resulting in information excesses and gaps

• Lack of centralized repository or management approach, leaving existing data inaccessible or unknown

• Lack of adequate collection program standards to address irregularities, inconsistencies, and incompleteness of data

Additionally, data mining is inhibited by:

• Not knowing what databases are out there

• Lack of standardized database formats

• Multiple levels of security among databases

• Lack of access to networks housing databases.[20]

Communication infrastructure in Afghanistan is immature. According to a 2010 ISAF Headquarters briefing, "The main communication challenge is the IT [information technology] Infrastructure in Afghanistan. Currently, the communications system in Afghanistan works via satellite (V-SAT) which is not powerful and not fast enough to fulfill all the communication needs of the country and also it is very expensive."[21] The ISAF Knowledge Management Office has also stated that "Afghanistan and Coalition forces cannot communicate effectively across Afghanistan."[22]

---

[19] Center for Army Analysis, 2010.

[20] Baranick et al., 2010, p. 5.

[21] ISAF Headquarters, 2010.

[22] ISAF Knowledge Management Office, 2010.

Data that are centrally assessed at the IJC or ISAF level are often inaccurate or incomplete due, at least in part, to database and data sharing issues. According to interviews with key assessment personnel, combat-experienced officers, and others with inside knowledge of the Afghanistan assessment processes, the SIGACT databases, including those found in the centralized Combined Information Data Network Exchange (CIDNE) database, are severely flawed, as discussed later in this chapter.

Despite ten years of effort and considerable expense, the U.S. military and ISAF cannot effectively capture, manage, and share the data necessary to conduct effective centralized assessment. Both operations and intelligence staffs place considerable data requirements on subordinate staffs, and these staffs generate thousands of reports each week; not all of these data feed assessment reports. Thus, it is worth considering whether additional requirements to collect, store, and share data for assessment are worth the output that centralized assessment delivers.

Problems with the Significant Activities Data Sets

SIGACTs are reported violent incidents ranging from threatening letters to key leaders to major assaults on coalition outposts. SIGACTs are captured at various levels of command in both Iraq and Afghanistan and then consolidated in a central database (CIDNE). Because SIGACTs are commonly used as key indicators of progress in Afghanistan, it is worth noting some expert opinions on the value of these data. Most of the conclusions regarding the Afghanistan SIGACT databases also apply to Iraq databases.[23] The following insights are drawn from interviews with subject-matter experts, comments from workshop participants, and official documents. Some of these points address database problems and not flaws in the data, but they are listed here to provide a holistic view of the issues pertaining to SIGACT data in Afghanistan (as well as Iraq).

- An expert on Afghanistan assessments and, specifically, SIGACT data, has stated, "SIGACTs are never accurate. The database isn't even complete. If you think that every incident that happens is entered into CIDNE you're kidding yourself."
- Furthermore, according to this analyst, "A lot of the fields in the SIGACT reports are listed as 'unknown,' so each report has varying levels of information."[24]
- A senior U.S. military officer with intimate knowledge of Afghanistan databases stated, "I'd say 10 to 20 percent of attack reports in CIDNE had misinterpretations of category. When you go through and read the narrative of the attack,

---

[23] This is not to imply that Iraq SIGACTs were sufficiently accurate to support accurate or meaningful centralized assessment. Just as there is no way to prove or disprove the accuracy of Afghanistan SIGACTs, there is no way to prove or disprove the accuracy of Iraq SIGACTs.

[24] These statements were made at the Allied Information Sharing Strategy Support to ISAF Population Metrics and Data Conference, Brunssum, Netherlands, September 1, 2010.

you find that it's very clear the wrong codes were plugged in to label the type of attack."

- This same officer noted, "Sometimes, reports are changed or updated eight or ten times after they are initially recorded."

- The same officer also noted, "There's stuff that's in the civilian casualty database [a subset of SIGACTs] that's not in CIDNE. The data that is in CIDNE is in text format in very large fields and is very hard to analyze."[25]

- A U.S. government official stated, "When [a DoS security contractor] is attacked, they don't report through the military; they report through the embassy. And when NGOs get attacked, they often don't report to anyone, but they keep their own databases."[26] In other words, there are entire sets of attack data that are never recorded in SIGACTs.

- A government-sponsored academic review of the assessment process in Afghanistan concluded, "As currently available in CONUS, CIDNE data provide limited support for empirical analyses owing to geographical or temporal gaps in coverage, and missing, incomplete and erroneous data. SIGACTs are limited in support of COIN-type security assessments; for example, those bearing on Afghan population security conditions outside areas of NATO force presence, non-kinetic threats, or attacks that do not involve forces with access to CIDNE reporting."[27]

- A U.S. military officer with two combat tours in Afghanistan stated, "We were pretty good about reporting SIGACTs. What bothered me was what happened to the reports after we submitted them. We would submit long narrative reports with lots of detail, and the CIDNE report would have almost nothing. Some clerk along the way must have cut out information he thought was unimportant."[28]

- Finally, according to a senior information manager in Afghanistan, "SIGACTs data reports above the ISAF level are off by about 30 percent."[29]

These anecdotal comments are not sufficient to call into question the validity of the entire SIGACT data set. Are SIGACTs generally accurate, and, if so, how can that accuracy be proven? Disproving the accuracy of a database that may contain more than 100,000 individual reports collected from thousands of sources over a ten-year period

---

[25] Statements by a senior U.S. military officer, Allied Information Sharing Strategy Support to ISAF Population Metrics and Data Conference, Brunssum, Netherlands, September 1, 2010.

[26] Statement by a U.S. government official, Allied Information Sharing Strategy Support to ISAF Population Metrics and Data Conference, Brunssum, Netherlands, September 1, 2010.

[27] Cabayan, 2010b.

[28] Statement by a U.S. military officer, Allied Information Sharing Strategy Support to ISAF Population Metrics and Data Conference, Brunssum, Netherlands, September 2, 2010.

[29] Senior information manager, interview with the author, Kabul, Afghanistan, May 2010.

presents many challenges.[30] Empirically disproving accuracy would require a separate and very costly study that might ultimately prove to be inconclusive. But if it is difficult to conclusively disprove the accuracy of the SIGACT data, it is also difficult to prove their accuracy. To enable such an analysis, it seems that the following would have to be made widely available:

- a single, complete database of incidents that could be examined for accuracy
- proof that consistent and accurate retroactive changes were made not only to baseline data but also to correct for lagging errors and changes in metric definitions (e.g., IED finds counted and then not counted as attacks)
- systematic means of backtracking a methodologically sound, representative sample of incidents to the originators of the reports
- with a system in place to check field reporting, a methodologically sound, peer-reviewed study showing the degree of accuracy of information on incidents in the centralized database.

If any part of NATO has made a systematic effort to validate SIGACT data by conducting field interviews, the report from this effort has not been sufficiently publicized. This process would require a team of trained field researchers to select an appropriate sample size from the CIDNE SIGACT database, identify the personnel involved in the incidents in question, locate these individuals, and then interview them to determine whether their initial tactical reports had been accurately reflected through multiple layers of re-reporting and aggregation. This would be a time-consuming and very costly effort. Just finding the persons involved in each incident would prove to be nearly impossible due to rotations, retirements, and outprocessing from the services. Since SIGACT data also include information from Afghan security forces, this research would have to study the accuracy of Afghan reporting as well: The team would require translators and interpreters to study these reports and conduct field interviews, assuming that the Afghans in question could be identified. There seems to be no reasonable way to prove the accuracy of the entire SIGACT database. This is problematic: SIGACT data not only are used in stand-alone assessment graphs, they are also figure prominently in holistic theater assessments, operational analyses, and research reports.[31]

There are many other problems with the SIGACT data, and there are some conceptual problems with the approach to tracking violent incidents in COIN that have

---

[30] This is an estimate; the actual number of reports is not specified in publicly available sources.

[31] See, e.g., Howard, 2007, or Ramjeet, 2008. Both sources present examples of detailed operations analysis research using SIGACT data. These kinds of analyses are common and often identify helpful patterns for tactical and operational analysis. There is no question that some of these reports have tactical and operational utility as long as the data are not taken at face value. However, they have far less utility when aggregated for theater campaign assessment.

yet to be adequately addressed in practice. Some of these problems were identified in Chapter Five, but other issues have plagued the validity and utility of SIGACT databases. For example, there is no sound method for differentiating between insurgency-related violence and other acts, like common criminal activity or the kind of intratribal, intertribal, and interethnic violence that is common in some areas of Afghanistan even during times of relative peace.[32] This means that there is also no accurate method for tracking trend shifts in non–insurgency-related violence, so it is not possible to *clearly* state that insurgency-related violence is rising or falling across the entire country at any one time. It would be possible to argue that this differentiation is unnecessary if the objective is to simply reduce overall violence, but SIGACT data are used to help staffs determine which type of forces to deploy in various areas (e.g., the army or police) and also to help determine whether specific areas are ready for a transition from military and paramilitary forces to normal police forces. They are also used to justify the continued need for U.S. military forces—forces that might not be needed to deal with a surge in criminal or intratribal violence.

With few exceptions, SIGACT data only reflect official reporting from U.S., coalition, or host-nation sources.[33] The government official quoted earlier stated that there are entire categories of reports from coalition officials (e.g., embassy employees, NGO staff) that are not counted in SIGACTs. There are other insurgent-related violent acts that that occur "off the grid," or out of sight of officials, and thus are not counted. If an insurgent attempts to intimidate a government official, that intimidation might be reported as an incident through official channels and eventually incorporated into SIGACTs and CIDNE. However, if insurgents intimidate local villagers in a remote area devoid of military or government officials, that incident most likely will never be reported. For the assessment staff, this unobserved incident—and the thousands like it—simply does not exist; these incidents are not reflected in time-series SIGACT reports. The same is true for IEDs that have been planted by insurgents but never discovered or detonated, direct-fire rounds shot at friendly vehicles that miss and are not recorded, insurgent hijackings of civilian vehicles at unobserved checkpoints, deliveries of night letters that go unreported, or attacks in which IEDs, rockets, mortars, rifles, machine guns, or other weapons misfire and the attack is canceled, to name a few examples. While these incidents are not reported, they do represent insurgent

---

[32] There may be fields in some versions of SIGACT report templates that allow the reporter to make these kinds of differentiations. However, very often, the reporter does not know the reason behind the incident. In both Afghanistan and Iraq, it is often impossible to differentiate between insurgents, criminals, and tribal members, because it is possible to be an insurgent criminal tribal member. Divining motivations for individual attacks would be difficult even with near-perfect intelligence. Further, data fields are often inaccurate or incomplete. Separating insurgent activity from criminal or intertribal activity is all but impossible through aggregated data analysis.

[33] Exceptions include efforts by data management teams to capture incidents from press reports that might not have been coded through the official SIGACT reporting process. However, press coverage may be even more scattered and inconsistent than military and government coverage.

capability and intent, and in many cases, they affect the population and conditions on the ground. Thus, SIGACT reporting is not synchronized with reality. There is no method for determining how many insurgency-related incidents go unreported across the country on a daily basis, so the significance of this additional gap between reporting and reality is unknown.

This gap undermines the validity of theater assessment that depends on time-series SIGACT data. Problems caused by this gap can be exacerbated when friendly forces move into previously uncovered areas and begin reporting a theretofore-unreported stream of insurgent-related incidents. Trend lines can jump, making it appear that there is a greater insurgent presence in a particular district or province when, in fact, coalition forces merely revealed the presence of insurgents who had been operating unnoticed. These kinds of jumps can also occur when military or other government forces instigate attacks by moving into an area where insurgents, criminals, or simply hostile tribal members were living or operating peacefully. And the number of actual events can drop in one area when government forces move in, only to increase in another area that is not under observation as insurgents set up shop away from prying eyes. As Kilcullen points out, "Violence tends to be high in contested areas and low in government-controlled areas. But it is also low in enemy-controlled areas, so that a low level of violence indicates *someone* is fully in control of a district, but does not tell us who."[34] The Vietnam case showed that insurgents often purposefully choose to refrain from violence in an effort to avoid detection or maintain freedom of movement, so the absence of violence can be and probably often is characteristic of a location where insurgents have a significant presence. In general, the presence or absence of violence, by itself, is not always meaningful for assessment.

Comparing SIGACT data to other variables is no less problematic than attempting to analyze the data on their own merits. A number of assessment staff officers and scholars have applied a variety of statistical methods to Afghanistan SIGACT data in an effort to find cause and effect between two variables. But at some point, statistical analysis tends to require (1) some quantity of known, accurate data, (2) a data set with generally known inaccuracies, or (3) strong evidence of correlation between two variables across a number of locations (e.g., violence always drops when soccer balls are distributed). However, this monograph argues that none of these requirements can be met through centralized assessment. It shows that analyses can be precise but inaccurate and that, in general, the completeness and quality of the data do not correlate with meaningful results. Any large-scale SIGACT study that claims to find a causal link between SIGACT data and any other variable should be carefully scrutinized.

As long as accuracy is not assumed, violent incident reporting may have value in context when local commanders can compare specific incidents to intelligence reporting and to their own (known) operations. They may still not be able to explicitly con-

---

[34] Kilcullen, 2009b, p. 5 (emphasis added).

clude cause and effect between their operations and a rise or fall in violence, but local commanders are best placed to interpret these incidents in a way that is meaningful for assessment. For example, Figure 7.1 shows operations analysis of SIGACT data at the RC level conducted for a UK military unit. It shows the number and types of reported attacks over a given period and at specific times of the day in two-hour increments. As long as this chart was used only to help analyze the time of day during which insurgents were more or less likely to attack coalition forces in RC South, then it could be useful. Any effort to extrapolate these data, or this chart, to show cause and effect for assessment would be inadvisable.

Theater-level aggregated SIGACT data should be used to assess the Afghanistan campaign only with the utmost caution, and then only when accompanied by stringent caveats.

Figure 7.1
Time-of-Day Attack Pattern Analysis for Regional Command South, Afghanistan

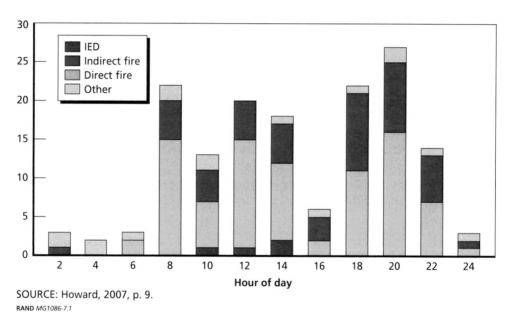

SOURCE: Howard, 2007, p. 9.
RAND *MG1086-7.1*

Analysis of the Afghanistan Assessment Process

The joint/combined force in Afghanistan has created a modified EBA process that generally mirrors the one applied in Iraq and is clearly evocative of the model applied in Vietnam.[35] As of early 2011, ISAF (which officially relies on NATO's Effects-Based

---

[35] See Thayer, 1985, 1975a, and 1975b.

Approach to Operations doctrine) framed both its operations and assessments in terms of effects.[36] There may be intent by senior leadership to emphasize commanders' analysis, but in practice, the various assessments of Afghanistan are derived primarily from quantitative and centralized assessment of individual data streams. Many millions of dollars have been spent improving data management systems, and no fewer than five interagency working groups are assigned to improving data collection and sharing in theater, across U.S. agencies, and among NATO partners. Efforts to improve assessment and further develop MOEs, MOPs, and indicators have incorporated work by mathematical modelers, social scientists, and EBA experts from the U.S. Army's operations research community. Both the ISAF and IJC staffs have aggressively pursued mathematical modeling processes in an effort to improve assessment design. Far from placing the planning and assessment process in the hands of the commander, strategic review and transition planning relies on input from several agencies, hundreds of staff officers, and hundreds of external experts.[37] The Afghanistan assessment model that Gregor accurately condemns as "inconsistent" is the U.S. military's and NATO's best effort at EBA.

Research conducted by the U.S. Naval War College and the Netherlands Defence Academy reinforces the notion that the modified version of EBA in use in Afghanistan has not met military or policymaker requirements for COIN. This research specifically addresses the failure of assessment to provide transparency and credibility. According to Stephen Downes-Martin of the U.S. Naval War College, "While no assessments process can be perfect or free of any criticism, the flaws in the currently used approaches are sufficiently egregious that professional military judgment on assessments is, rightfully, distrusted."[38] He finds that, "in the absence of a credible numbers-based theory for COIN, there is no objective numbers-based assessment. *At best one obtains precision without accuracy.*"[39] Downes-Martin also recorded complaints from the field:

> There is serious evidence that the collection capacity of regional commands' partner civilian organizations and major subordinate commands are overwhelmed by the number of metrics demanded, and that they rightfully do not trust the value of collecting on those metrics or the value of the assessments done using them. This evidence is in the form of *verbal admissions from senior officers that they make*

---

[36] To emphasize this point, most staffs at the RC level and below have "effects cells" designed to shape effects-based operational efforts.

[37] For example, from August 30 to September 3, 2010, at NATO Joint Forces Command, Brunssum, Netherlands, NATO held a conference of 150 officers and experts to examine data collection, data sharing, and metrics in Afghanistan.

[38] Downes-Martin, 2010b, preface. Downes-Martin was embedded with an operational assessment section both during its predeployment training and then for three months in Helmand Province in 2010. He has first-hand knowledge of the assessment process at both the RC and theater levels.

[39] Downes-Martin, 2010b, p. 7 (emphasis in original).

up what they do not have, and do not check the quality of what they do have, in their submissions. An additional reason given for not taking the metrics seriously was that they received no feedback from the requesting organization.[40]

Downes-Martin's observation that centralized quantitative assessment requirements generated falsified reports closely parallels the evidence of falsification described in the Vietnam case. Interviews and discussions conducted for the current study between November 2009 and February 2011 uncovered similar admissions from some officers who had served in Afghanistan. An operations researcher who had worked on a PRT in Afghanistan stated, "When we were asked to produce data we didn't have we just made it up."[41] I personally observed similar actions during two of three tours in Iraq.[42] Direct observation of the assessment process supports the conclusion that military officers often attempt to push back against requirements that they cannot logically or practically meet, but some do make up data to avoid irreconcilable friction and to allow them to focus on what they see as more important tasks. In Afghanistan, this behavior—seemingly common, judging by accounts from assessment staff, and observations by other researchers—is a direct result of what Downes-Martin refers to as "promiscuous metrics collection" driven by a core metrics list that cannot be relevant or collectible in all places across the country at the same time.[43]

Joseph Soeters of the Netherlands Defence Academy provides some insight into the EBA approaches attempted in Afghanistan in early 2009.[44] Soeters found that "any system that tries to cover all information needs—both operational and strategic—

---

[40] Downes-Martin, 2010b, p. 8 (emphasis added).

[41] Statement by an operations research analyst, conference working group, Allied Information Sharing Strategy Support to ISAF Population Metrics and Data Conference, Brunssum, Netherlands, September 1, 2010. Two military officers who overheard this statement concurred and added that they, too, made up data when mandated to report on indicators that did not exist.

[42] Commentary based on observations in Iraq should be taken as a professional, first-person perspective on such incidents. These observations do not qualify as empirical evidence of widespread falsification.

[43] Sometimes, these core metrics lists are included in CCIR lists and therefore address both assessment and intelligence requirements. Campbell, O'Hanlon, and Shapiro of the Brookings Institution have worked extensively with core metrics lists. They state,

If we were confident about which 10 or 15 or 20 metrics could best tell the story of the efforts in Iraq or Afghanistan, and had reliable data for those categories, we might have focused more narrowly on them. The truth is that the wars in both Afghanistan and Iraq have usually demonstrated an ability to confound short lists of metrics.

However, they go on to state, "In retrospect, some key metrics emerge with greater clarity" (Campbell, O'Hanlon, and Shapiro, 2009a, p. 4).

[44] In January 2009, Soeters led a team of researchers in Kandahar and Kabul, Afghanistan, on a study of the operational assessment process. The team conducted 63 semistructured interviews with the commander of RC South and the deputy chief of stability in Kabul, among others. They also attended assessment staff meetings and reviewed assessment documents. Soeters specifically focused on the use of EBO and EBA in COIN assessment, as opposed to the application of civilian business models in military operations.

becomes overly complex. . . . At the aggregate level of outcomes it is difficult, if not impossible, to attribute changes and results to certain inputs and activities." Forming standards of achievement is difficult when "results have multiple or even contradictory readings [or] when the environment is highly changeable."[45] Soeters's interviews revealed some of the underlying and rarely publicized frustrations with EBA from within the military. In addition, military officers and other officials have called the centralized EBA process "an illusion" that cannot react to a complex system and that assumes an unachievable level of predictability. One interviewee stated that EBA fails to deliver useful direction to subordinate staffs.[46] Soeters found that assessment staffs had poor tools to work with, because the "measures of effect and performance are unclear, change frequently and are difficult to measure. . . . The definitions these measurements should be based upon are often unclear, vague, and different."

Chapter Summary

Campaign assessment in Afghanistan is derived to a great extent from U.S. joint doctrine on EBA. On the surface, it looks a great deal like EBA in that it uses indicators to assess effects and to measure progress toward MOEs and MOPs. However, assessment in Afghanistan has typically been ad hoc and loosely coordinated, incorporating both EBA and pattern and trend analysis. In many cases through late 2010, assessment reports consisted of little more than republished time-series charts or statistical data that were devoid of meaningful context or explanation. The best analysts spend most of their time managing the vast quantities of centralized data necessary to feed EBA and pattern and trend reporting. To date, there has been no comprehensive effort to capture context from the field to shape time-series assessment reports, no common assessment methodology across the theater or over time, no method for systematically incorporating qualitative data into theater assessment, no common framework for pattern and trend analysis, and little or no comprehensive effort to incorporate caveats regarding data quality or the degree of subjectivity in assessment reporting.

As of late 2010, the point at which the focused research on the internal assessment process on Afghanistan ended, assessment of the Afghanistan campaign did not fare well against the assessment standards identified in this monograph. Table 7.2 applies the assessment standards described in Chapter One to the assessments produced by assessment staffs in Afghanistan, though not necessarily to the individual reports written by senior officers.

---

[45] Soeters, 2009, pp. 7–8.

[46] Soeters, 2009, p. 11.

Table 7.2
Performance of Afghanistan Assessment Against Assessment Standards

Standard	Afghanistan Assessment	Analysis
Transparent	No	Data aggregation and methods opaque, insufficient declassification of assessment reports
Credible	No	No holistic methodology, multiple and often contradictory reports, large data gaps
Relevant	No	While senior officer analysis is respected, assessments do not sufficiently support this analysis; this makes the process seem subjective
Balanced	No	Does not systematically incorporate qualitative or mixed data or subordinate reports
Analyzed	No	Analysis is not consistent across the theater, nor does it deliver a methodologically sound, overarching framework or report
Congruent	No	Does not reflect COIN joint/capstone doctrine
Parsimonious	No	Data requirements are extensive, poorly defined, and typically inconsistent from level to level

CHAPTER EIGHT

Why Does Centralized Assessment Fail in Counterinsurgency?

Our whole notion [is] that we can somehow develop a mathematical model that includes concrete achievements, factor in a time frame and voilà. Iraq doesn't work that way and Afghanistan doesn't work that way.

—*Ryan C. Crocker, U.S. ambassador to Afghanistan and former ambassador to Iraq*[1]

This chapter builds on the Vietnam case but specifically reflects the findings on assessment in Afghanistan and (to a lesser extent) Iraq. Before answering the question posed in the title of this chapter (Why does centralized assessment fail in COIN?), it is worth reiterating a point made in the introduction to Chapter Seven: A poorly conceived process—not the individuals executing the process or receiving its output—is the primary reason for the inherent weaknesses of U.S. military COIN assessment. More than a year of research, several years of personal involvement with the COIN assessment process, and numerous interactions with military and civilian assessment staffs from the battalion to the joint staff level revealed teams of assessment analysts that were both understaffed and overworked. Leaders and policymakers cannot adequately communicate requirements to collectors and analysts because COIN assessment is poorly defined and rarely addressed in the literature, professional education, or staff training. In many cases, operational assessment was a secondary or tertiary responsibility for officers and noncommissioned officers in combat theaters. There is little or no all-source analytic capability for campaign assessment at any level, and there is no doctrinal guidance on how one might conduct a *comprehensive* analysis of a COIN campaign. Doctrine and other available official guidance tend to be inherently contradictory, inadequate, and confusing. Almost all interactions with assessment staffs revealed hard-working people struggling to make the best of a bad situation. This observation of the current process seems to echo conditions during the Vietnam War.

---

[1] As quoted in Shadid, 2010.

Interviews with staff involved with assessment, subject-matter experts, and policymakers, along with a survey of the civilian and military literature, uncovered widespread dissatisfaction with the military assessment process as it is applied to Afghanistan. This dissatisfaction is evident from the company-grade officer level to the highest levels of policy (e.g., congressional and national security staff). In 16 months of research, interviews, conferences, workshops, and engagement with most of the producers and some of the official consumers of Afghanistan assessments (in total, well over 300 officers, civilian analysts, policymakers, and experts), no one was willing to characterize the current system as well-designed, transparent, or credible.

This chapter presents detailed findings matched to discussions of the seven standards framed in the beginning of this report and used to assess the Vietnam and Afghanistan cases: transparency, credibility, relevance, balance, analysis, congruence, and parsimony. It shows that centralized assessment does not meet any of these criteria and that the doctrinal and de facto approaches to U.S. COIN assessment are ineffective. This chapter also addresses some additional issues and provides examples of data challenges from the Iraq case.

Neither Transparent nor Credible

This section examines how EBA doctrine, pattern and trend analysis, and the modified versions of these approaches in Afghanistan have failed to deliver transparent or credible assessments to senior military officers and policymakers. The concepts of transparency and credibility are closely linked, and there is some necessary overlap between the two.

Transparency is a broad concept, but it is a helpful standard for reports that will be used to shape operations and support public review of the overall regional strategy of the United States. If democracies are to prosecute COIN, they must be able to openly justify to their populations the inherent risks and extended timelines associated with the typical COIN campaign.[2] Transparency allows policy staffs to dig into specific data to address any pressing concerns they might face, but without forcing an assessment staff to scramble to produce new reports. Transparent assessments help policymakers achieve depth of understanding when attempting to grasp the complexities of COIN. A classified but transparent assessment can also be fairly compared to an intelligence analysis product (in terms of sourcing), so transparency gives policy staffs a means to develop contrasting viewpoints to support decisionmaking. After some necessary declassification steps, including the removal of sensitive material, a transparent assessment can be used to help interested members of the public understand the complexities and challenges of the campaign. In this ultimate format, a transparent

---

[2] See the summary of findings in Connable and Libicki, 2010, for insight into COIN campaign dynamics.

campaign assessment can prevent creeping disillusionment with government policy, combat the impression that military assessment is overly optimistic, and reduce the impression that reports reflect unanchored subjective interpretations rather than in-depth reporting and analysis.[3] Ultimately, transparent assessment reports are widely releasable to official consumers and, ideally, unclassified and without caveat; the final iteration should be suitable for distribution on the Internet. Such reports reveal both the methods and data used at each level of assessment, from tactical to strategic, and allow for detailed and contextual analysis. Any and all subjectivity or justifications for disagreements between layers of command should be comprehensive, available, and understandable.

It is difficult to see how a centralized process can deliver transparency for COIN assessment. Centralized assessment is almost entirely reliant on quantitative data aggregation to produce assessment reports. At the highest levels of aggregation, these data are opaque to government and civilian consumers. They provide little opportunity for skeptical consumers to "see" into the assessment report to determine data or methodological validity. It would be possible to imagine opaque aggregation and transparent disaggregation on opposite ends of a continuum. Figure 8.1 shows transparency and disaggregated context on one end of a continuum and aggregation and opacity on another.

No matter how reports are presented or declassified, data aggregation is likely to undermine efforts to present assessments transparently. Because aggregated data tend not to have been analyzed in context at the level at which they are collected, accurate retrospective analysis of these data may not even be possible: It is difficult to find other sources of information that are sufficient to place such data in context well after

Figure 8.1
Relative Transparency of Aggregated and Disaggregated Data

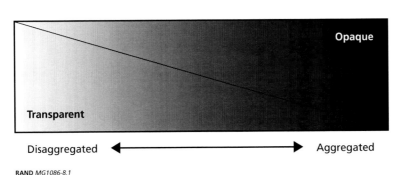

---

[3] This assumes, of course, that the transparent report meets other standards (e.g., it is credible, balanced, analyzed).

the situation on the ground has changed.[4] Aggregation essentially precludes effective analysis by anyone outside the combat theater. This analytic roadblock has likely led the press and subject-matter experts to be more critical of COIN campaigns than may be warranted otherwise.

Lack of transparency undermines credibility, but the credibility of assessment depends on a number of factors. *Credible* assessment reports must be transparent because opacity devalues any report to the level of opinion: If the data and methods used to produce the report cannot be openly examined and debated, consumers have no reason to trust the reporter. In a credible report, to the greatest extent possible, biases are revealed through iteration between commanders at various levels and top-level analytic staffs. Flaws in data collection, valuation, and data analysis are clearly explained. Methods are consistent and clear. Data are explained in context. The process used by commanders and staffs to select methods is also clearly explained. The report is both accurate and precise to the greatest degree possible.

It is unlikely that centralized assessment can meet these criteria. It is not possible to accurately or sufficiently account for data flaws and gaps because the degree to which aggregated data are inaccurate or incomplete cannot be known: COIN data are not collected using reliable scientific methods. The lack of control and consistency across data collection and management efforts prevents an assessment from showing even a reasonable estimate of bias and flaws in data collection. It is also not possible to determine how subordinate units valued any specific type of data over another once those data have been aggregated, so data valuation is typically unknown and often inaccurate.

It is possible to rectify the existing lack of overarching analytic methodology in the campaign assessment process. But, at best, structured analysis would provide dressing for an unappetizing array of dissociated and opaque sets of often inaccurate and gapped data. No core metrics list from any campaign has been widely accepted as credible and objective, so even expert analysis of core metrics is unlikely to produce a credible report.

The following two sections address additional considerations regarding efforts to develop credible assessment reporting.

Lack of Data Caveats and Citations Undermines Transparency and Credibility

Assessment staffs must take and present the quantitative data necessary to determine thresholds at face value. As of late 2010, it was rare to see a confidence rating attached to a quantitative source in an assessment report. Perception polls almost always carry statistical confidence bounds provided by the contracted polling organization, while other sources, such as attack reports, some types of economic data, and civil-servant

---

[4]   This supposition is based on efforts to conduct such analysis both from within a military intelligence staff in a combat theater and as a researcher.

employment rates, rarely do. This likely leads senior commanders and consumers to believe that assessment data possess a degree of accuracy that would be unachievable even in a peacetime environment.[5] Referring to Afghanistan, Anthony Cordesman from the Center for Strategic and International Studies identifies this problem:

> When numbers are applied to reality, they are always uncertain adjectives—not scientific quantification. This is especially true in conflicts like Afghanistan where accurate data collection is often difficult to impossible. Metrics often imply a false precision, particularly since they are rarely shown as full ranges or rounded to reflect uncertainty. The problem is made worse by the fact that many users do not really understand probabilities, statistics, or the overall process of quantification half as well as they understand words. In general, numbers that are unsupported by narrative are a remarkably poor method of communication.[6]

Not only are quantitative sources offered at face value, but the assessment reports themselves rarely contain sufficient citation; this is particularly true for unclassified reports released to the public. It is often impossible to find the original source of reporting used to build theater-level assessments. Sometimes, sources are provided, but these sources are often second- or third-hand aggregations (e.g., for reasons of classification, a theater-level assessment might cite national-level economic data or simply "CIDNE"). The absence of citation further degrades the transparency and credibility of the assessment process.

Weighting Metrics Reduces Transparency and Credibility

In Afghanistan, at least one of the analytic techniques that staffs apply to COIN data may do more harm than good. Centralized reports sometimes incorporate weighting of MOEs, MOPs, and indicators. FM 5-0, Appendix H, recommends a mathematical assessment process that includes both a weighting and a rating scheme.[7] This research did not reveal any verbatim adoption of the doctrinal model (perhaps a compelling critique of the model's relevance to COIN), but a number of other weighting schemes have emerged to support EBA at various levels of command in both Iraq and Afghanistan. Figure 8.2 shows another proposed EBA model that incorporates a weighting formula.[8] While this model was not designed for COIN and may never have been imple-

---

[5] And, when policymakers or their staffs discover errors or irregularities in the data, their confidence in DoD assessment reporting is further undermined.

[6] Cordesman, 2008, p. 3.

[7] See HQDA, 2010.

[8] The authors describe the thinking behind the formula:

> The mathematical mechanics involve an iterative process that repeats the similar steps for each level in the model hierarchy. At the lowest levels, each effect has a number (x) of MOEs associated with it, and each task has a number (y) of MOPs associated with it. In addition, we assign each MOE and MOP a weight reflecting relative

Figure 8.2
Effects-Based Mathematical Model with Weighting Scheme

Mission	$M_P = w_{(1)}O_{P(1)} + w_{(2)}O_{P(2)} + \ldots + w_i O_{P(i)},$ where $\sum_{n=1}^{1} w_n = 1$ Equation 4: Mission-level *performance* scores
Objective	$O_{P(i)} = w_{(i,1)}E_{P(i,1)} + w_{(i,2)}E_{P(i,2)} + \ldots + w_{(i,j)}E_{(i,j)},$ where $\sum_{n=1}^{f} w_{(i,n)} = 1$ Equation 5: Objective-level *performance* scores
Effect	$E_{P(i,j)} = w_{(i,j,1)}T_{P(i,j,1)} + w_{(i,j,2)}T_{P(i,j,2)} + \ldots + w_{(i,j,k)}T_{P(i,j,k)},$ where $\sum_{n=1}^{k} w_{(i,j,n)} = 1$ Equation 6: Effect-level *performance* scores
Task	$T_{P(i,j,k)} = w_{(i,j,k)_1}MOP_{(i,j,k)_1} + w_{(i,j,k)_2}MOP_{(i,j,k)_2} + \ldots + w_{(i,j,k)_9}MOP_{(i,j,k)_9},$ where $\sum_{n=1}^{y} w_{(i,j,k)_n} = 1$ Equation 7: Individual-task *performance* scores

SOURCE: Clark and Cook, 2008, Figure 5.
RAND *MG1086-8.2*

mented, it is emblematic of how difficult it can be to add weights to EBA or modified EBA, and of how weighting can undermine transparency and credibility.

This process does not pass the tests of transparency, credibility, or even understandability. Models that require specific skills and inside knowledge to understand and interpret degrade the value of assessments. In this formula, there is no clear, logical relationship between indicators and weights. The weighted combinations of metrics have no physical or logical meaning. There is no way to incorporate uncoded qualitative data into the formula, so the tens of thousands of narrative intelligence information reports, key leader engagement reports, civil affairs narratives, and other qualitative reports produced in theater would have no value in the model's output.[9] This model may or may not be helpful to a staff that is implementing the strictest interpretation of

importance. For each assessment period, we observe values associated with each MoE and MoP and input them into their respective models. [The first figure, not shown here] outlines the effects-scoring model; [the figure shown here as Figure 8.2] outlines the performance-scoring model. The MOE and MOP scores, between 0 and 1, indicate the level of a particular effect or task, respectively. A score of 1 indicates complete success—at least temporarily. This holds true for all scores at each level. (Clark and Cook, 2008)

[9] Bousquet (2009, p. 183) states, "Precise quantitative analysis must yield to a more qualitative understanding via the identification of more or less hidden patterns in the [complex adaptive] system's behavior."

EBA as presented in Chapter Four, but neither this model nor the model in FM 5-0, Appendix H, is relevant to contemporary COIN operations.

Even if this quantitative model could be adequately explained and justified, there is no way to avoid a high degree of subjectivity in this or any other weighting scheme (e.g., the one described in FM 5-0, Appendix H). At some point, someone has to make a decision about what is more important and what is less important to the course of the campaign.[10] Typically, a senior policymaker or military commander would make this decision, but subordinate staffs are also free to use their own internal weighting schemes; weighting could be added and hidden by anyone at any stage of the assessment process, at least as it is currently conceived and implemented. Because weighting requires subjectivity that is difficult to explain or justify, it is vulnerable to criticism. And because every consumer of assessment reports brings biases and opinions, consumers also apply their subjective interpretations to the weights (i.e., they might disagree that one thing is more important than another). Policymakers and senior commanders might feel justified in applying their own weights, thereby changing the entire basis of the assessment. This has happened: Staff officers in Afghanistan report that centralized weighting schemes were attempted but failed after senior officers could not agree how each effect should be weighted.[11]

For a weighting system to be accurate (at least by the standards of that specific staff) the staff must adjust weighting values to meet changing requirements and conditions. Early in the war, security might be the most important metric, but, later in the war, development might be more important. As importance shifts, weighting schemes would have to reflect that shift.[12] The mathematical modeling required to account for routine weighting shifts in a number of interrelated metrics over time would be complex to the point that it would be relatively meaningless to policymakers or senior military leaders—interpreting the model would become a subjective exercise. And, because weighting is highly subjective, every shift in weighting would also require subjective judgment. It would quickly become impossible to unravel who decided what was most important at each moment in time and why. If the likely end result of mathematical weighting is a return to subjective debate, then no staff should exert the time and energy required to apply weights to assessments.

---

[10] FM 5-0 (HQDA, 2010, p. B-35) recognizes that even in the course of action development process—a process that takes place before operations, typically in safe and static environments—weighting adds a great deal of subjectivity.

[11] This statement is based on discussions with coalition staff members in Kabul, Afghanistan, May 2–8, 2010.

[12] I observed two major shifts in weighting at the theater level in Afghanistan over a six-month period. These changes shifted weighting from one line of operation to another. Each shift should have created a large cascading effect throughout the models used to gauge subordinate MOEs and indicators.

Not Relevant to Policy Decisionmaking

Relevant reports should effectively and efficiently inform policy consideration of COIN. They should be sufficient to help senior military leaders and policymakers determine resourcing and strategy, and they should satisfy public demand for knowledge up to the point that they do not reveal sensitive information. The other six standards described here shape the degree to which assessment reports are or are not relevant. Chapter One presented a list of general requirements that policymakers might have for relevant COIN campaign assessment:

- a close approximation of ground truth upon which decisions can be based
- an appreciation for resource requirements (current and predictive)
- transparent and credible reporting that can be used to inform Congress and the public
- relevance to decisionmaking
- predictive analysis (sometimes).

Centralized assessment does not provide accuracy that can be substantiated through either an examination of the data or sound analysis, so policymakers can never be sure that it provides a close approximation of ground truth. Optimally, resource requirements are predicated on at least a relatively clear understanding of the threat, progress toward developing the host nation's government and security forces, the status of the population, and other relevant information. If this information is inadequate for strategy decisionmaking, it is also likely to be inadequate for resource decisionmaking. As established earlier, centralized assessment is neither transparent nor credible, so even if assessment reports appear relevant to decisionmaking, they are not sufficient to sustain a long-term COIN campaign that requires U.S. popular support. Thus, centralized assessment can provide some predictive analysis, but because it is not predicated on sound data or methodology, this analysis is insufficient for sound decisionmaking. The near-universal dissatisfaction with the COIN assessment process in policy circles emphasizes these findings.

Unbalanced

Balanced reports reflect information from all relevant sources available to military staffs and analysts, including both quantitative and qualitative data. Such reports should contain and reflect input from military commanders at all levels of command. They should also be broad enough in scope to incorporate nonmilitary information and open-source data. Finally, balanced reports include countervailing opinion and analysis, as well as data that both agree with and contradict overall findings.

The mathematical models recommended by the *Commander's Handbook* and FM 5-0, and employed in various forms in Afghanistan, rely on core lists of quantitative metrics.[13] While FM 5-0 tries to place equal value on qualitative and quantitative data, it also states, "Quantitative indicators prove less biased than qualitative indicators. In general, numbers based on observations are impartial."[14] This belief has been reflected at some theater-level staffs in Afghanistan at various times, but it is not correct in the context of loosely structured and nonscientific COIN assessment. COIN data collection and reporting are not objective efforts. Quantitative data reflect subjective bias that is hidden when aggregated. The processes of selecting MOEs and indicators and determining what to collect and report at the tactical level, the collection method, the reporting method, and the way in which data are displayed inherently involve subjectivity. Because staffs tend to collect a great deal of quantitative data but focus on a few key indicators, there is also subjectivity in the decision about what not to count or report. Quantitative and qualitative data are both subjective, but in different ways. Because COIN assessment is not controlled scientific research, the standards associated with scientific research will not be reflected in COIN data.[15]

Theater-level staffs continue to make modest efforts to incorporate qualitative information into assessments, but mathematical models, time-series charts, and core metrics lists require purely quantitative data. In practice, therefore, centralized assessment is driven by quantitative data.[16] It is possible to code some *qualitative* data for quantitative analysis, but coding requires trained analysts, expensive software, and a great deal of time. It is not feasible for even theater-level staffs to code the tens of thousands of qualitative reports fed into centralized databases every year.[17] According to an information analyst at a senior staff in Afghanistan, "We can't use any data that is not in a cohesive or productive format. Everything needs to be similarly coded [i.e., quantified] or it can't be inducted into a database." The analyst also stated that "the data must

---

[13] See HQDA, 2010, Appendix H. A senior information manager and analyst at IJC stated that incidents are run through a mathematical model to "determine causation" (interview with the author, Kabul, Afghanistan, May 2010).

[14] HQDA, 2010, p. 6-8.

[15] In any event, some scientists might claim that qualitative and quantitative data could be either more or less subjective, depending on how they are produced and obtained. There are ways to collect and analyze qualitative data that are highly structured, and there are various methods to reduce subjectivity in qualitative data. The assumption that one type of data is more or less subjective might apply when comparing a physical experiment to a perception poll, but it does not necessarily apply in the study of a social problem. For this reason, the statement in FM 5-0 is misleading.

[16] An examination of the mid-2010 version of the IJC District Assessment Framework Tool, which checks subordinate commanders' assessments against an objectives metrics list, showed that the majority of the approximately 80 metrics required percentage ratings, binary responses (e.g., present or not present), or some other form of quantitative representation.

[17] Even if these reports could be coded, the process of coding would inject an additional layer of subjectivity and a layer of opacity to the finished assessment report.

be pre-coded before it reaches the our level. Our analysts don't necessarily have enough context to code raw reporting from the RC-level and below."[18] The RCs do not have the capacity to do this.[19] Therefore, in practice in Afghanistan, qualitative data are not *systematically* incorporated into theater-level assessments.[20] This also proved true in Iraq.[21]

Centralized mathematical models, and all centralized assessments for that matter, reflect only a subset—perhaps 40–60 percent—of all available information in theater.[22] This means that centralized assessment provides a picture that, at best, reflects about 60 percent of the events, insights, and reflections that are readily available to analysts. Some aspects of COIN that are crucial to success simply cannot be usefully quantified, even through coding. Unless one places absolute reliance on public opinion polling, it is not possible to accurately quantify insurgent motivations, civilian satisfaction with the government, or government legitimacy. Efforts to quantify the personal capability of key leaders, the potential of a military unit to fight and win, or the degree to which a government institution is prepared to accept responsibility for its portfolio are likely to fall short or mislead. Quantitative metrics are generally incapable of describing the kind of political agreements and development that tend to be the keystone of lasting government stability. An unbalanced approach to assessment cannot hope to provide a policymaker with a reasonably accurate and holistic understanding of the COIN environment or the campaign.

Not Sufficiently Analyzed

> There are a lot of things in the calculus of factors affecting progress of the war, there's no way we know of yet to measure it.
> —*General Creighton W. Abrams*[23]

---

[18] Interview with the author, Kabul, Afghanistan, May 2010.

[19] In the case of at least one RC, the entire assessment staff consists of two people.

[20] According to FM 3-24 (HQDA, 2006c, p. 4-6), "COIN operations often involve complex societal issues that may not lend themselves to quantifiable measures of effectiveness."

[21] This finding is based on direct observation of and participation in the assessment process in Iraq over the course of three tours from 2003 to 2006.

[22] These percentages are based on a generalized assumption regarding the availability of quantitative versus qualitative data in COIN. They are based on direct observation of COIN from within a staff headquarters over three tours, research on the Afghanistan campaign, and a study of 89 COIN cases for *How Insurgencies End* (Connable and Libicki, 2010). This percentage does not, of course, take into account the fact that a great deal of relevant information is not readily available to analysts due to the limitations posed by the COIN environment.

[23] Sorley, 2004, p. 363.

Analyzed, finished reports should reflect analyses of all available data to produce a unified holistic assessment. This process should be objective and predicated on at least a general methodological framework that can be modified to fit changing conditions and, if necessary, challenged by consumers. The requirement for holistic analysis is commonly voiced in the assessment literature. But assessment doctrine does not clearly prescribe or define theater-level *analysis* of information, only a working group approach to assessment and the application of weights and thresholds. Analysis of raw data and subordinate commanders' assessments is limited by manpower constraints, the constant demand of external data and briefing requirements, and the inability of assessment staff members to travel outside the theater command's forward operating bases, even infrequently. Theater-level staffs in Afghanistan are particularly prone to resource limitations and demands for data. This section draws on research conducted in Afghanistan in 2010 and 2011.

Analysts at top-level staffs in Kabul spend a great deal of time rectifying bad or missing data, retroactively adjusting time-series graphs, responding to directive requirements for raw data from congressional and policy staffs, and trying to convince parallel staff sections to provide relevant data to feed centralized assessment. This was also true in Vietnam and Iraq. The mechanics of centralized assessment take up a great deal of time, and the absence of transparent and credible analysis contributes to a constant demand for raw data by senior policymakers. In the words of an assessment officer at a theater-level command in Afghanistan, "A lot of what we do is superficial."[24] This officer's staff was overwhelmed with requirements for briefing slides and had little time to immerse itself in all-source information.[25] Another analyst whose job it was to "red-team," or provide alternative analysis to RC assessments, was equally overwhelmed. This analyst, responsible for tracking and red-teaming an entire RC, stated, "We're doers, not thinkers. We're collecting so much information that we don't have time to analyze it. We're just a reporting agency. We put our labels on subordinate products."[26]

When analysis does occur, it is typically conducted in an ad hoc and limited fashion. Staffs publish a recurring set of time-series reports intended to show progress or lack of progress along (typically) three lines of operation: governance, security, and development. There is no analytic or statistical method used to compare or combine these reports in a way that would produce a clear quantitative analysis of the entire campaign (but no method could hope to show strong correlation between multiple aggregated variables in COIN).[27] Instead, analysts examine individual variables (e.g.,

---

[24] Information analyst, interview with the author, Afghanistan, May 6, 2010.

[25] Sometimes the staff has only six people, depending on rotations and leave. Since this research was conducted, it is possible that there has been an increase in these levels.

[26] Information analyst, interview with the author, Afghanistan, May 6, 2010.

[27] The exception to this rule was the NC3A effort to produce a combined mathematical report derived from a regression analysis. That effort was reportedly halted in mid-2010.

number of attacks reported) and attempt to explain trends through subjective analysis, or they conduct limited comparative studies to show patterns in activity. For example, one staff produced a report that compared time-series attack data with weather and seasonal data to show the effects of weather and seasons on insurgent activity. This report demonstrated a definable pattern, but this pattern could, at best, be considered little more than an analytic tool for subjective assessment of other time-series trends, or an operational planning tool.

The mathematical EBA formulas recommended in the *Commander's Handbook* and FM 5-0 have yet to be accepted by theater-level staffs: In 2009, the ISAF commander rejected the most comprehensive EBA analysis model developed by his staff. There is general agreement among military staff officers interviewed for this study that there is no mathematical solution to COIN. Because it is, in all practicality, a mathematical model, EBA does not offer a useful analytic process for holistic COIN assessment. Pattern analysis and time-series trend analysis are not holistic analyses. Both pattern analysis and trend analysis are methods that might (*with known data sets*) be used to develop information to feed holistic analysis, but these efforts do not effectively stand on their own to deliver holistic reports. Any effort to aggregate or conduct regression analysis on quantitative patterns and trends would succumb to the same fate as mathematical EBA solutions. Efforts to analyze centralized aggregated data through less structured approaches (e.g., subjective analysis of aggregated data) are insufficiently transparent and credible to meet policymaker requirements. As of early 2011, there was no proven centralized analytic method for producing a transparent and credible assessment derived from aggregated quantitative data.

This monograph recommends the incorporation of all-source fusion methodology from the military intelligence field to improve assessment analysis. This recommendation is derived primarily from JP 2-0, *Joint Intelligence*, but it also relies on other military intelligence publications, intelligence community publications, and the professional literature on intelligence analysis.[28] It is worthy of noting here that the application of all-source analytic methodology to campaign assessment can improve holistic analysis but it cannot account for all of the inherent flaws in centralized assessment. All-source analytic methods would be best matched with layered contextual assessment.

Incongruent with Doctrinal Understanding of Counterinsurgency

COIN assessment theory should be congruent with U.S. joint and service understanding of warfare, of the COIN environment, and the way in which COIN campaigns should be prosecuted. Appendix E references a 2008 memorandum from the com-

---

[28] This recommendation is explained in greater detail in Chapter Nine.

manding general of U.S. Joint Forces Command that officially removes EBO and its associated themes and concepts from the joint lexicon. But the joint doctrine that shaped this assessment process remains centralized, quantitative, and reliant on effects-based theory and language.[29] The Marine Corps continues to cleave to the principles of maneuver warfare but refers to many Army publications that retain effects-based language. At first glance, U.S. Army doctrine seems to include some internal contradictions on EBO and EBA. The 2008 version of FM 3-0 clearly states that "Army forces do not use the joint systems analysis of the operational environment, effects-based approach to planning, or effects assessment" and that effects-based theory and language are introduced in doctrine only to familiarize soldiers with joint doctrine. (FM 3-0 was published several months before Mattis issued his letter.)[30] However, FM 5-0 prescribes the use of EBO in planning, operations, and assessment.[31] Appendix H of FM 5-0 clearly spells out an effects-based process for assessment without offering an alternative.[32]

It might or might not be possible to apply a mathematical assessment model like the one prescribed in FM 5-0 or the *Commander's Handbook* to a conventional warfare environment. However, it is internally inconsistent to simultaneously accept complexity and chaos (FM 3-24 and FM 5-0), conceptually reject SoSA, ONA, and EBO (FM 3-0), and also dictate the use of EBO and EBA models in COIN (FM 5-0). An examination of the hierarchy of Army doctrine as described in the *2007 Army Modernization Plan* clearly gives FM 3-0 ("capstone doctrine") higher precedence than FM 5-0 (a "reference" document).[33] In the Army's hierarchy of doctrinal publications, all subordinate publications are to mesh with the capstone publications; thus, the rejection of effects-based theory as articulated in FM 3-0 appears to be the Army's official

---

[29] JP 2-0 (U.S. Joint Chiefs of Staff, 2007) clearly spells out methods to examine nodes and links within a SoSA model and describes ways to support an effects-based assessment process.

[30] HQDA, 2008, p. D-2. An April 29, 2011, document issued by the U.S. Army Combined Arms Center provides what is termed "unofficial guidance" to training developers on Army terms that have been added, rescinded, retained, or revised. This spreadsheet of terms rescinds "effects based operations," "effects cell," and "effects coordinator," describing each as "non-doctrinal terms."

[31] For example, "By identifying the possible emergence of unintended consequences or threats, commanders consider exploitable opportunities to create effects that reinforce the desired end state" (HQDA, 2010, p. 3-11).

[32] FM 5-0 also states,

It is often difficult to establish a link or correlation that clearly identifies actions that produce effects beyond the physical domains. The relationship between action taken (cause) and nonphysical effects may be coincidental. Then the occurrence of an effect is either purely accidental or perhaps caused by the correlation of two or more actions executed to achieve the effect. (HQDA, 2010, p. 6-7)

[33] HQDA, 2007.

doctrinal position on EBO and EBA.[34] It seems that Army doctrine rejects EBO while retaining EBO language to address some standing joint requirements in the absence of a viable alternative.[35]

Joint COIN doctrine remains largely effects-based, despite the absence of explicit EBO language in standing publications. JP 3-24 calls for a systems approach to understanding the COIN environment and an effects-based approach to planning, operations, and assessment:

> Identifying desired and undesired effects within the OE [operating environment] connects military strategic and operational objectives to tactical tasks. Combined with a systems perspective of the COIN environment, the identification of desired and undesired effects informs a holistic view of the OE. Counterinsurgents plan joint COIN operations by developing strategic and operational objectives supported by measurable strategic and operational effects and assessment indicators.[36]

JP 3-24 provides more suggestion than direction for assessment, and like some service doctrine, it proposes EBA while recognizing that the complexity of the COIN environment requires modifications to the approach: "[T]he complex nature of COIN operations makes progress difficult to measure."[37] In the same section, it recommends the use of the Interagency Conflict Assessment Framework, but that is a loosely structured guide for preoperational assessment and not a campaign assessment tool or methodology.[38] The joint Army/Marine Corps publication FM 3-24 skirts the issue by presenting EBA at face value and then cautioning against overreliance on EBA, without providing an alternative.[39] In total, the 2006 version of FM 3-24 devotes three pages to COIN campaign assessment.

According to U.S. Army and Marine Corps doctrine, as well as nearly all the literature on COIN examined for this study, COIN is best practiced as a decentralized type of warfare predicated on "mission command." Decentralization and mission command necessitate a loosely structured, localized approach to prosecuting war.

---

[34] According to the plan (HQDA, 2007):

The Army has two capstone manuals regarding doctrine: Field Manual 1, *The Army*; and FM 3-0, *Operations*. FM 1 contains the Army's vision. FM 3-0 provides the principles for conducting full spectrum operations, and describes the operational role of linking tactical operations to strategic aims. It details how Army forces conduct operations in unified action. These two doctrinal publications establish the foundation in preparing the Army to dominate land warfare in Joint operations.

[35] ISAF is a NATO (not U.S.) command. As of mid-2010, NATO doctrine incorporated the "effects-based approach to operations" (EBAO), a modified version of EBO that still relies on basic effects-based concepts.

[36] Joint Chiefs of Staff, 2009c, p. IX-4.

[37] Joint Chiefs of Staff, 2009c, p. X-17.

[38] Reconstruction and Stabilization Policy Coordinating Committee, 2009.

[39] HQDA, 2006c, p. 5-27.

This approach allows commanders at the tactical level to assess local conditions and tailor the ways in which they address local challenges. This means that there are few if any generalizable or consistent activities across a COIN theater of operations. In COIN, the impact of decentralization of operations on assessment is exacerbated by the complexity and uncertainty inherent in the environment. It is difficult—if not impossible—to develop a practical centralized model for COIN assessment because complex COIN environments cannot be clearly interpreted through a centralized process that removes data from their salient context. The incongruence between decentralized and complex COIN operations and centralized, decontextualized assessment has led military staffs to rely on ad hoc assessment methods that leave policymakers and the public dissatisfied with U.S. COIN campaign assessments. FM 3-24 applies to both Army and Marine Corps COIN operations:

> *Mission command* is the conduct of military operations through decentralized execution based upon mission orders for effective mission accomplishment. . . . Mission command is ideally suited to the mosaic nature of COIN operations. Local commanders have the best grasp of their situations. . . . Thus, effective COIN operations are decentralized, and higher commanders owe it to their subordinates to push as many capabilities as possible down to their level.[40]

If it is truly the case that "local commanders have the best grasp of their situations," then local commanders are also best positioned to write assessments; at the very least, their assessments should be prominent in the overall theater process. JP 3-24 states, "Subjective assessment at all levels is essential to understand the diverse and complex nature of COIN problems,"[41] and FM 3-0 concurs: "Generally, the echelon at which a specific operation, task, or action is conducted should be the echelon at which it is assessed."[42] Centralized COIN assessment runs counter to not only these doctrinal statements but also more generally to U.S. joint and capstone doctrine on both warfare and COIN. Further, it is written in a way that seems intended to confound military officers attempting to create an assessment process. It simultaneously recommends the use of detailed mathematical models and recommends against reliance on mathematical precision and accuracy. As a result, military staffs preparing to deploy to combat tend to find assessment a confusing and unhelpful distraction rather than a means for

---

[40] HQDA, 2006c, p. 1-26 (emphasis in original).

[41] Joint Chiefs of Staff, 2009c, p. X-17.

[42] HQDA, 2008, p. 5-17.

supporting decisionmaking.[43] At least in part because of this confusion, they tend to devote only fleeting attention to assessment until they are in theater.[44]

Data Aggregation and the Counterinsurgency Mosaic

FM 3-24 and nearly all of the literature on COIN cited in this monograph reinforce the idea that insurgency tends to be a local phenomenon. National, regional, or ethnic issues might provide a common theme to insurgency, but individuals and small groups are the insurgency. With some possible exceptions (e.g., the Chinese communist insurgency), the root causes and motivations that feed insurgent recruitment and support tend to be localized.[45] This dynamic has proven to be particularly salient in Afghanistan, a country with a population broken into many distinct ethnic and tribal groups, mostly spread out across thousands of small and often isolated villages. Each village, district, province, and region has unique physical and human terrain, levels of violence, degrees of economic development, and governance. With few exceptions, there are no blanket solutions to any COIN problem: Local concerns tend to require local solutions.[46] As demonstrated in places like Vietnam, Afghanistan, Iraq, the Philippines, and Malaya, COIN is a battalion commander's war.[47]

The events that transpire in each battalion's area of operations are also unique. An attack in one district probably will not have the same meaning as an attack in another district. For example, the attack might be tied to an intertribal contracting dispute in one district (one tribe wants the other's government contract), while it might be clearly tied to a significant Taliban presence in another. Shops opening in one market might indicate increasing stability and economic development, while, in another market, this may reflect the success of the Taliban shadow government in the absence of Afghan government presence. Following the guidance in FM 3-0, this information—whether quantitative or qualitative—has clear value when assessed at the level at which it is collected. The battalion staff should be able to explain that the intertribal attack does not

---

[43] This last observation was made over the course of two years spent preparing U.S. Army and Marine Corps units from the division to the battalion level for deployment to Afghanistan through structured discussions, briefings, seminars, and workshops.

[44] This finding is derived from direct observation of numerous tactical units during their predeployment workups and deployments to Afghanistan and Iraq.

[45] Kilcullen, 2009a. However, this is not always the case. The example of the Chinese Revolution stands this assumption on its head. As with nearly all assumptions regarding COIN, this one should be considered generalizable only with caution and with the understanding that each COIN case is unique.

[46] See Connable and Libicki, 2010, pp. 151–156.

[47] In some cases, it is a company or platoon leader's or platoon commander's war, depending on force lay-downs and a range of elements in the operating environment (e.g., physical terrain). This monograph focuses on battalion-level operations with the express intent of building staff processes. Writing in the mid-1960s, David Galula (2005, pp. 110–111) describes a "basic unit of counterinsurgency warfare" at the battalion level during earlier phases and down to the squad level during transition.

reflect increased insurgent activity and that the market openings can be either positive or negative. As data are aggregated, however, that context disappears. When applied to a core metrics list, more attacks can simply mean that things are getting worse and that all shop openings are beneficial.[48] Multilayered data aggregation without embedded analysis leads to an exponential loss of context and concurrent decrease in relevant meaning from level to level. Aggregation without embedded analysis is the primary cause of opacity and loss of credibility in assessment.

Figure 8.3 offers a notional example of the kinds of data aggregation challenges that can develop within a regiment's (or brigade's) area of operations. It shows that each of four battalions has reported something different about school attendance, but, taken as an aggregate across the regimental area of operations, the judgment is that a large number of children go to school and that attendance reflects support for the government. But these data may be painting a misleading portrait. They are not compared to the total number of children in the area of operations, because no accurate census data are available. Therefore, it is not possible to know what percentage of children overall attend school. The Taliban clearly affected attendance in 2nd Battalion's zone and probably in 3rd Battalion's, where children are paid to go to school. It appears that no one goes to school in the 4th Battalion's area, but in reality, attendance is unknown; it could be zero, 50,000, or some number in between. In this notional case, the aggre-

Figure 8.3
How Aggregation Hides Context

How many children attend school?

1st Battalion:
- 25,000 attend every day across all districts
- Positive view of commander, Afghans
- Commander: "School attendance reflects success"

2nd Battalion:
- 25,000 attend some days in most districts
- Want to attend but Taliban prevents them
- Commander: "Insurgent violence reduces attendance"

3rd Battalion:
- 50,000 attend 3 days/week across all districts
- Do not trust or support the government
- Commander: "We pay them to attend school"

4th Battalion:
- 0 attend schools in any district
- No data available *but they might attend*
- Commander: "We don't know how many attend"

1st Regiment:
- 100,000 attend school in area of responsibility
- School attendance reflects support
- Commander: "Attendance is high and positive"

NOTE: The data in the figure are notional.

RAND *MG1086-8.3*

---

[48] See Kilcullen, 2009a, for a lucid explanation of the impact of aggregating attack incident data. See Downes-Martin, 2010b, for another critique of data aggregation in the DAM. While he agreed that context is important, the director of AAG stated that detail at the theater level is not important, and that "more attacks means things are getting worse" (interview with the author, Kabul, Afghanistan, May 2010).

gated assessment does not reflect operationally critical context at the battalion level: The commander is making an assessment that fails to interpret the data in context.

Unparsimonious, Costly, Risky, and Contentious

> What I don't know I make up, what I do know I don't bother checking . . . because the metrics system is junk, the scoring mechanism is broken, and we get no feedback.
>
> —*Senior military officer on an RC staff, 2010*[49]

COIN assessment should be parsimonious. Collection and reporting requirements for assessment should be carefully considered in light of the demands and risk they may leverage on subordinate units. Assessment should rely to the greatest extent possible on information that is generated through intelligence and operational activities without requiring additional collection and reporting.

As Chapter Five showed, experts on assessment generally agree that core metrics lists should be parsimonious and that efforts to collect data from subordinate units should be frugal. This expert opinion coincides with U.S. doctrine that states, "Excessive reporting requirements can render an otherwise valid assessment plan onerous and untenable."[50] This simple advice does not go far enough to shape assessment or data collection for assessment. Both the Vietnam and (to a lesser extent) Afghanistan cases show that centralized COIN assessment requires the collection, reporting, and management of enormous amounts of quantitative data. Requirements for data leveraged upon subordinate units rarely reflect a thorough and iterative effort to determine the cost and risk that might be associated with the requirement. As a result, subordinate units sometimes ignore requests for data, fabricate data, or endure cost and risk to collect data that are not used to produce relevant assessments. It is possible for assessment staffs to improve the collection requirements process to include considerations of cost and risk, but centralized assessment cannot function without large quantities of aggregated quantitative data. Not all of the data demands for centralized assessment can be met with data produced for intelligence and operations functions, so a considerable amount of recurring collection and reporting will always be necessary to feed these demands.

Because centralized assessment is likely to place onerous demand on subordinate units, it is also likely to create some amount of disharmony between subordinate and senior staffs. While this can be alleviated to some extent through clear guidance and iterative discussion, combat tempo and the distributed nature of COIN opera-

---

[49] Statement by a senior coalition military officer, Afghanistan, 2010, as quoted in a briefing delivered by Stephen Downes-Martin at the U.S. Naval War College, Newport, Rhode Island, October 19, 2010. Downes-Martin interviewed the officer in mid-2010.

[50] HQDA, 2010, p. H-3.

tions make this type of process difficult and unlikely. In practice, core metrics lists are created through a process that assessment staffs and combat officers tend to refer to as "BOGSAT," or "Bunch Of Guys Sitting Around a Table."[51] The subjectivity of core metrics selection is inherent in the process, and tactical-level officers understand that core metrics selection is subjective. There is an ingrained suspicion of core metrics at the tactical level, and this suspicion is unlikely to be overcome even in the best of circumstances. Centralized assessment is likely to produce staff friction that will lead either to cost and risk of questionable value or to disingenuous reporting.

Additional Factors Contributing to the Failure of Centralized Assessment in Counterinsurgency

This section addresses a number of additional factors that contribute to the failure of centralized assessment in COIN. They include overoptimism, the inability to adequately cope with missing data in statistical trend analysis, and the subjectivity inherent in time-series thresholds. It also provides a number of examples of data inaccuracy and inconsistency, and it describes the district assessment process to demonstrate how color-coding and aggregation can undermine transparency and credibility.

Overly Optimistic

> Listen, the president didn't ask for a "situation" report, he asked for a "progress" report. And that's what I've given him—not a report on the situation, but a report on the progress we've made.
>
> —*Robert W. Komer, as quoted in* Vietnam: A History[52]

> The post–World War II American [military reporting] system was receptive only to the recording of sunny hours. All reports were by nature "progress reports."
>
> —*Neil Sheehan*[53]

---

[51] I participated in a number of such "BOGSAT" sessions between late 2009 and early 2011. All of these efforts were well-intentioned and professional but ultimately unhelpful. Some of these sessions were unstructured or at low levels of authority, while others were structured and coordinated by high-level authorities. No matter how the sessions were structured, because they were removed from the military planning staff and process, they necessarily boiled down to debates over the partially informed opinions of those present for the discussion. The sessions I attended never reflected iteration with field units, rarely referred to operational planning documents, and never considered the risk of collection through structured and realistic analysis. The best of these efforts relied on the participation of more than 100 top experts and military offers over a weeklong series of offsite sessions. This group had access to all relevant documentation, and it included members of the theater and component staffs. Its findings were not adopted.

[52] Karnow, 1984, p. 515.

[53] Sheehan, 1988, p. 287.

"Progress reports" are neither unusual in military operations nor necessarily always optimistic. Military operations are conducted to make progress toward objectives; sometimes they succeed and sometimes they fail or struggle to make progress. These two quotes (above) reflect perceptions about the assessment process that are relevant to the way in which assessments are viewed by consumers and the public. Komer's comment may reflect pressure from senior policymakers to provide good news through "objective" analysis, while Sheehan's comment reflects the cynicism that perceived overoptimism in military assessment reports sometimes elicit.

Overoptimism in military campaign assessment reporting can undermine transparency and credibility and also contribute to the existing distrust of military assessment reports in some policy circles and among the general public. Sometimes, this overoptimism is self-generated, and sometimes it results from pressure within the policy or military hierarchy. Gelb and Betts describe how assessment staffs during the Vietnam War were subjected to similar pressure to deliver positive assessment reports:

> Most [assessment analysts] were under varying degrees of pressure, both explicit and implicit, to produce certain *kinds* of assessments. Whether they biased their reports in a certain direction because someone else told them to do it or because they knew this would be best for them in the long run . . . is not particularly important in this context; the fact is that the pressure was there nonetheless.[54]

This analysis of Vietnam-era assessment is reinforced by the accounts of those who were involved in the assessment process during this period, as cited throughout this monograph. Testimony in the *Westmoreland v. CBS* case by retired Marine Corps colonel, COIN expert, and then-Congressman Paul N. McCloskey, Jr., on MACV briefings to visiting policymakers offers specific insight into the way in which quantitative data were used to engender positive perceptions:

> There was heavy stress on numbers, i.e., body count, crew-served weapons captured, strength of VC units, and particularly the favorable *trends* in those numbers in every category as compared with three months earlier, a year earlier, etc. I do not recall a single unfavorable trend reported to us, and there was a consistent and strong expression that there was "light at the end of the tunnel," that our "nation building" program was succeeding, that the VC strength was steadily eroding, and that in due course we would be able to return to an advisory status to a strong and stable South Vietnamese government. . . . [I saw a manual] with a title along the lines "Standard Operating Procedure for Handling Visiting CODELs [Congressional Delegations]." The manual explicitly outlined the requirement that CODELs were to be provided only with facts favorable to MACV's perfor-

---

[54] Gelb and Betts, 1979, p. 309 (emphasis in original). They provide additional insight into this dynamic on pages 319–321.

mance and directed withholding facts that would make "MACV's mission more difficult."[55]

It is clear that Vietnam-era analysts were under considerable pressure to deliver positive reports.[56] Research into the Iraq and Afghanistan campaigns revealed similar pressures on analysts in both theaters. Downes-Martin also observed overoptimism in Afghanistan assessment reports generated at both the RC and theater levels.[57] Observation and interviews for this study showed that the majority of this observed and reported pressure in Iraq and Afghanistan was implicit and probably nowhere near as egregious as during the Vietnam campaign. This kind of pressure is typically nuanced, and it is sometimes unintentional or self-generated. In their discussion of assessments, Gelb and Betts point out that military officers are trained and educated to be optimistic and forward-leaning and that delivering bad news could represent a "personal failure."[58] This last claim is an overgeneralization of military culture, but it is sometimes accurate.

As with Vietnam-era assessment reports, overoptimism in Iraq and Afghanistan assessment reporting is most apparent in the emphasis that is sometimes placed on positive information rather than in an imbalance in the published reports. In other words, reports from both eras often show positive and negative information, but the ways in which the reports are presented may emphasize an optimistic outlook that the report does not necessarily warrant.[59] As McCloskey's testimony shows, oversell was taken to an extreme during the Vietnam War. Karnow also provides a number of examples in which Robert S. McNamara presented overly optimistic public assessments that he then countered with pessimistic but private reports to President Johnson.[60] This oversell backfired when the Pentagon Papers were leaked to the public.

To a great degree, overoptimism is an issue of perception. In the absence of precise and accurate determinative assessments, only subjective judgment can say whether reports are overly optimistic rather than objectively optimistic. Sheehan judged that Vietnam-era progress reports were overly optimistic based on his perceptions and research of the available data. Although overoptimism may be inherent to some degree,

---

[55] Westmoreland's memorandum in opposition to CBS's motion to dismiss, October 19, 1982 (emphasis in original), in *Westmoreland v. CBS*, 1984. McCloskey ran the COIN school for reservists preparing to deploy to Vietnam and served in Vietnam from 1964 to 1965, prior to being elected to Congress.

[56] McNamara testified in the same case that "General Westmoreland was not under any pressure to deliver good news on the war."

[57] Downes-Martin, 2010b, pp. 2–7.

[58] Gelb and Betts, 1979, p. 309.

[59] This is a subjective judgment based on nine years of experience reading COIN assessments from Afghanistan and eight years of experience reading COIN assessments from Iraq, as well as a more intensive examination of assessment reporting between November 2009 and March 2011.

[60] Karnow, 1984, pp. 341, 357, 440.

the best way to counter the perception of overoptimism from the public or policy-makers is to create depth and transparency in reporting.

Inability to Adequately Address Missing Data in Time-Series Trend Analysis

This section addresses a significant concern in the application of statistical methods to data sets with unknown inconsistencies and gaps. Some amount of accurate (or valid) data is necessary for any kind of statistical analysis. This is certainly true for the kind of time-series analysis that is common in COIN assessment. But nearly every scientific study is based on the assumption that a percentage of data will necessarily be missing. A study of 300 scientific articles that had been published within a three-year period showed that a full 90 percent of the studies described in the articles were missing data and that the average amount of missing data across these 270 studies "well exceeded" 30 percent.[61] Nonscientific military assessment of COIN campaigns must also assume that there will be sizable gaps in sets of information and that the unstructured nature of COIN assessment may mean that gaps will be larger than in a controlled scientific study because it is generally not possible to collect data from all relevant areas due to violence or threat. It will also be very difficult, if not impossible, to provide a sound estimate of the amount of data that is missing because the assessment analysts do not control and typically cannot inspect collection measures in the field.

There is an entire field of research on the phenomenon of *missing data*, a statistical term referring to gaps in data sets due to improper or incomplete collection or data corruption. Statisticians have mathematical techniques (e.g., data creation algorithms) designed to fill in these gaps for scientific analysis. The theory and terminology from missing data analysis is informative. In very simple terms, ignorable, or "missing-at-random," data are the kind of data that can be made up for logically and systematically through statistical inference.[62] "Nonignorable data" is also somewhat self-explanatory, but it is very hard to make up for nonignorable data using missing data analysis.

This chapter argues that each of the individual tiles in the COIN mosaic may be important to the campaign in their own way. In practice, it is typically the people or events that are ignored that turn out to be the most dangerous to the government. It is not enough to state that "nearly all of the tribe supports the government" if 5 percent of the tribe is able to destabilize an entire district. It is not enough to state that an opinion poll covered "most of the country" if the poll-takers were prohibited from entering the

---

[61] McKnight et al., 2007, p. 3. Since this analysis describes missing data problems in scientific studies, it makes sense that missing data problems in loosely structured COIN assessment would be even more significant. However, there is no available empirical evidence to support that conclusion.

[62] *Inference* is used in the broadest sense here. Missing data analysis can be based on imputation, regression analyses, or other approaches.

most dangerous—and potentially most important—areas due to security concerns.[63] Efforts to bridge data gaps probably could not account for the damage that *might* be caused by inaccurate or missing reports.[64] The inability to safely jump data gaps poses a serious challenge to researchers trying to apply statistical methods to COIN analysis. Social scientist Paul D. Allison at the University of Pennsylvania (the author of *Missing Data*) states that missing data methods

> depend on certain easily violated assumptions for their validity. Not only that, there is no way to test whether or not the most crucial assumptions are satisfied. The upshot is that although some missing data methods are clearly better than others, none of them really can be described as good. The only really good solution to the missing data problem is not to have any.[65]

The inability to control data collection and data management, the lack of reliable data collection methods, and the fluid nature of requirements means that it is impossible to accurately track, describe, or account for missing data in a way that would satisfy basic requirements for validity. This means that the most common technique in centralized assessment—time-series analysis—is of questionable utility when integrated into a COIN campaign assessment.

Time-Series Thresholds Are Subjective and Unhelpfully Shape Behavior

As with all other aspects of centralized assessment, the lack of context confuses the meaning and undermines the value of time-series thresholds. In some cases, thresholds are selected through behind-the-scenes discussion or by the commander's preference. In a few cases, thresholds are transparently (and commendably) tied to national objectives, but selection remains a subjective judgment call. For example, in Iraq, the coalition set a threshold for electricity that matched prewar levels: If it could achieve this level of service, then it could show that it had put Iraq back on its feet. But this threshold proved elusive as demand rose above prewar levels and was never met; frustration continued unabated as expectations changed. Coalition planners did not factor into the equation the possibility that the removal of sanctions coupled with more plentiful electricity might lead Iraqis to purchase more air conditioners, refrigerators, and tele-

---

[63] This is the most common criticism of poll results in both Iraq and Afghanistan, based on my discussions with consumers of polls in DoD and other government agencies. The second most common complaint seems to stem from a lack of transparency.

[64] For example, what if a host-nation military unit exaggerates an incident report to show that it was attacked by 100 insurgents when, in fact, it was attacked by ten? This single report could drastically alter both quantitative and qualitative analysis. What if the same host-nation unit was actually attacked by 100 insurgents but failed to report the attack or the report was lost before it could be officially recorded?

[65] Allison, 2002, p. 2.

vision sets than they had prior to the war. So, while the original threshold was met, it proved to be relatively meaningless in terms of effect.

Assessment of Iraq's mobile-phone network provides another example of the failure of subjective thresholds to capture actual progress. As violence ebbed in Iraq, Iraqi cell-phone purchases skyrocketed, but the overall number of phones hid the fact that many Iraqis had to purchase three or more phones because there were no available roaming services from region to region.[66] The subjective threshold proved to be misleading.

Aggregated threshold analyses are susceptible to Simpson's paradox: They can show what appears to be forward progress while, according to the same indicator, many of the key areas in the country are sliding backward.[67] For example, a security threshold that sets a maximum acceptable number of violent attacks across the country (e.g., 50 attacks per day) might reveal a lack of progress in a specific district that is suffering a relatively high five attacks per day, positive progress at the province level because that same province might be suffering only five attacks per day (all in the aforementioned district), and then a lack of progress at the national level because the total number of attacks may be well over 50. Without detailed analysis, the national threshold can be misleading.

The use of thresholds shapes behavior in ways that are often unforeseen. This was proven in the Vietnam case: Military officers changed their behavior to increase body counts or show positive HES reporting data because policymakers and senior military leaders had set the performance thresholds for both programs. Strict quantitative thresholds encourage dishonest reporting. And while thresholds might help focus military units on key tasks, so do traditional military mission statements and objectives; the addition of thresholds for operations is superfluous and akin to micromanagement.

Thresholds are poor assessment devices because the data needed to show accurate quantitative progress do not exist in COIN environments. In the absence of a mathematical solution to COIN, it is difficult to envision a systematic, transparent, and credible means of establishing mathematical solutions to specific aspects of COIN operations. There seems to be no way for a staff to adequately defend the selection of a time-series threshold as anything but a subjective guess as to what level of progress or suppression of activity might be necessary to effect a meaningful change.[68] Time-series thresholds may seem to be the most direct and commonsense method for showing

---

[66] This information is based on my experience working with development metrics in Iraq and is derived, in part, from discussions with experts on Iraq assessment between 2004 and 2011. For further information on electricity and cell-phone usage in Iraq, see Brookings Institution, 2010.

[67] For an explanation of Simpson's paradox and its effect on aggregated assessment, see Connable and Libicki, 2010, p. 139.

[68] This might not be the case in a conventional campaign designed to degrade a known conventional military force, and it might not apply to some MOPs that are focused on internal U.S. performance. For example, it may be possible to set a defensible threshold for reducing the number of days it takes to move troops into theater in

progress in COIN campaigns, but in practice, they are misleading and often counter-productive. To the credit of commanders and staffs in Afghanistan and Iraq, the use of thresholds appeared to be waning in 2011.

Incomplete and Inaccurate Data: U.S. Department of Defense Reports on Iraq

Most data on contemporary COIN campaigns (Iraq and Afghanistan) remain classi-fied or otherwise restricted.[69] Even if most of these data were available for unclassified research, an empirical study of data from a complex insurgency like Iraq or Afghani-stan would take months or years and would require extensive fieldwork and interviews with members of tactical units. It is possible, however, to gain insight into data quality by other means.[70] A commonly referenced source for official, publicly released data are the 9010- and (now) the 1230-series reports produced by DoD for Congress. These reports are very lightly sourced because they cannot show restricted data, a serious flaw not in the way the report is written but in the structure of the assessment process. These reports typically refer to the command that provided the data (e.g., "MNF-I" for Multi-National Force–Iraq or "USFOR-A" for U.S. Forces–Afghanistan) rather than the original source of the information. This makes it nearly impossible for someone outside the U.S. government to check the validity of the data.

It is also possible to identify major inconsistencies in historical 1230-series reports that bring into question the overall quality of the data. For example, the November 2006 version of the 9010 report on Iraq shows that there were approximately 1,000 sectarian killings there in October 2006, but the December 2007 report shows that there were approximately 1,700 such killings in October 2006.[71] There is no explana-tion offered to account for the additional 700 killings; it could be a result of lagging reports or simply a correction of erroneous reporting. There is no way to tell whether either of these reports is accurate, due to the opacity of the sourcing.[72]

---

order to improve efficiency. This kind of initiative has little to do with holistic COIN campaign assessment, however.

[69] Some aggregated data reports are unclassified, and, in some cases, specific data sets are unclassified. However, to clearly determine quality, it would be necessary to study all types of data from multiple sources that had been captured over an extended period. Even with sampling techniques, a thorough study of this type would require access to restricted data.

[70] For example, the September 2009 report lists hundreds if not thousands of discrete elements of information but contains only 35 footnotes referencing a total of three sources, while the March 2010 report on Iraq also jams a staggering amount of data into the narrative text (at least 35 individual pieces of information or conclusions on page 53 alone) but offers only 24 footnotes referencing three sources throughout the entire report. DoD, 2009b, p. 38; DoD, 2010b, p. 77.

[71] DoD, 2006, p. 24; DoD, 2007, p. 17.

[72] The classified versions of these reports are more adequately sourced but still not adequate to sustain objective credibility. The most commonly read reports—and the only ones available to the public—are the unclassified versions provided on the DoD website.

Incomplete and Inaccurate Data: Civilian Casualties

The rate of civilian casualties is a commonly used metric in COIN. The thinking behind the use of this metric is that the more civilians who are killed, the less likely the remaining civilians are to trust or support the government because of resulting instability, fear, and a perception that the government is either ruthless or incompetent. Civilian casualty data are often disputed, however. Jonathan Schroden states that, as of 2009, there was no official total civilian casualty figure for Iraq and that

> even if we could consistently separate civilians from insurgents, the actual number of such casualties still may not be easily determined with any degree of accuracy. . . . One effort, called the "Iraq Body Count," has compiled a running tally via open-source reporting that currently stands around 90,000 [as of 2009]. A second effort by The Brookings Institution relied on a combination of data sources, and gave a total closer to 104,000. A third effort, by researchers at the Johns Hopkins Bloomberg Institute of Public Health, generated a substantially different count of over a half-million.[73]

While the Iraq Body Count (IBC) and Brookings numbers are fairly close to each other, they are still estimates and not actual counts. Schroden points out that official Iraqi statistics are corrupt and inaccurate. He also compared these three databases with three others, but none were in agreement.[74] Some of the media sources that fed the IBC database were controlled by insurgents or reflected inaccurate circular reporting. For example, in 2004, I witnessed an incident that resulted in a false report of 40 civilian casualties, later tallied as accurate by IBC.[75] This false report was re-reported in various media outlets and is still part of the IBC total; IBC staff had no consistent way of checking their sources in the field. A 2011 DoD report identified 21,000 civilian deaths by IEDs in Iraq between 2005 and 2010, but the DoD spokesman discussing the report stated that the military was "unable to quantify the claim" and that it reflected U.S. reporting only.[76] The Congressional Research Service lists six different sources that estimate total civilian deaths in Iraq; at the time of that review, these totals ranged from 34,832 to 793,663.[77] Despite these critical flaws in civilian casualty data, they are still highlighted as accurate centralized quantitative metrics rather than rough estimates in the 1230-series reports.[78]

---

[73] Schroden, 2009, p. 720.

[74] Schroden, 2009, p. 721.

[75] Connable, 2010.

[76] Vanden Brook, 2011.

[77] Fischer, 2008, p. 3.

[78] See, e.g., DoD, 2010b, p. 29.

Figure 8.4 shows a graph from a Multi-National Force–Iraq briefing to Congress in 2008. It shows casualty statistics, presented in time-series trend format, from January 2006 through March 2008. Two separate data sets are depicted in the graph. The first, in purple, is the combined total reflected in Iraqi and coalition data, while the second, in blue, reflects coalition data alone. The graph shows the degree of potential disparity between data sets that address the same category of data. In December 2006, for example, the coalition recorded approximately 1,250 civilian casualties, while the combined figure was approximately 3,700 casualties. This means that Iraqi tabulations of total civilian casualties were about two times higher than the coalition's overall total—just for this category of data. This disparity throws both estimates into question: Were the Iraqis inflating their figures, or did they have better data than the coalition? Were the coalition forces underreporting casualties, or were they more discerning than the Iraqis? There is no way to tell from the graph or the accompanying report. One or the other figure might be accurate, or neither might be accurate. The graph is instructive not only for an examination of civilian casualty data but also for a broader understanding of how host-nation or third-party reporting can skew centralized data sets. Iraqi and Afghan incident reporting and other types of reporting are often rolled up into coalition data sets without differentiation in time-series presentations for assessment.

Figure 8.4
Civilian Casualty Data, Iraq, 2007 and Early 2008

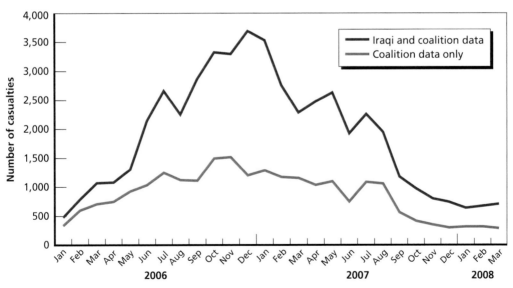

SOURCE: Multi-National Force–Iraq, 2008, p. 2.

Misleading Metrics: Insurgent Recidivism

U.S. and coalition forces hold and often release insurgent prisoners in Iraq and Afghanistan as part of routine COIN operations. Some widely publicized problems with the treatment of prisoners early in the wars revealed inadequate prisoner handling practices, poor evidentiary procedures, and a lack of forethought and supervision in the way in which prisoners were processed prior to release. These problems led to a series of reforms, and a number of programs were implemented in Iraq to improve processes and to reconcile and educate prisoners in an effort to help reintegrate them into society upon release. Military journal articles and official reports have trumpeted the improvements wrought by these programs, and it is widely recognized that many of these improvements were genuine and effective in remedying concerns about prisoner treatment. But their contribution to the COIN campaign is less clear.

Overall improvement of the prisoner handling and reintegration process was, and is, typically measured through the recidivism metric. This metric showed that very few prisoners who had been released have been reclassified as recidivists. In early June 2009, the official recidivism rate for released Iraqi prisoners was 0.1 percent, compared to 1.2 percent for all of 2007 and 7.7 percent for the three prior years combined.[79] According to the DoD semiannual report on the COIN campaign in Afghanistan, the official 2010 recidivism rate for Afghanistan was 1.2 percent.[80] The U.S. national average for criminal recidivism tends to hover above 60 percent.[81] At first glance, it appears that the prisoner programs in both Iraq and Afghanistan have seen remarkable, if not miraculous, success. These metrics are intended to show that significant progress has been made in reducing participation in the insurgency, thereby indicating some degree of overall success in the campaign.

But there are some complexities in recidivism statistics that belie the simplicity of these types of reports. Strictly defined, recidivism means a relapse into a former activity or behavior. By this definition, to qualify as a recidivist, a former insurgent would only have to commit a new insurgent act; rearrest, reconviction, or reincarceration would not be necessary. The U.S. National Institute of Justice states, "Recidivism cannot accurately be measured just by using arrest data because not all crime is discovered."[82]

---

[79] Brown, Goepner, and Clark, 2009, p. 45.

[80] DoD, 2010b, p. 61.

[81] Langan and Levine, 2002, p. 1. Official recidivism studies tend to rely on three years of data collection and additional time for processing and analysis. This report shows that the recidivism rate for prisoners released in 1994 was 67.5 percent over the subsequent three years, while the 1983–1986 study revealed a recidivism rate of 62.5 percent.

[82] U.S. National Institute of Justice, 2008. To account for this gap in data, the U.S. government recommends that criminal behavior studies rely not only on arrest data but also on interviews with offenders, analysis of crime events, and various statistical methods. For practical purposes, though, most studies tend to equate recidivism with rearrest, reconviction, or reincarceration. The rearrest rate for U.S. criminals in a 1994–1997 study was 67.5 percent, and the reincarceration rate for U.S. criminals in a 1994–1997 study by the U.S. National Institute

By this standard, it would be impossible to determine the actual recidivism rate in either Iraq or Afghanistan because only a few of the hundreds of insurgent attacks that take place each week could be clearly tied to individual insurgents.[83] In fact, the commander of Combined Joint Interagency Task Force–435 (CJIATF-435), which runs the prison system in Afghanistan, made it clear that he tracked only rearrest (or "recapture") rates, drawing a clear distinction between rearrest and recidivism:

> And we refer to it as recapture, not recidivism. Of this year, 550 [prisoners] have been released. There have been four individuals recaptured on the battlefield, so less than half of—about half of one percent, or less than one percent, of a recapture rate. Now, I can't give you recidivism, because maybe someone went back to the fight and was killed and we can't recapture that. So he wasn't recaptured.[84]

However, the DoD report on progress in Afghanistan does not account for any of these nuances or caveats. In contrast to the comments by the CJIATF-435 commander, it describes the CJIATF-435's success in achieving a 1.2-percent "recidivism rate."[85] It explains the success of its programs by tying this very low rate to its reintegration programs: "CJIATF-435 assesses that reintegration programs are working to prevent previously detained individuals from rejoining the insurgency." The report does not say how this conclusion was reached, so there is no way to determine correlation between this improperly labeled statistic and the success or failure of any program.

Recidivism statistics are often presented in a way that would allow the reader to assume that the military had the means to know the actual rate at which released prisoners returned to committing insurgent acts (an unknowable figure), or perhaps the theater-wide rearrest rate. Even the rearrest or recapture rate would be a difficult figure to obtain from remote tactical units or Afghan units that might not report all arrests or obtain accurate biometric data before releasing prisoners at the local level.

---

of Justice was 25.4 percent. Even by the strictest standard for recidivism (reincarceration), a 0.1- or 1.2-percent recidivism rate is remarkable.

[83] It would be possible to argue that criminal behavior and insurgent behavior cannot be equated, and therefore these rates reflect success against the insurgency rather than an aberration in normal human behavior. This would be a difficult argument to make, considering the degree to which criminal activity and insurgent activity blend together, but there does not appear to be any empirical evidence equating the two types of behavior for statistical analysis. It might also be possible to argue that Iraqis or Afghans are dramatically less prone to criminal activity than Americans. However, there does not appear to be empirical evidence that would support this argument.

[84] DoD, 2010d. He goes on to state that this success is indicative of the success of the task force's reintegration program, a correlation that reflects his informed opinion but is neither clear nor proven empirically. He does not state that insurgents could recommit insurgent acts and escape both capture and death, as happens on a daily basis in both theaters. He does not state whether these statistics reflect only theater-level prison operations, or also tactical arrests by infantry battalions in rural areas that might not result in a transfer to central prisons due to lack of evidence or adequate transportation.

[85] DoD, 2010b, p. 61. This percentage accounts for a greater number of released prisoners than the 550 cited in the CJIATF-435 commander's quote.

In reality, these statistics might only indicate the rate at which released prisoners were reincarcerated at province- or theater-level prisons. The rate of reincarceration might reflect the success of reintegration programs, but it also might show that formerly incarcerated insurgents became better at evading capture, that they were improperly identified upon detention, that they were killed while conducting attacks, that patterns of coalition operation had changed, or any other number of factors. A low rearrest or reincarceration rate might actually mean that security forces are incapable of keeping up with insurgent activity. When carefully analyzed, the official Afghanistan recidivism statistic takes on much less (or different) importance for campaign assessment than at first glance.

Iraq statistics are similarly opaque. A 0.1-percent recidivist rate for any type of behavior seems to be farfetched at best. It it more likely that this number reflects a rearrest or (even more likely) the theater-level reincarceration rate. Because it is presented as a recidivism rate without explanation or caveat, this statistic has the potential to be misleading.

Aggregation and Color-Coding: The District Assessment Model

This section explores the DAM as it was developed in late 2009 and early 2010 and deployed through late 2010. It refers to this best-known Afghanistan assessment tool to address broader concerns with color-coding and context.

Figure 8.5 shows an example of a DAM map. The map shows the color-coded ratings for "key terrain districts" and the extensive amount of space unaddressed by comprehensive assessment. The color codes run from green (good) to red (bad), with white areas showing up as "not assessed." In this example, the gray areas are key terrain districts that were not assessed in this particular reporting period. The rest of the map area should also be considered white, or not assessed. The data in the figure are notional, but they reflect general coverage of Afghanistan assessment during early 2010.

Former Congressman John V. Tunney, a critic of pacification assessment systems in Vietnam, said in 1968, "Since pacification is based on the dubious premise that Government control results in political loyalty, reports of progress in pacification [in Vietnam] continue to be misleading for both the American people and policymakers in Washington."[86] The DAM is an improvement over the HES in that it tries to assess loyalty by examining security, development, and governance. As of mid-2010, the district assessment was delivered in several formats: a consolidated color-coded map, a briefing containing a series of color-coded regional maps, and an operations plan assessment report (a narrative version of the district assessment maps). Downes-Martin's research has produced findings on the mapped district assessments that are very similar to those presented in later RAND research. It is therefore appropriate to refer to his insights to summarize concerns with district assessment color-coding:

---

[86] Tunney, 1968, p. 2.

Figure 8.5
The ISAF Joint Command District Assessment

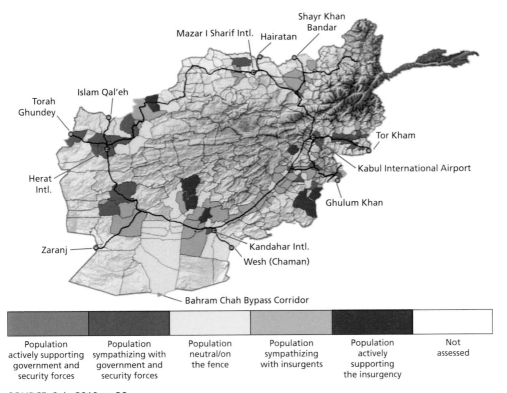

Population actively supporting government and security forces	Population sympathizing with government and security forces	Population neutral/on the fence	Population sympathizing with insurgents	Population actively supporting the insurgency	Not assessed

SOURCE: Sok, 2010, p. 36.
NOTE: The data in the figure are notional.
RAND *MG1086-8.5*

A color-coded map hides information. The single color-coding is an average (*not* a summary) of a large number of underlying factors. It is not only possible, but also likely, that an average (i.e. color on the color-coded map) stay the same as some factors improve and others degrade. The color code tells us nothing useful about this effect, and so one must give narrative explanations about the improving and degrading factors. Since smart staff often does this anyway, the color-coded map becomes pointless at best and misleading at worst.[87]

---

[87] Downes-Martin, 2010b, p. 6 (emphasis in original). He also points out,

> Finally there is the observed tendency among some RCs to "average colors". They present separate colors for their Region for the three LOOs [lines of operation] (Security, Governance and Development), then provide an overall assessment color which happens to be the color of the average point on the color bar chart of the three LOO colors. This is not coincidence, since RCs have been observed during briefings to have difficulty explaining in operational terms why they have given the color to the overall assessment. (Downes-Martin, 2010b, p. 5)

Color-coding is the most aggressive form of data aggregation, since the color scheme forces the assessment into only a few very narrow categories (e.g., red, yellow, blue, green). As with thresholds, color-coded assessments are susceptible to Simpson's paradox. Color codes are, in fact, codes: They represent a narrow set of numbers (usually three, or five at most) and demand simplification and averaging of complex problems.

Figure 8.6 illustrates this situation. While the figure uses notional data, it represents a very real problem that RC-level commanders and their subordinates face when trying to assign color codes to key terrain districts, a process that I observed firsthand while at IJC during the final part of a six-week district assessment cycle. It seems clear that this notional district does not contain a population that would merit the official rating definition level of "neutral or on the fence," but color-coding demands averaging.[88] Taken together, Figures 8.7 and 8.8 reveal additional concerns about data aggregation in the color-coding process. Each smaller square represents a village within a district (also using notional information). In Figure 8.7, color-coding rules essentially suppress information, hiding contradictory data. Figure 8.8 shows how color-coding can suppress critical information, e.g., the most important tribe is on the fence, or the most populous district in the province favors the insurgents.

As of early 2011, ISAF was deeply immersed in building a transition model to set the stage for the 2011 strategic timeline. The flaws in the district assessment process

Figure 8.6
The Effect of Averaging Color Codes in the District Assessment Model

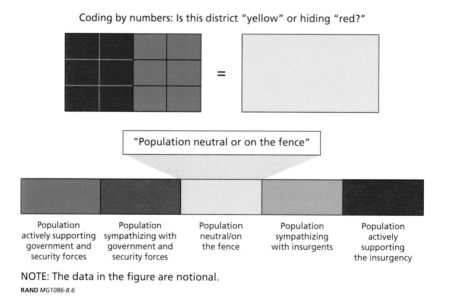

NOTE: The data in the figure are notional.
RAND MG1086-8.6

---

[88] See DoD, 2010b, p. 35.

Figure 8.7
The Effect of Aggregating Averaged Color Codes

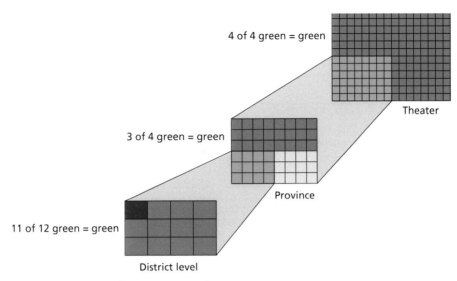

4 of 4 green = green

Theater

3 of 4 green = green

Province

11 of 12 green = green

District level

NOTE: The data in the figure are notional.
RAND MG1086-8.7

Figure 8.8
Analysis of the Effect of Aggregating Averaged
Color Codes

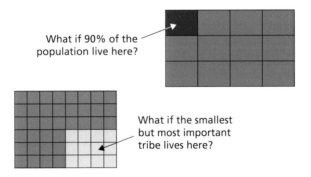

What if 90% of the
population live here?

What if the smallest
but most important
tribe lives here?

NOTE: The data in the figure are notional.
RAND MG1086-8.8

might carry over to the transition process. Downes-Martin's analysis of the incorporation of the district assessment maps into transition planning seems accurate:

> The end state [from transition] is, loosely speaking, a Region (or District, or the entire Country) that is suitable for transition to full GIRoA [government of the Islamic Republic of Afghanistan] control, where "suitable" means there is some

good chance that GIRoA will be able to keep the Region (or District, or the entire Country) stable and secure. But unless we have a credible theory that links the level of active support for an Insurgency to the likelihood of GIRoA successfully running an area, then we have no connection between the rolled-up color-coded assessment and the desired end state. In this case the assessment does not provide senior leaders tasked with judging the suitability of a Region or District for transition with a credible assessment of suitability.[89]

The senior staff at IJC attempted to make allowances for problems with data quality and aggregation, but even thoughtful iteration between the IJC and RC staffs could not make up for these pervasive flaws. By the time the district assessment was delivered to higher headquarters, it was aggregated, averaged, and opaque (to at least some degree). While FM 5-0 exhorts assessment staffs to record justifications for assessments, these justifications were not systematically recorded and were not clearly available to policymakers or the public. It was therefore not possible for senior leaders or journalists to either disaggregate the quantitative models behind the assessments or determine how and why judgments were levied from the battalion to the theater level. It was also not possible to see or understand changes in core metrics requirements, changes in metrics definitions, changes in theory or models, or adjustments to baseline data (most of which remained unavailable to the general public). Transparency and credibility simply were not achievable with the EBA district assessment process as it existed in 2010.

An April 29, 2010, *New York Times* article shows how a lack of transparency, or at least the absence of embedded contextual narrative, can lead to erroneous interpretations of theater assessments. The article refers to the IJC district assessment and states, "The number of districts sympathetic to the insurgency or supportive of it increased to 48 in March 2010 from 33 in December 2009."[90] This statement does not accurately reflect the assessment. According to the head of the IJC assessment section, the IJC added approximately 20 districts to the review between the December and March cycles.[91] No clear indication of this change was reflected in the briefing slide that contained the color-coded map. To the reporter, who was reporting from Kabul, it looked like things had suddenly taken a turn for the worse when, in fact, the reporting parameters had changed.

By the time this monograph is published, the district assessment may no longer be in use in Afghanistan. It is already being supplanted by products like the District Deep Dive, an effort to provide in-depth contextual narrative for transition assess-

---

[89] Downes-Martin, 2010b, p. 7.

[90] Rubin, 2010a.

[91] Senior officer, IJC Information Dominance Center Assessments Section, interview with the author, Kabul, Afghanistan, May 2010.

ment.[92] Because the District Deep Dive is a relatively new process, an examination of its efficacy or applicability to the overall assessment process was outside the bounds of the current research effort. But despite this probable shift in practice, this examination of the DAM will remain relevant: The dangers of aggregation and color-coding will persist, and the importance of context in COIN is immutable. Color-coding seems to be a default technique for assessment staffs and commanders, so future generations of counterinsurgents should be aware of inherent pitfalls. Policymakers faced with this kind of confusion express skepticism and frustration with operational COIN assessments.[93]

Chapter Summary

By the definitions proposed in this monograph, the U.S. military has yet to produce a relevant, transparent, or credible COIN assessment methodology. This is not a failure of effort or intellect; in most cases, energetic and intelligent people draft and review the operational assessments delivered to policymakers. Rather, it is a failure of effects-based theory or centralized pattern and trend analysis to replace or improve upon the fundamental and broadly accepted theories of warfare. While some of the findings in this chapter may not be technically relevant after early 2011, the underlying issues will persist as long as the COIN environment remains uncertain and U.S. doctrine calls for distributed, mission-type operations.

It is possible to create a relevant, transparent, and credible assessment process. This monograph proposes options designed to meet the seven standards for assessment articulated in Chapter One. However, these standards are offered to help frame *any* solution to the challenge of COIN assessment. Based on the findings presented in this chapter, prospective solutions that lie outside the closed framework of centralized assessment are the most likely to prove successful in COIN.

---

[92] Perceptions of the District Deep Dive expressed by a group of assessment officials during a December 2010 workshop in The Hague, Netherlands, were generally dismissive. Several officials saw the effort as rushed, inaccurate, and incomplete.

[93] For example, see Abramowitz, 2007, and Walker, 2007.

Conclusions, Recommendations, and Options

Sound assessment blends qualitative and quantitative analysis with the judgment and intuition of all leaders. Great care must be applied here, as COIN operations often involve complex societal issues that may not lend themselves to quantifiable measures of effectiveness. Moreover, bad assumptions and false data can undermine the validity of both assessments and the conclusions drawn from them. Data and metrics can inform a commander's assessment. However they must not be allowed to dominate it in uncertain situations. Subjective and intuitive assessment must not be replaced by an exclusive focus on data or metrics. Commanders must exercise their professional judgment in determining the proper balance.

—*FM 3-24,* Counterinsurgency[1]

Throughout the operations process, commanders integrate their own assessments with those of the staff, subordinate commanders, and other partners in the area of operations.

—*FM 5-0,* The Operations Process[2]

Military campaign assessment is an arcane process that receives little attention until it fails. The assessment process failed in Vietnam, thereby contributing to the poor policy decisionmaking that led to America's first major COIN defeat. Assessments of the Afghanistan campaign suffer from many of the same drawbacks as the Vietnam-era assessments, but the Afghanistan campaign is ongoing. As of this writing, the United States does not intend to fully transition the leading role for the security of Afghanistan to the Afghan government until 2014. In any event, assessment is likely to continue after this transition. This leaves sufficient time for the military to improve COIN assessment to support the policy decisions that will shape the end of U.S. involvement in the war. Comprehensive improvements to COIN assessment theory and practice in Afghanistan might also help reshape U.S. assessment doctrine, training, and education over the long term.

---

[1] HQDA, 2006c, p. 4-6.

[2] HQDA, 2010, p. 6-1.

This chapter presents conclusions and a summation of the critical analysis presented in this monograph, as well as recommendations and options intended to help guide efforts to improve COIN assessment in Afghanistan and the U.S. military staff process in general.

Conclusions

Making effective and lasting improvements to the assessment process will require revisiting basic assumptions. This monograph addressed a number of assumptions—some official, some implied in the military literature—that underlie the doctrinal and de facto U.S. military COIN assessment process. They include, but are not limited to, the following:

- Assessment is a centralized and quantitative process.
- The COIN operating environment is a system that can be understood through systems analysis.
- For EBA: It is possible to deliver precise and accurate campaign assessments.
- For pattern and trend analysis: Generalized pattern and trend analyses are sufficient to support policy decisionmaking in COIN.
- Quantitative data and methods are inherently objective and therefore preferable to mixed or qualitative data and methods.

These assumptions have not provided a foundation for effective assessment. Policymakers attempting to penetrate the complexity of COIN are poorly served by centralized and highly quantitative military campaign assessments. Such assessments either attempt to portray accuracy that is beyond the grasp of even a theoretical military collection capability or unintentionally convey accuracy because they are presented through quantitative charts without sufficient caveat. Policymakers can be enticed into making overly determinative decisions based on information they believe to be complete and accurate when, in fact, centralized assessment reports are typically neither complete nor accurate. Centralized EBA and pattern and trend assessment do not account for as much as 40–60 percent of available information, because these processes do not systematically incorporate qualitative data. This problem is inherent in a centralized quantitative assessment process and, in the absence of a massive and quite expensive coding effort at multiple layers of command, it will remain an inherent problem. At best, centralized quantitative assessments are informed by qualitative reports, but because assessment does not reflect dedicated all-source analysis, this tends not to be the case. The absence of qualitative input into the centralized process corrodes both the transparency and credibility of centralized assessment.

The COIN environment can be loosely analogized to a complex adaptive system. But COIN campaigns cannot be *effectively* understood by applying systems analysis methods (e.g. EBA) to assessment because COIN is not, in fact, an interconnected system that generates obtainable and understandable feedback. Instead, the COIN environment is an unbounded cloud of individuals and groups, perceptions and actions, emergence, and happenstance, all of which are unpredictably whipsawed by chaotic violence and uncertainty. At best, assessment staffs gain an imperfect sense of general patterns and trends from aggregated COIN data, but these patterns and trends do not provide policymakers with a holistic understanding of the environment or the campaign.

Chapters Five through Eight explored the subjectivity inherent in all aspects of assessment, as well as the degree to which aggregated quantitative data used to build patterns and trends are incomplete and erroneous. The levels of completeness and quality for nearly all categories of aggregated COIN data are unknown and cannot reasonably be known. This inability to gauge completeness and accuracy unanchors aggregated data sets from ground truth to the point that COIN data do not provide a credible basis for centralized analysis. Compounding this problem is the lack of a credible methodology that could correlate all available quantitative data to produce a holistic centralized assessment; EBA has not proven credible in COIN. As a result, centralized assessment briefings tend to look more like an assembly of dissociated quantitative data reports than holistic assessment. No matter how these reports are packaged, they provide senior military leaders and policymakers with the kind of vague impressionistic understanding of the campaign described by Donald Rumsfeld in Chapter Two.

Figure 9.1 provides a generalized schematic of the impressionistic decisionmaking process derived from centralized quantitative assessment. In the figure, the theater-level assessment staff produces a report containing primarily quantitative data and summaries of subordinate reports, but these data and reports are not systematically analyzed to form a holistic picture. Data quality is unknown, and patterns and trends are open to subjective interpretation without clear correlation between data sets. The commander provides insight, which is holistic and informed by the theater assessment report. The decisionmaker sees an array of dissociated, loosely associated, or falsely associated and decontextualized data, patterns, trends, and finished reports (e.g., intelligence reports), as well as the commander's report. The decisionmaker gains an impression from this mass of information, compares that impression to that of the commander, and makes an impressionistic decision.

This process places undue burden on the assessment staff, the military commander, and the policymaker. The staff must scramble to piece together what will necessarily be an incomplete picture from inaccurate and incomplete data, so it never delivers an optimized product. Assessment staffs have almost no opportunity to succeed. The commander provides a unified picture to the policymaker in the form of a professional assessment, but the value of this assessment rests almost entirely on the

Figure 9.1
Centralized Impressionistic Decisionmaking

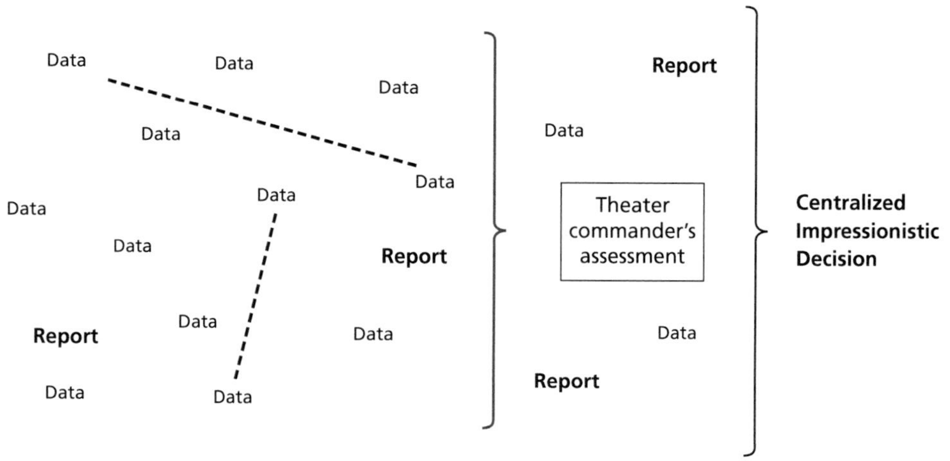

NOTE: "Data" are raw figures without context or analysis; "reports" are finished but narrowly focused products from subordinate commands, intelligence centers, PRTs, or other organizations.
RAND *MG1086-9.1*

commander's position (billet) and reputation because there is no *structured* contextual analytic depth to the assessment. This, in turn, leads to a situation in which an experienced and trusted military commander's analysis and *coup d'oeil* are regularly challenged by policymakers, lawmakers, and the public. The policymaker is forced to make decisions that tend to be more impressionistic than necessary and is placed in the position of defending an opaque, less-than-credible assessment to a suspicious public. No matter how capable the military commander or how expert the advice of assessment staffs or intelligence agencies, decisions based on this flat analytic process will *appear to be* highly impressionistic to the public.

All policy decisions are impressionistic to some extent. Even peacetime decisions based on highly accurate quantitative information require some impressionistic analysis. Policymakers try to take in all available input and make what is ultimately a subjective decision: They have the evidence in front of them, but the final decision rests with their judgment and intuition. The purpose of assessment and analysis is to reduce the window of uncertainty to help policymakers narrow the range of possible conclusions and options. COIN assessment should support decisionmaking that is *relatively less impressionistic and subjective* than an uninformed or poorly informed decision: The more transparent and credible the report, the less impressionistic the decision. Few if any strategic COIN decisions should be deterministic.[3]

---

[3] Some operational and tactical decisions differ from strategic decisions because better data are available. Decisions that might be deterministic include logistics decisions in which data are obtained from known sources and

Any military assessment process for COIN should reflect the military's understanding of the COIN operating environment. For the ground combat services most commonly associated with population-centric COIN (the Army and Marine Corps), this understanding is reflected in capstone military doctrine: MCDP-1, *Warfighting*, and FM 3-0, *Operations*.[4] Both of these doctrinal publications describe the environment as chaotic and unpredictable, and both services acknowledge that the fog of war can be reduced but never eliminated. Assessment of COIN campaigns should also reflect COIN doctrine: FM 3-24/MCWP 3-33.5, *Counterinsurgency*, and JP 3-24, *Counterinsurgency Operations*.[5] COIN doctrine is less clear and is sometimes contradictory. In places, it coincides with capstone doctrine; for example, it acknowledges the chaotic nature of the COIN environment, it recommends the use of mission-type orders and distributed operations, and it recommends that assessment be conducted at the lowest possible level. FM 3-24 states, "Subjective assessment at all levels is essential to understand the diverse and complex nature of COIN problems."[6] However, it also recommends the use of SoSA and EBA.[7] The weight of capstone doctrine and the near-universal focus on mission-type orders and distributed operations in other publications on irregular warfare suggest that COIN assessment should reflect the more traditional (maneuver warfare) understanding of COIN as opposed to an effects-based approach.[8] COIN assessment should reflect and embody the enduring realities of chaos, friction, uncertainty, and unpredictability, and it should also reflect operations that are (at least in theory) predicated on mission-type orders and distributed command and control.

This monograph argues that the centralized and quantitative EBA approach recommended in doctrine is not applicable to COIN, at least not in a way that would effectively support policy decisionmaking. EBA demands "mathematical rigor," but COIN is more likely to produce immeasurable uncertainty.[9] The EBA process was developed to assess conventional warfare, and it is framed in a way that is antithetical to distributed COIN operations. In EBA, battalion-level assessment is little more than

---

outcomes are predictable, operational force disposition decisions, communication decisions, and any other decision in which accurate data are known to be available and the problem is sufficiently bounded.

[4]   Headquarters, U.S. Marine Corps, 1997b; HQDA, 2008.

[5]   HQDA, 2006c; U.S. Joint Chiefs of Staff, 2009c.

[6]   HQDA, 2006c, p. 5-26.

[7]   Comparing JP 3-24 with JP 2-0 shows that joint COIN doctrine depends on determining and measuring effects through the use of joint intelligence preparation of the environment (JIPOE), as described in JP 3-24. JIPOE is an effects-based process predicated on systems analysis and precise node and link identification and targeting. See U.S. Joint Chiefs of Staff, 2009c.

[8]   Some EBO advocates state that there is no difference between effects-based and traditional, or maneuver, warfare operations, and some doctrine makes a similar point. This finding reflects the findings of this research effort, which recognizes that there is an unsettled debate between EBO advocates and critics. The finding addresses COIN assessment only.

[9]   HQDA, 2010, p. H-3.

movement and combat reporting (e.g., "We have crossed phase line x," or "We inflicted 50 casualties").[10] This system does not allow tactical commanders to inject meaningful insight into theater-level campaign assessment. While EBA is recommended in FM 3-24, EBA also contradicts (or is contradicted by) the same manual, which states, "Local commanders have the best grasp of their situations."[11] EBA also contradicts (or is contradicted by) FM 5-0, which states, "Generally, the echelon at which a specific operation, task, or action is conducted should be the echelon at which it is assessed."[12] Table 9.1 compares EBA to the nature of COIN as described both in U.S. doctrine and in the literature.

This monograph also argues that pattern and trend analysis, while more realistic for COIN than EBA, cannot effectively support policy. When they are not based on exceptionally poor data or generally misleading, patterns can inform operational planning and intelligence analysis. They can also inform assessment, but they cannot be central to campaign assessment because they cannot be predictive in the absence of other analyses; they do not *necessarily* show change over time, nor do they necessarily show a shift in conditions. None of these centralized analyses can show defensible causation to support strategic decisionmaking. Trend analysis, and particularly time-series analysis, is of greater interest to campaign assessment because it purports to show measurable change over time. But unlike pattern analysis, trend analysis tied to thresholds does require a degree of accuracy that the COIN environment cannot provide. Trend analysis that purports to deliver even "reasonable accuracy" without the use of thresholds is dependent on data control that typically does not exist in COIN. Both pattern

Table 9.1
Effects-Based Assessment Compared to the Nature of Counterinsurgency

Effects-Based Assessment	Nature of COIN
Precise	Imprecise
Systematic	Chaotic
Certain	Uncertain
Centralized	Distributed
Conventional	Irregular
Effects-based	Mission-based

---

[10]　U.S. Joint Chiefs of Staff, 2007, p. IV-20.

[11]　HQDA, 2006c, p. 1-26.

[12]　HQDA, 2010, p. 6-1. This is particularly confusing because FM 5-0 also recommends a highly centralized EBA process that relies on a mathematical formula to produce a single assessment number for a campaign.

and trend analysis can deliver findings that are misleading, or precise but inaccurate, but neither captures context. Pattern and trend analysis contradicts the same key statements in FM 3-24 and FM 5-0 as EBA. Neither pattern analysis nor trend analysis is sufficient to provide or even to support an effective holistic campaign assessment.

Centralized assessment places undue burden on subordinate units in the form of core metrics collection and reporting requirements. Centralized lists of "the right things to assess" can become more important than military objectives because quantitative indicators appear to be tangible. It is tantalizing to believe that a commander or assessment staff might be able to show measurable progress in a military campaign that is inherently chaotic and essentially immeasurable. Demand for quantitative data, or even qualitative data that have no bearing on local conditions or objectives, can inadvertently steer tactical units (e.g., an infantry battalion) toward metrics-driven operations and away from the distributed, commander-driven operations called for in U.S. military doctrine. It might be possible to overstate the effect of core metrics lists on operational behavior, but in a worst-case scenario, the military could get caught up in a metrics-driven campaign that produces copious data but not victory (as in Vietnam). While metrics-driven operations could conceivably be productive—indicators and MOEs in doctrine are supposed to be tied to objectives—it is not possible to equate core metrics indicators with military objectives in *COIN*. Centralized indicators cannot account for variations in tactical conditions or tactical and operational mission-type objectives.

Even when commanders and staffs do not conduct metrics-driven operations, they are still faced with collecting and reporting reams of quantitative data for centralized assessment that, in their own words, provide little in the way of useful feedback or support. Collection of some types of assessment data entails physical risk, and most collection efforts require the expenditure of resources (e.g., patrol time, vehicle fuel, maintenance time, administrative time). Core metrics lists levy a "tax" on tactical units that must be paid in time, resources, and, in some cases, life and limb—or be ignored or falsified. These costs and risks affect tactical military units (and also PRTs) that are embroiled in complex and dangerous combat operations. If the centralized assessment process were effective, this tax might be justifiable rather than an undue burden, but it is not. A colloquial U.S. military aphorism describes the tactical-level reaction to centralized assessment: "The juice ain't worth the squeeze."

Table 9.2 compares EBA and pattern and trend analysis to the standards for assessment introduced in Chapter One. The findings in the table reflect the findings from a broad examination of both methodologies and a review of their implementation in Vietnam, Afghanistan, and (to a lesser extent) Iraq. The standards, as discussed earlier, are derived from a detailed literature review, direct observation, interviews, a study of 89 insurgency cases, and participation in the assessment process in two separate COIN campaigns. They are intended to provide a framework for the critical analysis of existing practice and for determining recommendations and options for improvement.

Table 9.2
Comparison of Centralized Assessment Methods to Assessment Standards

Standard	Effects-Based Assessment	Pattern and Trend Analysis
Transparent	No	No
Credible	No	No
Relevant	No	No
Balanced	No	No
Analyzed	No	No
Congruent	No	No
Parsimonious	No	No

The table shows that neither EBA nor pattern and trend analysis compares well with the seven standards for effective COIN campaign assessment.

An effective assessment process would address each of these seven standards and possibly others not considered in this study. It would provide policymakers with a relevant, transparent, and credible assessment report derived from all relevant and available data; it would provide contextual understanding of the campaign and the operating environment; it would reflect trusted commanders' insights but also internalized checks and balances; it would provide useful feedback and continuity to operational and tactical-level units; it would not place undue collection and reporting burdens on these units; it would contain dedicated and objective analysis; and it would be the product of a whole-of-government approach. If these standards were to replace the assumptions listed at the beginning of this chapter, military campaign assessment would be greatly improved.

No assessment process offers a panacea. Assessing COIN will always be a challenging and laborious process that will deliver imperfect results. Tactical units will always be responsible in some way for helping senior military leaders and policymakers understand the environment and the contextual aspects of the campaign. These senior leaders and policymakers will never receive a complete picture from a campaign assessment; it will be incumbent upon them to build a more complete picture of the strategic situation by incorporating top-level intelligence analyses, input from trusted advisors, and other reporting into their decisionmaking processes. However, an effective campaign assessment can serve as a platform for sound decisionmaking. A transparent and credible assessment can help sustain popular support for COIN campaigns that historically have lasted about ten years and can also help policymakers identify a failing strategy. Effective assessment can help win COIN campaigns, while ineffective assessment can undermine strategy. Assessment can only be effective if the assessment methodology accounts for the realities and complexities of war. The current system does not, but there is sufficient time to put an effective process in place to help shape a successful conclusion to the war in Afghanistan.

The following sections present recommendations and options for improving the assessment in Afghanistan and in the U.S. military staff planning process in general. The section titled "Options" proposes an alternative to centralized assessment derived from the seven standards for assessment. Chapter Ten presents this process in detail, while Appendixes A and B provide a template and example, respectively.

Recommendations

This section offers recommendations for improving COIN assessment doctrine and for improving the military's approach to incorporating COIN assessment into education and planning. It also recommends a review of the interagency approaches to COIN assessment.

Conduct a Joint and Service Review of Military Assessment Doctrine

The evidence presented here strongly suggests that the military would benefit from a review of COIN assessment doctrine, which would provide impetus for change. Although doctrine is intended to guide rather than explicitly instruct military planning and execution, a more detailed explanation of the role of assessment and the various approaches to COIN assessment would be helpful for the operational, training, and education communities in each of the services. A review of COIN assessment doctrine would also support improvements in assessment across the full spectrum of military operations (including, e.g., conventional war, noncombatant evacuation operations).

Reviews of military assessment doctrine—joint, Army, and Marine Corps, at a minimum—should focus on developing a theory of assessment that better aligns with doctrinal understanding of warfare and of COIN, specifically. Once this theory is established, the review should focus on finding ways to improve assessment processes, methods, and integration. This monograph offers a framework of standards that could be used to guide the development of a theory of assessment as well as a process and methods. However, this framework merely constitutes a starting point for a comprehensive doctrinal review. In addition, the review should address ways to meet policymaker requirements, incorporate qualitative data, capture context, minimize data collection requirements, provide holistic analysis, integrate and support commanders' analyses, capture interagency information, and produce adequately sourced and contextual reports.

Incorporate Counterinsurgency Assessment into Training and Education

Some military training and education programs incorporate aspects of COIN assessment into their curriculum. For example, the Marine Air-Ground Task Force Staff Training Program instructs Marine Corps staffs on how to assess COIN as part of predeployment training packages for Iraq and Afghanistan. This monograph did not examine the military assessment training and education programs, but a review of the way in which COIN assessment is taught might be beneficial. To maximize effectiveness, training and education on assessment would be seamlessly integrated into existing training and education at all levels of staff functionality. Both a captain at a battalion staff and a colonel at a corps staff should understand the practical need for useful assessment and the underlying theories of assessment doctrine.

Conduct an Interagency Review of Counterinsurgency Assessment

Research conducted on Afghanistan COIN assessment between 2009 and 2011 revealed little formal and no comprehensive collaboration on assessment between the military and civilian agencies. Any effort to develop an interagency assessment process will necessarily be fraught with bureaucratic complexities, equity issues, and differences in mindset. Currently, policymakers receive separate assessments from each agency (primarily DoD and DoS). However, according to U.S. government and military doctrine, COIN should be prosecuted by a joint military-civilian team working in harmony toward a unified strategic objective. JP 3-24 and the *U.S. Government Counterinsurgency Guide* are the best efforts to pull together interagency doctrine on COIN, but neither sufficiently addresses assessment. Campaign assessment is mentioned only briefly in one paragraph of the U.S. government guide.[13] A review of the interagency assessment policy would benefit all agencies as well as policymakers.

Incorporate All-Source Analytic Methodology into the Campaign Assessment Process

Holistic analysis is generally absent from campaign assessment, at least in part because there is no clearly defined assessment analysis process. Campaign assessment is not intelligence analysis, but all-source intelligence methodology provides an existing and well-established framework for assessment that could be used to both improve and structure campaign assessment reports. All-source analysis for intelligence is best defined as a function of intelligence fusion:

> Fusion is the process of collecting and examining information from all available sources and intelligence disciplines to derive as complete an assessment as possible of detected activity. It draws on the complementary strengths of all intel-

---

[13] U.S. Government Interagency Counterinsurgency Initiative, 2009, pp. 47–48. The term "assessment" is used throughout the guide in other contexts. For example, it is used to describe assessment of a *prospective* COIN campaign. See also U.S. Joint Chiefs of Staff, 2009c.

ligence disciplines, and relies on an all-source approach to intelligence collection and analysis.[14]

The process of intelligence analysis is semistructured. In practice, an intelligence analyst is given an analytic task or a general intelligence requirement. Sometimes, the analyst can generate collection requirements (or the analyst's command does so) to create tailored information to inform this specific analytic task. The analyst then acquires and reads and gauges the quality and veracity of as many relevant reports as possible. This process allows the analyst to build knowledge of the subject and to frame analysis to perform the task or meet the requirement. The analyst then writes the report in a narrative format, incorporating evidence from all sources of information available, as well as iterative discussions with fellow analysts and operators.

All-source analysis is particularly suited to complex and chaotic environments like COIN. It is structured enough to provide a consistent framework across an entire theater but flexible enough to respond to changes in the type and availability of data that occur on a daily basis in COIN. An experienced all-source analyst is capable of seeking out, analyzing, and incorporating any type of information that might be available, including attack data, human intelligence reports, economic data, open-source press reports, geospatial data, NGO reporting, and host-nation reporting. This kind of analysis is particularly suited to the proposed alternative to centralized assessment presented in the following chapter. However, it could also benefit centralized assessment efforts.

Analytic methodology is shaped by guidance in joint doctrine, service doctrine, and intelligence community directives. All-source analysis is semistructured, but these guidelines help bound analysis and improve quality and credibility. Many of these guidelines could be easily applied to assessment. For example, Intelligence Community Directive 203 establishes "core principles of the analytic craft." These standards require that analysis be objective, remain independent of political considerations, be timely, be based on all available sources of intelligence, and exhibit proper standards of analytic tradecraft. These tradecraft standards are exacting, and each of these standards could also be applied to, or modified for, campaign assessment.[15]

---

[14] U.S. Joint Chiefs of Staff, 2007, p. II-11. All-source analysis is specifically described as analysis that incorporates "all sources of information, most frequently including human resources intelligence, imagery intelligence, measurement and signature intelligence, signals intelligence, and open-source data in the production of finished intelligence." (U.S. Joint Chiefs of Staff, 2007, p. GL-5).

[15] Office of the Director of National Intelligence, 2007, pp. 2–4. Tradecraft standards in the directive are as follows:

    1. Properly describes quality and reliability of underlying sources

    2. Properly caveats and expresses uncertainties or confidence in analytic judgments

    3. Properly distinguishes between underlying intelligence and analysts' assumptions and judgments

    4. Incorporates alternative analysis where appropriate

The greatest benefit of applying all-source fusion methodology to assessment would be in providing a semistructured means for military staffs to create a holistic assessment rather than a disconnected set of patterns, trends, and raw data. All-source fusion analysis will not provide a formula for assessment that produces a clear quantitative finding, but this research found that the state of warfare and technology in early 2011 precludes reliance on any such formula. It also will not substitute for a layered, contextual assessment; all-source analysis, by itself, will simply provide an improved version of the flat impressionistic decision reports described earlier.

There are ample resources available to help modify all-source methodology for assessment, including the Sherman Kent School at the CIA, the service intelligence schools, and other organizations that teach or apply all-source fusion methodology. The adoption of all-source fusion methodology should not preclude the application of scientific methods or scientific rigor to the assessment process, with the understanding that assessment is not and cannot be conflated with scientific research. The semistructured nature of all-source analysis leaves considerable room to incorporate other methods and standards.

Options

This section first provides an option for improving the existing processes and methods of military COIN assessment that, while available to policymakers and the military, is not supported by this research. It is presented here in the interest of exploring all relevant options to improve COIN assessment. The second option presented here, contextual assessment, is described in the following section and then explained in detail in Chapter Ten.

Option One: Improve the Existing Centralized Assessment Process

This option assumes that centralized assessment as it currently exists in doctrine and practice will be retained. No matter how assessment is changed and improved, some policy consumers will continue to require the production of time-series graphs and pattern and trend analyses. However, any improvement to centralized assessment would be marginal because it would not address the inherent inconsistency between centralized assessment and COIN. The following recommendations could improve to some extent the transparency and credibility of a centralized assessment process—either

---

5. Demonstrates relevance to U.S. national security

6. Uses logical argumentation

7. Exhibits consistency of analysis over time, or highlights changes and explains rationale

8. Makes accurate judgments and assessments

EBA or pattern and trend analysis—should that process be retained. This is a suggested and not necessarily exhaustive list:

- Create a single narrative theater assessment process and deliver a single report from a single theater-level staff to the relevant combatant command.
- Publish this report without classification or caveat, or at least publish a well-cited version of the report without classification or caveat.
- Provide checks and balances on the theater report by
  - conducting a thorough red-team, b-team, devil's advocate, or other competing analysis[16]
  - developing a structured alternative assessment process within the intelligence community and at a selected level in DoD above theater.
- Restrict the practice of raw data dissemination outside of theater to the greatest extent possible.
- Design and implement a single and broadly promulgated set of assessment standards that could be translated and applied from the policy to the tactical level.
- Select or clearly describe how to select MOEs, MOPs, and indicators. Then,
  - clearly explain how MOEs and MOPs are tied to campaign objectives and, in turn, how they will show strategic success or failure
  - define the MOEs, MOPs, or indicators in both simple terms and in great detail to ensure that all possible differences in field conditions are accounted for
  - set a method of data reporting from the tactical to the theater level for each MOE and MOP
  - describe how to set a baseline of data for time-series analysis (or define the baseline)
  - set an effect threshold or explain in clear terms how to set an effect threshold.
- Provide a simple means for recording disagreements between various levels of command to meet doctrinal standards. Incorporate these reports into the final theater report.
- Eliminate all weighting schemes from the assessment process.
- Eliminate from the assessment report all color-coded maps and graphics (i.e., those in which a color equals a number, like in the ISAF DAM and MACV strategic maps). If these reports are required, attach appropriate caveats.
- Provide clear and detailed citations for all data, where possible. Where specific citations are restricted, provide a full list of citations to appropriate government consumers.
- Provide a reliability rating for all data, including data sets (such as polls) that already incorporate a reliability rating from a contracted polling agency.

---

[16] See *A Tradecraft Primer: Structured Analytic Techniques for Improving Intelligence Analysis* (U.S. Government, 2009).

- Ensure that data collection, reporting, transfer, and management practices deliver and sustain databases that are generally complete and accurate. Find a method to determine the completeness and accuracy of the database, if possible. Explain to consumers the degree of completeness and accuracy of data sets to the greatest extent possible.
- Clearly define and describe the analytic process that the theater staff will use to compare and contrast available data to produce the assessment report. This process should meet clear methodological standards as described by either the intelligence literature or literature from a selected set of scientific fields.

The continued existence of a quantitative, centralized process is likely to make the inclusion of other methods (e.g., all-source analysis) in theater assessment irrelevant. A combined approach will probably do little to improve transparency and credibility, as decisionmakers and the public tend to rely on what *appears* to be the most concise and accurate reporting.

Option Two: Build a Decentralized and Bottom-Up Assessment Process
If centralized assessment is untenable, then a decentralized, or bottom-up, process should be considered for COIN. The following list of assumptions should be considered as a potential underpinning for a decentralized approach to COIN assessment. They could replace the assumptions identified in the introduction to the section "Conclusions," earlier in this chapter. These assumptions are drawn from U.S. military doctrine on warfare, doctrine on COIN, the literature on COIN, and observations and interviews conducted for this study. This option is explored in greater detail in Chapter Ten.

- Wars are fought in a complex adaptive environment marked by chaos, friction, disorder, and the fog of war. Largely because COIN campaigns are conducted on both physical and human terrain, the COIN environment lies on the far end of a spectrum of complexity and chaos in warfare.
- A complex and chaotic environment undermines data collection and analysis and reduces the ability of a military staff to perceive the operating environment. This effect is exacerbated when staff are far removed from the battlefield.
- To make up for these limitations, effective COIN campaigns involve distributed, or mission-type, command and control, allowing maximum leeway in operations at the battalion level.[17]
- Distributed operations succeed because they are fought in ways appropriate for the local context: Tactical commanders have maximum flexibility to find and

---

[17] In this case, *distributed operations* is meant to broadly describe the kind of dispersed operations common to COIN. It is not intended to directly reflect the concept of distributed operations drawn from defense transformation theory.

describe solutions that will work in their area of operations but perhaps not in others.

- Local operations and local solutions require local assessments that defy aggregation into simplified theater assessment reports.
- Because COIN is complex and local assessments are not easily or effectively aggregated, theater-level assessments cannot be easily or effectively simplified.
- Because COIN defies simplification through quantitative analysis and similar analytic methods, centralized assessment is not usefully applicable to COIN.
- If centralized assessment is not usefully applicable to COIN, some form of decentralized assessment might be.
- Decentralized COIN assessment should be transparent, credible, relevant, balanced, analyzed, congruent, and parsimonious, as defined earlier in this monograph.

Some members of the community of experts on Afghanistan COIN assessment believe that a bottom-driven, contextual process is called for. Freek-Jan Toevank, a Dutch operations research analyst tasked with studying assessment in Afghanistan, developed recommendations to improve COIN assessment. Toevank determined that assessments are best conducted at the lowest possible level. In addition, "Assessment should provide authoritative claims on robustness of assessments, including processes, anonymized sources, and methods."[18] According to one former military commander and Afghanistan policy analyst,

> There is nobody in the province that has a better understanding of what's going on in the province than the USAID, State, USDA [U.S. Department of Agriculture], representatives, and maneuver battalion commanders. We need to keep metrics simple and as few as possible. Once you get into the nitty-gritty of what is taking place in the provinces, you need to let battalion and PRT commanders develop their own metrics that will help them develop their own assessments.[19]

The only way to build a bottom-up, transparent, and credible alternative to centralized assessment is to develop a process that is, in its ultimate manifestation, unclassified and without caveat. This is an inherent risk: Transparency *of process* must be absolute. This does not mean that intelligence reports, sensitive operations reports, or other sensitive communications need be revealed. It is not necessary to open all the books to the public, so to speak. There are three ways to maximize transparency without compromising security. First, assessments are produced on a time cycle that allows for a thorough security review; the kinds of "running estimates" delivered on a daily or

---

[18] Toevank, 2010. Toevank conducted empirical research over two separate periods between 2004 and 2008 in Kabul and Tarin Kowt, Afghanistan.

[19] Statement by a former military commander during a conference briefing, 2010.

weekly basis through operations channels (distinct from campaign assessments) would remain at the necessary level of classification. Second, data can be aggregated *at the battalion level* and still be kept essentially in context; it is not necessary to describe individual events or people in detail. Third, for data that are retained in the report, the time between data collection and assessment reporting (within the assessment cycle—at least a matter of weeks) reduces the tactical value of the information to the insurgents while retaining the operational and strategic value of the assessment.

Transparency and the development of a comprehensive, written framework for assessment will contribute to the credibility of the bottom-up option. However, at least the classified versions of bottom-up contextual assessment will have to contain precise and accurate citation that will allow senior policy staffs to identify sources of information. Reports that are both transparent and credible will be relevant if they also deliver a close approximation of ground truth. In theory, a bottom-up approach might fulfill this need more consistently than a centralized approach. Bottom-up assessment should reflect all types of data in context, it should be congruent with the fundamental understanding of warfare and COIN as expressed in literature and doctrine, and it should be parsimonious to the greatest extent possible. The following section examines the practicality of bottom-up assessment as it relates to the development of accurate reporting.

Practicality of a Bottom-Up Approach

Justification for the development of a bottom-up contextual assessment process rests on the notion that some level of assessment can and should be produced by battalion staffs. In current practice, battalion commanders periodically write a narrative assessment of their battlespace (sometimes, the intelligence or operations summary suffices), while their staffs routinely send up brief reports and often reams of raw quantitative and qualitative data to higher headquarters.[20] Interviews revealed an impression at some higher military and policy staff levels that battalion commanders write their narratives based on their personal opinions and that data play little or no role at this level.[21] In reality, battalion commanders are immersed in data on a daily basis. They receive a steady stream of reports on violent incidents, civil affairs actions, key leader engagements, human intelligence collection, biographies, friendly operations, ANSF partnering, local economics, and local NGO operations. Proficient commanders also

---

[20] For example, infantry atmospherics reports from local villages, key leader engagement reports, civil affairs reports, specific and aggregated attack reports, operations reports on friendly movement and actions, progress reports on the development of local security forces, intelligence information reports, and so on.

[21] This was not a universal assumption at IJC or ISAF, and, in fact, the IJC commander clearly wanted to rely on commanders' narrative assessments. However, the IJC assessment process rests on the concept that subordinate commanders' assessments must be checked against an objectives metrics list. IJC uses the District Assessment Framework Tool to conduct a quantitative check on subordinate assessments.

conduct frequent battlefield circulation: They go out and see firsthand what is happening in their battlespace. Good commanders are therefore able to put data in immediate and relevant context. A battalion staff is made up of field- and company-grade officers and senior NCOs who are similarly immersed in data.[22] These personnel typically have the same type of training and experience as those of the same rank assigned to the theater staff.

Effective battalion commanders analyze data in context, incorporate the advice of their staffs, take into account personal observations gathered from battlefield circulation, and then write their assessments. Thus, battalion assessments are also holistic assessments. Because commanders are both interpreting data and reporting on their own performance, some level of subjectivity is always present. But because the battalion-level assessment incorporates staff input and data analysis, it reflects a blend of all types of information. Because battalion staffs should be more capable of putting data in context than theater-level staffs ("Local commanders have the best grasp of their situations," according to FM 3-24), battalion staffs are better positioned than the theater staff to assess COIN in their area of operations.

In combat, battalion staffs tend to be shorthanded, overworked, and likely to focus only on those tasks that are going to prove useful to their primary mission. They are unlikely to adopt a complicated assessment reporting format that requires the collection and reporting of data that are not relevant to their operations or area of operations. Nor are they likely to put effort into filing narrative and contextual reports that do not effectively inform their immediate superiors about ongoing situations and near-term requirements. However, they tend to be both capable of and willing to produce reports that make sense to them and that show an impact on their near-term and long-term situations. Battalion staffs are accustomed to filling out narrative operations reports and narrative intelligence reports that explain what has happened in their area and why, and what is likely to happen in the near term. They are also accustomed to producing requests for resources in narrative format. Therefore, battalion staffs are well positioned to write narrative assessment reports that make sense to them, that are relevant to their mission, and that inform their immediate superiors (brigade or regimental staff and the commander) about their requirements. A bottom-up, contextual assessment can capitalize on this existing predilection and capability.

---

[22] This synopsis of battalion-level staff functions is based on my experience with an infantry battalion while serving in the Marine Corps, service on an infantry regiment staff as the assistant operations officer, debriefings of infantry battalion commanders, observations of battalion staffs in combat over three tours in Iraq, and interactions with battalion staffs during peacetime training during more than 21 years of active military service.

Deflation, Inflation, and Inherent Subjectivity

The advantage of context at the battalion level does not mitigate the very real problem of deflation and inflation of assessments. Although I am not aware of any published empirical information to support this assumption, there is a strong sense among assessment personnel that commanders tend to assess events negatively when they first arrive in theater and in a better light when they leave. This dynamic is not necessarily disingenuous; it could simply reflect increasing levels of knowledge and comfort with local leaders and events over time. In many cases, the improvement might reflect reality.[23] However, a senior military officer with experience in Afghanistan stated,

> When you come in to command, you're fired up and ready to go. I think that during the first two months . . . when you're trying to figure out what's going on you tend to temper your expectations. I've never heard of a gaining commander looking back and saying, "This guy has it all figured out, I don't need to make any changes."[24]

Figure 9.2 is a simplified and somewhat amplified depiction of this effect: The assessment tends to be negative when a staff first arrives, then rises as the deployment continues. When units perform a relief-in-place transfer-of-authority, the new commander in this notional scenario arrives with a negative bias and then follows the same assessment trajectory as the previous commander. While the widespread belief in a

Figure 9.2
Subjectivity in Commanders' Assessments

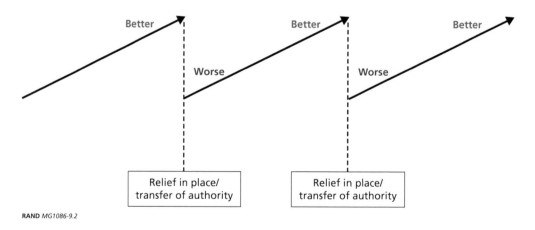

RAND *MG1086-9.2*

---

[23] This effect was recognized by the IJC and AAG assessment staffs, by members of a 70-person working group on assessments in March 2010, and in many other gatherings of experts.

[24] Statement by senior military officer, Allied Information Sharing Strategy Support to ISAF Population Metrics and Data Conference, Brunssum, Netherlands, September 1, 2010.

Manichean division between subjective and objective assessment described throughout this monograph is false, it is fair to say that a single assessment written by a commander and staff with a vested interest in the outcome cannot be considered strictly transparent or credible. It is not enough to trust battalion, regimental, or even regional assessments without additional analysis. The proposed alternative to centralized assessment explored in Chapter Ten is designed to address subjectivity in assessment reporting, but no assessment process can ever eliminate subjectivity.

While command outlook and perspective may change considerably during a relief-in-place/transfer-of-authority period, the functional aspects of assessment are likely to reflect continuity. Typically, new units pick up and continue the practices of departing units for a period—say, one or two months—before implementing major changes in process and operations.[25] This is not always the case, but the saw-tooth changes in subjective opinion reflected in Figure 9.2 should not necessarily be conflated with a similarly abrupt change in assessment methods and practice.

---

[25] This analysis is based on direct observation of numerous such transitions in Iraq between 2003 and 2006, as well as discussions with U.S. military officers on the subject between 2003 and 2011.

CHAPTER TEN

A Proposed Alternative to Centralized Assessment

This chapter explores in greater detail the alternative to centralized COIN campaign assessment proposed earlier in this monograph. Tentatively called contextual assessment, or CA, this concept was developed over the course of this research effort to meet the specific needs of COIN campaign assessment; it is not necessarily appropriate for other types of military assessment. The process presented here is derived from a number of sources reviewed for this study and is framed on existing models of military reporting. Specifically, the model is based in the description of warfare found in MCDP 1, *Warfighting*, and FM 3-0, *Operations*, as well as in the literature on the nature of warfare, such as Carl Von Clausewitz's *On War*.[1] It is also based on the description of COIN in JP 3-24, *Counterinsurgency Operations*, FM 3-24, *Counterinsurgency*, and the literature on COIN.[2]

The CA model is intended to address the seven standards for assessment identified in this monograph (see Chapter One), and it is specifically designed to capture relevant context in a layered format that also reflects critical internal analysis. The specific framework is based on a survey of both the operational and intelligence reporting formats used by the U.S. military in Vietnam, Iraq, and Afghanistan. These formats include daily operational reports filed by operations staffs in infantry units, intelligence summaries filed by intelligence staffs, advisor reports (see Appendix C), operational narratives produced at operational-level military staffs, and intelligence analysis fusion products developed by operational and theater-level intelligence staffs. The process described here is intended to fit within the observed "battle rhythm" of a contemporary U.S. military force deployed to execute a COIN campaign in a coalition environment. The template and example provided in Appendixes A and B, respectively, are framed to generally align with the format and style of writing identified in the survey of operational and intelligence reports.

The remainder of this chapter describes the step-by-step process for COIN CA in Afghanistan or for a prospective COIN campaign. CA is a framework intended to

---

[1] See Headquarters, U.S. Marine Corps, 1997b; HQDA, 2008; and Clausewitz, 1874.

[2] See U.S. Joint Chiefs of Staff, 2009c, and HQDA, 2006c.

draw from the traditionalist approach to COIN—what this approach would imply for assessment to better align assessment with capstone military doctrine and COIN operational doctrine. This chapter also further refines the concepts of both CA and mission-type metrics (MtM). MtM are a subset of CA or a tool intended for use in such an assessment. It is predicated on the idea that there are no "good" or "bad" metrics, only contextualized metrics. In COIN, it is not possible to have a centralized list of the "right" things to count, as the *Commander's Handbook* suggests.[3] Proposed definitions are as follows:

- *Contextual assessment (CA):* A transparent, bottom-up method of assessing progress in a COIN campaign that captures embedded contextual data—both qualitative and quantitative—at each layer of assessment, from the tactical to the strategic.
- *Mission-type metrics (MtM):* Relevant qualitative and quantitative indicators that exist,[4] can be collected, and can be contextualized at the level at which they are collected. MtM are indicators that show progress immediately against mission or campaign objectives. MtM are identified by staffs at each level of assessment for each assessment period; *they may or may not carry over from one assessment period to the next.*

MtM have value only in one place and only as long as they provide contextual insight into the success or failure of the COIN campaign. They are not intended to feed a centralized core metrics list in COIN or other irregular warfare missions that depend on distributed operations. *Mission-type* is drawn from *mission-type tactics* or *mission-type orders* (the same as "mission orders"), the process by which commanders issue flexible guidance to subordinates, particularly in distributed operations.[5] In this process, commanders and staffs at each level of command would view MtM as an intelligence analyst would view intelligence sources: They are either available or unavailable, reasonably or unreasonably collectible, and relevant or not relevant. This flexibility allows the commander and the assessment staff to look at all available information without prejudice and without the need for unnecessary or onerous data collection and reporting at the company level.

---

[3] "In today's increasingly dynamic operating environment, [commanders] can only gain sufficient situational awareness and adapt their current operations, future operations, and future plans if the staff is assessing 'the right things' in the operational assessment" (Joint Warfighting Center, 2006, p. IV-18).

[4] The stipulation to collect only those indicators that exist is intended to ward off what Downes-Martin refers to as "promiscuous metrics collection."

[5] "In the context of command and control, also called *mission command and control*. Mission tactics involves the use of *mission-type orders*. Mission-type order: 'Order to a unit to perform a mission without specifying how it is to be accomplished.' (Joint Pub 1-02)" (Headquarters, U.S. Marine Corps, 1997b, p. 105; emphasis in original)

While any and all information should be considered for assessment and analysis, commanders are free to dictate the consistent collection of like information over time at their level of responsibility. Local commanders are best positioned to direct the collection of information over time for several reasons: (1) they understand the immediate cost and risk of that collection; (2) they and their staffs can analyze that information in context; and (3) they can adjust collection and reporting to meet current local conditions and context. If a battalion commander determines that it would be valuable to know how many attacks were reported over time in the battalion's area of operations, that commander could assess the cost and risk of collection with input from immediate subordinates. If conditions changed, or it became obvious that the information was no longer relevant, the commander could quickly adjust the collection efforts.

There is already a similar process in place for the collection and reporting of mission-relevant information. Both joint and service doctrine calls for CCIR collection and reporting to feed operational planning and decisionmaking. Such information is necessary for commanders to effectively prosecute their operations and accomplish their missions. The requirements are flexible and change as often as necessary to meet changing conditions in the field. According to JP 5-0, *The Operations Process*:

> CCIRs comprise information requirements identified by the commander as being critical to timely information management and the decision-making process that affect successful mission accomplishment. CCIRs result from an analysis of information requirements in the context of the mission and the commander's intent. The two key subcomponents are *critical friendly force information and priority intelligence requirements*. The information needed to verify or refute a planning assumption is an example of a CCIR. CCIRs are not static. Commanders refine and update them throughout an operation based on actionable information they need for decisionmaking. They are situation-dependent, focused on predictable events or activities, time-sensitive, and always established by an order or plan.[6]

Joint doctrine already closely links CCIRs with assessment. JP 5-0 states, "The CCIR process is linked to the assessment process by the commander's need for timely information and recommendations to make decisions."[7] But while joint doctrine recommends a commander-centric, localized, mission-focused, and flexible data collection process for assessment, it goes on to recommend the application of a highly centralized, rigid, and process-centric assessment analysis framework: EBA. The first part of the joint doctrinal recommendation for information collection and reporting—the description of CCIRs—is congruous with COIN doctrine, is broadly accepted by

---

[6] U.S. Joint Chiefs of Staff, 2006, p. III-27 (emphasis in original). See also HQDA, 2010, p. B-9. CA would expand the concept of CCIRs to encompass all aspects of the campaign, including civilian and economic reporting.

[7] U.S. Joint Chiefs of Staff, 2006, p. III-57.

the military operations and intelligence community, and is also generally accepted and used by tactical units. It would seem logical either to create an assessment collection and reporting process that is analogous to CCIRs or to use the CCIR process to feed assessment. The MtM process could be used in addition to CCIRs, it could be used to modify CCIRs, or it could be deemed sufficient to meet assessment requirements without the addition of a new process. Since, by definition, CCIRs are designed to collect and report data that "affect successful mission accomplishment," either the second or third option would conceivably be sufficient to meet assessment information requirements.

CCIRs provide a means for local commanders to shape their own collection requirements and activities, but, in practice, they also tend to reflect the CCIR requirements of superior commanders. (For example, a battalion CCIR list might contain CCIRs from a brigade or regimental commander.) This could conceivably provide a pathway for the reintroduction of core metrics for assessment. It might be necessary to separate MtM from CCIRs to ensure that the MtM primarily reflect local context. Because CA is an untested, hypothetical process, it would be necessary to resolve these kinds of detailed procedural issues during iterative development. *Preferably, no standing metrics requirements would be issued even at the tactical level: Assessments would be fed with available information and with limited, short-duration collection requirements that addressed concerns most relevant to the situation at that time.*

By dictating the use of bottom-up CA and MtM, the theater command would be exercising prototypical maneuver warfare command and control. In the framework presented in this monograph, CA would begin at the battalion level of command, the first level at which a cohesive military staff exists (although input from the platoon and company levels could easily be incorporated in formal reports). The assessment would build up from the battalion to the theater level in a step-by-step process. *Every step of the process builds a layered assessment designed to ultimately address theater-level strategic objectives.* For convenience, this chapter adopts the terminology used to describe current conditions in segments of a COIN theater similar to Afghanistan: districts, provinces, and regions. In other campaigns, these levels may not be directly analogous, but the idea of layered assessment should still hold true. The CA process assumes that mission analysis, planning, and deployment are complete and that clear strategic objectives exist. Each commander will have determined both specified and implied objectives and tasks and will have written a clear mission statement, and commander's intent. In other words, everyone involved will understand his or her role in the assessment process. The theater command will have issued its overall instructions for the CA process (including reporting periods), and thus each battalion, while retaining contextual flexibility, will file reports that are comparable in format.

The CA process is founded on the finding that COIN assessment should be bottom-up, and should reflect local context and layered analysis. Figure 10.1 presents the outline of a layered contextual assessment process for COIN.[8]

This building-block approach is reflected in the following step-by-step process. Appendix A offers a series of templates for CA that can be modified for use at each level of assessment and reporting.

Figure 10.1
Contextual Impressionistic Decisionmaking

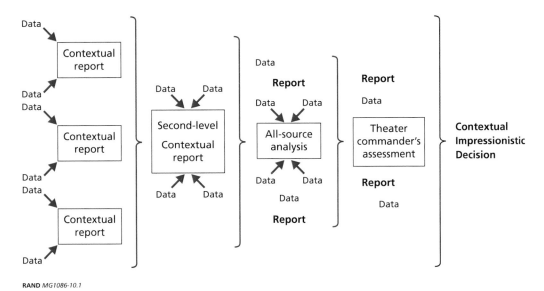

RAND *MG1086-10.1*

The Contextual Assessment Process

Step 1: Form an Assessment Working Group

The battalion staff (and staff through the RC level) establishes an assessment working group (AWG).[9] At the battalion level, this is a principals' working group made up of (at least) the intelligence officer, operations officer, logistics officer, plans officer, communication officer, civil affairs officer (if available), and host-nation security force officer (or equivalent, if available). Because the working group comprises battalion staff, these officers should already interact regularly during the course of normal business. It should be easier to provide incentive to key staff to participate in assessments if they control the assessment process at their level of effort and have a clear role in shaping

---

[8]   To see the striking contrast with traditional centralized assessment, see Figure 9.1 in Chapter Nine.

[9]   HQDA, 2010, pp. 6-9–6-10.

the theater-level assessment. The battalion commander should choose a leader for the AWG to ensure consistent recording and information management. If necessary, the AWG could consist of lower-level officers, but at some point, a principal's group should meet to review and discuss the assessment.

The battalion commander should carefully choose a staff member to write the report. This person should have good writing ability and, if possible, should have a holistic perspective of the battlespace. In some cases, the commander will want to write the report. At the regiment/brigade and RC levels, the commander might choose a civilian advisor (e.g., the DoS representative) to write the report. These advisors tend to have a more objective view than some of their military counterparts and are typically trained to write narrative reports.[10] It is necessary to keep in mind that because the AWG consists of staff members who work full time for the commander, they are, in essence, grading their boss's performance. This dynamic exists both in CA and in EBA. Superior commanders and the theater analytic team will have to keep this relationship in mind as they analyze subordinate assessments.

Step 2: Shape the Mission-Type Metric Requirements and Acquire Information

The commander and staff determine collection requirements for assessment in a structured, step-by-step process designed to keep the collection process parsimonious and targeted to the battalion mission. The first step would be to conduct a mission analysis and an operational analysis to determine which types of information might inform mission success. This list would reflect both mission requirements and the context of the operational environment (i.e., not only what is important but also what matters here and now). The next step would be to determine the availability of existing information through standard reporting (e.g., logistics reports, intelligence reports, civil affairs reports) to eliminate redundancy in collection. The third step would be to develop a draft list of collection requirements tied to the mission and the situation. The fourth step would be to iterate this draft with staff sections at the battalion, as well as with company commanders, to determine feasibility, cost, and risk of collection for each item on the list. The fifth step would be a commander's review of the list and implementation. Once the list is in place, it would be up to staff sections and subordinate commanders to find appropriate ways to produce the information. This information would be collected at the battalion level, along with all other relevant and available information.

This process would be repeated at least once per reporting cycle (e.g., monthly) and also as often as deemed necessary by the commander. The commander retains the prerogative to change the MtM list (like the CCIR list) at any time without going

---

[10] The recommendation to assign the writing task to the civilian advisor (when one is available) was offered by Colin F. Jackson of the U.S. Naval War College at a working group of assessment analysts at the U.S. Naval War College in Newport, Rhode Island, on October 19, 2010.

through this process, but establishing a pattern of staff analysis over time would help refine these changes once they were made. *However, the commander and staff should keep in mind that it is not necessary to create a structured MtM list.* Instead, the staff could use available information from operations and intelligence reports based on existing CCIR requirements, as well as information from miscellaneous reporting (e.g., a one-time key leader engagement report), to build the assessment. This would allow for greater flexibility at the company and platoon levels, and would preclude situations in which information requirements might unintentionally drive operations in a direction that did not support the commander's intent.

Step 3: Write the Battalion-Level Assessment

The AWG meets and discusses the available data and the current situation and then briefs the commander on its initial assessment. If possible, company commanders and leaders of attached organizations (e.g., a USAID representative or a special operations team) should participate in the AWG meeting to help the AWG ascertain ground truth. The commander can then write an assessment or have the AWG or a desig-nated staff officer write it for the commander's signature; each commander must have leeway in this internal process. There are no limits to the length of the battalion-level assessment because the document is intended to capture contextual detail, but it must include, at a minimum, the following elements:[11]

- a contextual narrative that incorporates both quantitative and qualitative infor-mation, explaining how and why the information shows progress or lack of prog-ress toward the battalion's mission
- a justification for the assessment that includes information about the battalion's context and explains the thought process behind the commander's assessment (in a form written or approved by the commander).

This second element also provides an opportunity for the commander to offer predictive analysis similar to that offered in operations summaries.[12]

Step 4: Assess Battalion Assessments at the Brigade or Regiment Level

The process of forming an AWG is replicated at each level of command. Each bat-talion sends its assessment—containing all required contextualized information and justifications—up to the AWG at the brigade (Army) or regiment (Marine Corps)

---

[11] At this stage, it might be appropriate to place distribution restrictions on the document to allow for higher-headquarters security review. Eventually, the full report should be released without caveat.

[12] The battalion-level assessment could also include written input from companies, if available and applicable, and whole informational reports if necessary and appropriate to support the contextual narrative.

level.[13] The AWG reviews each battalion assessment and then writes a mandatory assessment of the assessment. This gives the higher commanders and staffs an opportunity to agree or disagree with the assessment, expound on the assessment using additional information, and provide any other pertinent comments.

Step 5: Shape the Brigade or Regimental Mission-Type Metrics

Brigade and regimental AWGs follow the same process as the battalion AWG in selecting MtM. They incorporate only information that (1) exists,[14] (2) can be collected at reasonable cost and risk in relation to the urgency of the requirement, (3) is relevant to the brigade or regiment, and (4) can be contextualized. Contextualizing information becomes increasingly challenging as MtM lists are established at successive layers of command. Staffs and commanders at these higher headquarters have to absorb and understand at least twice, and often three or four times, as much information as at the battalion level, and battlefield circulation becomes more problematic and less contextual. MtM information at this level might include NGO reporting, PRT reporting, or other information that might not exist at the battalion level; this remains in the hands of the AWG. Brigade and regimental collection assets might be used to obtain some of this information.

Commanders at this level should also keep in mind that it is not necessary to create a formal MtM list. Because brigades and regiments have more robust staffs than battalions, this is an appropriate level at which to implement an all-source assessment analysis process. If implemented in accordance with intelligence literature and doctrine, the process should function with information available from CCIRs and miscellaneous reporting, supplemented by some directed MtM collection. This process should preclude the need for structured assessment information requirements that might unintentionally replicate core metrics.

Step 6: Write the Brigade or Regimental Assessment

This process also parallels the battalion process. All requirements that apply at that level apply at each successive level of command. However, an additional requirement applies: All battalion assessments, in their entirety, must be included in the single assessment document that is sent to higher headquarters. The document must also include the written assessment of each battalion's assessment report, or the overall brigade or regimental assessment should address differences. Some intelligence staffs in Afghanistan (e.g., RC Southwest) already embed subordinate assessments. Thus, incorporating subordinate reports at this stage reflects existing practice.

---

[13] Or NATO equivalent.

[14] Or that can be collected with assets at this level, depending on the commander's judgment.

Step 7: Repeat the Process at the Regional Command Level

Each step of this process is repeated at the RC level with the same stipulations. The RC-level assessment that is sent to the theater level will contain each battalion and brigade/regimental assessment from the RC area of operations, all justifications, all assessments of assessments, and any whole information reports that might prove critical in supporting contextual analysis. At no point is any information separated from the single holistic assessment unless it is deemed a security concern. Removal of information from subordinate reports should be documented in the final serialized theater report.

At the RC level and above, the assessment should routinely incorporate input from nonorganic military units and civilian organizations in the region. These inputs might come from PRTs, special operations units, nongovernmental aid organizations, host-nation elements, and coalition partners. It would be preferable to provide the entire text of these inputs in the final report because it may be difficult to retrieve nonorganic data once the report is submitted (e.g., PRT reports might never be serialized or retained in military databases). The AWG should include a brief description of each report, its valuation of the report, and its analysis of how the report fits into the holistic assessment.

Step 8: Conduct the Theater Review

The assessment staff reads and reviews all battalion, brigade/regimental, and RC assessments, as well as all additional information contained in the single document passed up from each RC.[15] Because, in this construct, the assessment staff is not overly burdened with producing briefings or slides and has the opportunity not only to read operations and intelligence reporting but also to travel in theater, staff members are well positioned to *analyze* the subordinate assessments.[16] Analysis is the assessment staff's primary purpose. It must be able to leverage its knowledge and general familiarity with each region of the theater to be able to see gaps or weaknesses in subordinate assessments.

---

[15] Approximately 40–60 officers and NCOs from all staff sections and varied specialties would make up the theater-level assessment staff (they should fall immediately under the director of assessments or equivalent). Each staff section would have a permanent representative on the assessment staff, and principal staff officers would make up the theater-level AWG (separate from the assessment staff). Members of the assessment staff would have carte blanche to travel within theater, and no fewer than 50 percent of them should be on the road at any one time. The purpose of the assessment staff visits would not be to inspect subordinate commands but to engage with subordinate staffs and gain firsthand context and understanding of battalion, brigade/regiment, and RC-level assessment processes.

[16] It may be not necessary to produce additional reporting to feed the theater assessment provided that subordinate assessments contain adequate information and the staff reads the daily operations and intelligence summaries that are produced across the theater.

Step 9: Write the Theater Analysis

After conducting the theater review, the theater-level assessment staff writes a theater-level analysis of the entire assessment. This report does not need to address each battalion report individually, but it should address the overall situation and any significant differences with subordinate assessments. The theater analysis report does not supplant any subordinate reports; it is added as the next layer in the single assessment document.

The report should be framed on all-source analysis methodology. The theater analysts would have a thorough, layered, and contextual report upon which to focus their analysis. They would conduct all-source analysis to provide a separate and perhaps competing view. This analysis would rely on additional reporting not covered in the assessment (e.g., host-nation institutional reporting, theater-level special operations reporting, coalition reporting) and on the reports cited in the assessment. The analysts would attempt to ascertain bias in the selection of sources and the interpretation of data, as well as in the commanders' assessments. This need not be a contentious process, and it would be most successful if the analysts were able to travel to subordinate commands to engage with commanders or at least able to ask questions of subordinate staffs.

Step 10: Write the Theater Commander Assessment

Once the theater-level analytic team has produced its report, the theater commander writes an assessment of that assessment explaining the theater-level "theory of assessment" points of agreement or disagreement with subordinates and the theater analysis. The commander then offers an overall assessment of the campaign in narrative format. There should be no limit to the length of this narrative; each campaign offers varying levels of complexity, and each commander must be given the leeway to write as much or as little as necessary.

Step 11: Deliver the Report

After writing the assessment of the theater-level assessment, the commander signs off on the final version of the overall assessment. This report includes each battalion, brigade and regimental, and RC assessment; the theater analysis; and the theater commander's own assessment. Quantitative data are embedded in narrative text at each level of assessment, and all graphics are clearly contextualized at the level at which the information was collected. The report is then delivered to higher headquarters. It will eventually form the backbone of the national-level assessment that is delivered to policymakers and the public.

Aggregation of Contextual Reports and Layered Summaries

Figure 10.2 shows the layering of reports from battalion to theater level. It depicts the theater analytic assessment as covering all reports up to the commander's. At the first level, the battalion report is written and sent to the brigade or regiment. At that level, four battalion reports are consolidated and then sent up to the RC with the brigade or regiment report attached. This process is repeated from the regional to the theater level, where the theater report, theater analysis reports, and commander's assessment are written. The figure assumes a notional number of battalions per brigade or regiment, brigades or regiments per region, and regions within the theater.

Figure 10.2
The Contextual Assessment Reporting Process

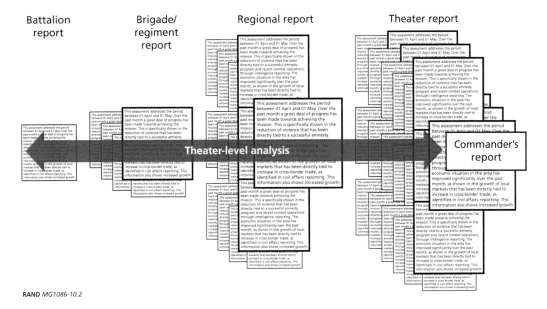

RAND MG1086-10.2

Contextual Assessment in Practice

> [C]onceptual frameworks must stay open to change on the basis of new information from the external world and avoid at all cost closing on themselves, interpreting all new information through the prism of rigid and untouchable schemata.
>
> —*Antoine Bousquet*[17]

---

[17] Bousquet, 2009, p. 193.

The formatted but loosely directed step-by-step process presented here would be the entire assessment process for the theater. It is intended not only to meet the seven standards for assessment developed for this study but also to give structure to commanders' assessments. Instead of continuing to treat commanders' assessments as purely subjective opinion and using quantitative data as a means to "check" commanders, this process should help put commanders' analyses in relevant context. Commanders at each level would be able to anchor their analyses in the layered reports, and the senior-subordinate assessments at each layer would place these analyses in further context. Instead of a distant assessment staff attempting to check a battalion commander's assessment with inaccurate, incomplete, and aggregated data, the inherent subjectivity in subordinate commanders' assessments would be addressed transparently through contextual data presentation and superior staff assessment. The theater commander's assessment would rest upon a solid foundation of layered, contextual, and carefully analyzed assessment reporting rather than a set of dissociated and often inaccurate time-series charts and tables.

If the theater commander is able to deliver a report with this kind of granular contextual information and analysis, the demands for raw, aggregated, and decontextualized data might diminish. With the existence of a thorough CA report, there should be no *need* to build a core metrics list or leverage time-consuming (or risk-inducing) data requirements on subordinate commands.[18] Requirements for the theater staff to create executive summaries and briefings—documents that tend to have little context, that suffer from data aggregation, and that are usually poorly sourced—might also diminish because each commander's assessment at each level is, in effect, a well-sourced executive summary. The final report would contain detailed narrative and data, as well as several layers of summary assessment. Commanders would continue to produce and receive operations and intelligence reports in accordance with standing orders.

CA might produce a report that is hundreds of pages long. No staff will expect a senior officer or policymaker to read the entire report; that is not its purpose. The purpose of CA would be to ensure that policymakers, senior officers, or staffs are *able* to read the entire report or desired sections, that they are able to find sources of information, and that they understand that summaries are methodically built on contextual data and narrative. Nothing will preclude staffs from excerpting summaries from specific levels of reporting and providing briefs with scalable detail.

An effective staff should have no need for a formal MtM list. The relevant recommendation in this chapter incorporates such a list primarily to help ease the transition from a more centralized process to a truly decentralized but structured process. Envisioning a best-case scenario, staffs would simply collect and analyze data as needed or

---

[18] As discussed earlier, collection requirements can induce risk because it may be necessary for military personnel to physically expose themselves to danger to collect data. For example, if a collection requirement calls for a unit to count the number of water wells in a district, patrols will have to be sent out to count the wells. Members of the patrol could be killed or wounded while conducting this seemingly mundane mission.

as the data presented. In this case, MtM would represent a concept rather than a tangible list of collection requirements.

This system as described is hypothetical and therefore unproven. There are potential benefits, costs, and risks associated with implementing CA, as discussed next.

Potential Benefits and Constraints

This section describes the potential benefits of CA, along with some potential downsides associated with each benefit. The following section examines some of the risks and costs inherent in the CA process. The benefits of the process proposed here include the following, with constraints on these benefits in italics:

- CA and MtM reflect the kind of flexibility envisioned by the developers of maneuver warfare theory as interpreted in joint and service doctrine.[19] Increased flexibility in collection and assessment should enhance the ability of tactical and operational units to conduct COIN. *Conversely, centralized, decontextualized quantitative analysis will be more challenging once CA and MtM are implemented because centralized quantitative data will be less readily available.*
- Transparency is nearly absolute for policymakers, and public reports can be highly transparent. This should dramatically improve the credibility of the theater and national assessments and, possibly, the credibility of U.S. regional policy. *Transparency carries risks, not all of which can be completely mitigated through a security review.*
- Because staffs at each level produce their own detailed and contextual assessments, tactical and operational units should benefit from being freed from an assessment process that currently "fails to deliver useful direction to subordinate staffs."[20] *The benefits of CA to tactical units will still derive in great part from higher-level analysis. If no analytic framework is applied to assessment then these benefits may still not be realized.*
- MtM reduce the impact that core metrics lists have on operational behavior.
- The process places data in context, and a detailed, contextual assessment better supports executive decisionmaking and resource allocation. *Detailed assessment requires time to develop and, to some extent, to read. Even a short executive summary of a narrative report is longer than a single briefing slide.*
- Data collection and management requirements are both reduced and redirected to address information that is more relevant and therefore more valuable to assessment. *It may not be possible to reduce demands for centralized data in current operations.*

---

[19] See, for example, Headquarters, U.S. Marine Corps, 2005.

[20] Soeters, 2009, p. 11.

- A reduction in data management requirements might produce some cost savings, although, as stated, all operations and intelligence information channels must still be maintained. *There is a risk that some useful data might not be collected or reported.*
- Contextualized data assessment could reduce the kind of assessment shopping and raw data analysis currently practiced by consumers who are dissatisfied with EBA products. *However, this may not be the case if CA or a similar process is not properly and clearly implemented.*
- CA could improve accountability for campaign progress while ensuring that tactical commanders have a venue to address issues that might be of strategic concern. *The availability of a comprehensive document might invite interpretation and exploitation of data by those with an agenda (e.g., those opposing U.S. policy in the region). If not carefully reviewed for sensitive data, CA could compromise operations.*
- CA and MtM could be quickly implemented because neither requires special training, new data networks, or overly complex formatting: Units from the battalion to the RC level already write assessments of some kind. Improved methods could be introduced over time. *Quick implementation will inevitably result in inconsistencies in product and push-back from some units. A slower, more comprehensive rollout of the process would be preferable.*

Potential Risks or Shortcomings and Mitigating Factors

The CA process proposed here is not perfect, nor is it applicable to all theater campaigns. Prior to implementing a new process, decisionmakers will need to consider the potential risks and shortcomings of CA, including the following (mitigating factors are presented in the nested bullets):

- Producing an unclassified assessment of an ongoing military campaign risks exposing sensitive information to hostile actors.
  - ISAF is already creating unclassified assessments. The ultimate DoD assessments (the 1230- and 1231-series reports), which should reflect all subordinate assessments, are unclassified and officially posted on the public Internet.
  - Assessments should not reveal ongoing or future operations. This information stays within operations channels.
  - Data can be stripped of sensitive detail, such as names or specific locations.
  - Commanders will have to exercise judgment at each level to determine risk.
  - Classifying an assessment makes it appear that the military has something to hide.
- Eliminating core metrics lists and centralized data collection places considerable trust in the hands of subordinate commanders. This could undermine the credibility of the assessment.

- This is a "trust-but-verify" system: Each level has built-in checks, and the red team, expanded in size and given increased authority to travel and communicate, adds a top layer of alternative assessment.
- The current system, based on core metrics lists and centralized data collection, already suffers from a lack of credibility. Some of the benefit in this system is relative.
- Transparency of CA and MtM methods and data should offset the loss of centralized data analysis.
- MtM information cannot be baselined and cannot be adequately tracked in time-series analysis, the principal method of showing campaign progress. What will happen when units turn over and conduct a relief-in-place?
  - The core metrics lists at ISAF and IJC have been fluctuating, in some cases quite dramatically, over the course of the Afghanistan campaign. When a core metrics list fluctuates, it sends ripples across the entire theater and generates its own wave of cascading effects in ISAF and subordinate commands. MtM affect only the staff processes and data collection that are internal to one subordinate command.
  - Doctrine calls for routine adjustments to the assessment process, measures, and metrics to match operations (FM 3-0, FM 5-0, FM 3-24).
  - Neither CA nor MtM will prevent subordinate units from baselining and using time-series assessments at the level at which the data are acquired.
  - However, it is not clear that any assessment staff has accurately baselined a data set in Afghanistan or provided methodologically sound justification as to how the baseline was defined and the thresholds were established.
  - It is impossible to eliminate all fluctuations in assessment that result from turnovers. Some of these fluctuations will be alleviated by the consistent production of a formatted battalion-level assessment; lack of consistency in the current process undermines continuity. Some of these fluctuations are also beneficial, however. Barring the existence of indefatigable commanders, it is important to get a new perspective on the problem periodically. The (inevitable) resultant shifts in the MtM list are easier to process in CA than in a centralized, effects-based process.
- Consumers have come to expect raw data, executive summaries, aggregated time-series briefings, and "objective" centralized data. They will question why the reports they have been receiving for the past nine years are inadequate.
  - Providing an effective, transparent, and credible product should help mitigate these concerns.
  - There may be a false expectation at the theater level and below that all senior policymakers and staff officers are demanding simplified assessments. Interactions with senior staffs at DoD and DoS and in Congress indicate that many U.S. decisionmakers want more, not less, detail from the field. Currently, these

requirements manifest in requests for raw data in the absence of a trusted holistic assessment.

- Even if this system is implemented, senior commanders and staffs will continue to demand raw data collection and reporting, extensive graphic briefings, and the development of stand-alone executive summaries. CA might simply add to the workload of subordinate staffs.
  - Sustained demands for raw data for centralized assessment, graphics-laden briefings, and stand-alone summaries would indicate that the system is not being successfully implemented.
  - Ideally, these requirements would be actively restricted at the theater level.

Inevitably, CA will reveal many of the flaws identified here. Battalions will report unevenly, and friction might occur as reports are assessed at each level. Some commanders and policymakers will not like the idea of a distributed, flexible indicator process and will continue to demand raw data collection and reporting. These problems and others should be anticipated and planned for: They already exist in most places. RCs already report in accordance with their own capabilities and circumstances, and there is no single, clear assessment method for any command below the theater level in Afghanistan.

CA does not require additional or specialized training; the process could be implemented quickly, while all-source analytic capability could be added more gradually. The description of CA offered in this chapter can and should be modified to meet specific conditions, but not to the point that transparency and credibility are lost. Based on observation of mid-level and senior staffs in both Iraq and Afghanistan, CA may be counterintuitive to some officers accustomed to centralizing raw data and producing stand-alone briefings. A measure of command influence probably will be necessary to ensure that EBA does not creep back into the assessment process once CA has been implemented.

Chapter Summary

CA offers the military and policymakers a method for producing a credible report that should inform good decisionmaking and also help sustain public support for policy through transparency. However, applying CA in half-measures will not remedy an overall process that remains mired in effects-based theory and aggregation. If policymakers continue to demand EBA, centralized pattern and trend analysis, and aggregated raw data, they should not expect satisfactory results from military operational assessments.

Contextual Assessment Template

This appendix contains a template for CA and MtM. As with the recommendations in Chapter Nine, the template is intended to serve as a springboard for discussion rather than a fixed recommendation. Each command and each campaign will bring specific requirements that will reshape reporting formats.

Report templates are offered here from the battalion through the theater level. The first template is also the most simple: It gives overworked and lightly staffed battalions an opportunity to produce a simple and straightforward report with context and citation. Each report builds in complexity, with commanders at each level required to review reports one level down. The theater AWG writes the theater assessment prior to the production of the theater-level analysis. The theater analytic team then reviews all assessments, including the theater assessment. At this point, the theater commander writes an overall assessment.

The template headings are in bold text, while discussion of each section is provided in bracketed and italicized text. The assessment documents are narrative text reports, but they may contain graphics. The report should not be produced in a graphic presentation format, such as PowerPoint. Graphic visualization programs constrict narrative, encourage cut-and-paste replication of subordinate reporting, and are difficult to search in contemporary data networks.

Battalion Assessment

Assessment code: [*Each assessment report will be coded for tracking purposes only, e.g., "Report #2-10."*]

Area assessed: [*should be defined using both a description and a map*]

Period of assessment: [*from-to, information cutoff date, date of report*]

Previous assessment code with comments: [*code from the last assessment for this area; should note whether the area of operations or unit has changed*]

Unit reporting: [*unit identification*]

Objectives and commander's intent: [*What are the unit's objectives and mission during the period covered by the report? This is included to ensure that the report is tied directly to objectives and so the consumer can understand what the unit was trying to achieve. What is the commander's intent, drawn from the operations order? How did the commander intend to achieve these objectives?*]

Commander's comments and outlook: [*The commander provides insight and then looks forward through the next reporting period, identifying probable trends and possible major events or tipping points. There is no limit to the length of this section.*]

Assessment rationale, including description of MtM: [*explains which approach was taken to assess the AO and why; explains why certain data or types of data were examined; can be written by the commander or staff*]

Contextual assessment incorporating MtM: [*The staff places any information in context within the overall assessment. The format of this section will depend heavily on theater command direction. For example, it could be broken down by lines of operation, or it could be broken down by district. Either way, it should consist of contextual narrative, and all data should be placed in context. Any graphics should also be contextualized, and relevant caveats should be included in the graphics themselves. Endnote or footnote references identify sources of data. There is no limit to the length of this section.*]

References and data gaps: [*All citation is identified in full. Data gaps are clearly explained. For example, if there has been no friendly presence and no reporting from a particular village, the situation in that village should be identified as "unknown." It would also be appropriate to insert requests for intelligence collection in conjunction with the identification of data gaps. These requirements would not be published in an unclassified report.*]

Brigade/Regimental Assessment

Assessment code: [*Each assessment report will be coded for tracking purposes only, e.g., "Report #2-10."*]

Area assessed: [*should be defined using both a description and a map*]

Period of assessment: [*from-to, information cutoff date, date of report*]

Previous assessment code with comments: [*code from the last assessment for this area; should note whether the area of operations or unit has changed*]

Unit reporting and deployment period: [*unit identification and from-to dates of presence in the specified area of operations*]

Assessment codes for all included subordinate assessments: [*self-explanatory*]

Objectives and commander's intent: [*What are the unit's objectives and mission during the period covered by the report? This is included to ensure that the report is tied directly to objectives and so the consumer can understand what the unit was trying to achieve. What is the commander's intent, drawn from the operations order? How did the commander intend to achieve these objectives?*]

Commander's comments and outlook: [*The commander provides insight and then looks forward through the next reporting period, identifying probable trends and possible major events or tipping points. There is no limit to the length of this section.*]

Assessment rationale, including description of MtM: [*explains which approach was taken to assess the AO and why; explains why certain data or types of data were examined; can be written by the commander or staff*]

Contextual assessment incorporating MtM: [*The staff places any information in context within the overall assessment. The format of this section will depend heavily on theater command direction. For example, it could be broken down by lines of operation, or it could be broken down by district. Either way, it should consist of contextual narrative, and all data should be placed in context. Any graphics should also be contextualized, and relevant caveats should be included in the graphics themselves. Endnote or footnote references identify sources of data. There is no limit to the length of this section.*]

Subordinate assessment comments and justification: [*The commander comments on each subordinate assessment one level down; in this case, to the battalion level. Comments reflect concurrence or nonconcurrence and explain any differences of command analysis. Any disagreements should be clearly spelled out. If necessary, this could be handled by a red-team element.*]

References: [*All citations are identified in full.*]

Regional Command Assessment

Assessment code: [*Each assessment report will be coded for tracking purposes only, e.g., "Report #2-10."*]

Area assessed: [*should be defined using both a description and a map*]

Time period of assessment: [*from-to, information cutoff date, date of report*]

Previous assessment code with comments: [*code from the last assessment for this area; should note whether the area of operations or unit has changed*]

Unit reporting and deployment period: [*unit identification and from-to dates of presence in the specified area of operations*]

Assessment codes for all included subordinate assessments: [*from battalion up*]

Objectives and commander's intent: [*What are the unit's objectives and mission during the period covered by the report? This is included to ensure that the report is tied directly to objectives and so the consumer can understand what the unit was trying to achieve. What is the commander's intent, drawn from the operations order? How did the commander intend to achieve these objectives?*]

Commander's comments and outlook: [*The commander provides insight and then looks forward through the next reporting period, identifying probable trends and possible major events or tipping points. There is no limit to the length of this section.*]

Assessment rationale, including description of MtM: [*explains which approach was taken to assess the AO and why; explains why certain data or types of data were examined; can be written by the commander or staff*]

Contextual assessment incorporating MtM: [*The staff places in any context within the overall assessment. The format of this section will depend heavily on theater command direction. For example, it could be broken down by lines of operation, or it could be broken down by district. Either way, it should consist of contextual narrative and all data should be placed in context. Any graphics should also be contextualized, and relevant caveats should be included in the graphics themselves. Endnote or footnote references identify sources of data. There is no limit to the length of this section.*]

Subordinate assessment comments and justification: [*The commander comments on each subordinate assessment one level down; in this case, to the brigade or regimental level. Comments reflect concurrence or nonconcurrence and explain any differences of command analysis. Any disagreements should be clearly spelled out. At this level the command can also comment on battalion-level assessments if it chooses to do so. If necessary, this could be handled by a red-team element.*]

References: [*All citations are identified in full.*]

Theater Analysis

Assessment code: [*Each assessment report will be coded for tracking purposes only, e.g., "Report #2-10."*]

Period of assessment: [*from-to, information cutoff date, date of report*]

Previous assessment code with comments: [*code from the last assessment for this area; should note whether the area of operations or unit has changed*]

Theater analysis director: [*Identify the leader of the analytic team at the time the report was written.*]

Assessment codes for all assessments addressed in this report: [*from battalion up*]

Theater analysis comments and outlook: [*The analytic team offers general comments on the overall theater assessment, summarizes points of agreement and disagreement, and offers either a concurring or competing outlook for the coming reporting period.*]

Analysis of theater assessment: [*This section analyzes the theater-level assessment in detail, addressing all points of concurrence and nonconcurrence and including reference to data that might support a contrasting viewpoint. This section is written after the theater assessment is written but before the theater commander's assessment is written. There is no limit to the length of this section.*]

Analysis of selected subordinate assessments: [*Once the analytic team has reviewed all subordinate assessments, it can choose to comment on selected assessments down to the battalion level, comment on points of disagreement between subordinate units, or offer an alternative analysis for any (and, if necessary, all) assessments. There is no limit to the length of this section.*]

References: [*All citations are identified in full.*]

Theater Assessment

Assessment code: [*Each assessment report will be coded for tracking purposes only, e.g., "Report #2-10."*]

Period of assessment: [*from-to, information cutoff date, date of report*]

Previous assessment code with comments: [*code from the last assessment for this area; should note whether the area of operations or unit has changed*]

Commanding officer: [*identifies the theater commander*]

Assessment codes for all included subordinate assessments: [*from battalion up, including the theater analysis report*]

Objectives and commander's intent: [*What are the unit's objectives and mission during the period covered by the report? This is included to ensure that the report is tied directly to objectives and so the consumer can understand what the unit was trying to achieve. What is the commander's intent, drawn from the operations order? How did the commander intend to achieve these objectives?*]

Commander's comments and outlook: [*The commander provides insight and then looks forward through the next reporting period, identifying probable trends and possible major events or tipping points. The theater commander should also comment on the analysis of the theater assessment. There is no limit to the length of this section. This section is included in the report only after the theater analysis team has written its report.*]

Assessment rationale including description of MtM: [*explains which approach was taken to assess the AO and why; explains why certain data or types of data were examined; can be written by the commander or staff*]

Contextual assessment incorporating MtM: [*The staff places any information from the MtM list or any other information in context within the overall assessment. The format of this section will depend heavily on combatant command and national direction. Any graphics should also be contextualized, and relevant caveats should be included in the graphics themselves. Endnote or footnote references identify sources of data. There is no limit to the length of this section.*]

Subordinate assessment comments and justification: [*The commander comments on each subordinate assessment one level down; in this case, to the regional command level. Comments reflect concurrence or nonconcurrence and explain any differences of command analysis. Any disagreements should be clearly spelled out. At this level, the command can also comment on battalion-level assessments if it chooses to do so. If necessary, this could be handled by a red-team element. This is an opportunity to produce classified annexes to the*] *report, if needed, but the separation of subordinate material for security reasons should be justified in writing.*]

References: [*All citations are identified in full.*]

Notional Example of a Brigade/Regimental-Level Contextual Assessment

This appendix contains a basic example of a brigade/regimental level contextual assessment report. This specific level of reporting was chosen because it provides sufficient opportunity to show detail while also showing analysis of a subordinate assessment. While the report is modeled on the Afghanistan case, *data in this example are strictly notional and have no basis in U.S. or NATO military operations, assessment, or official documentation of any kind*. This example is intended to inform rather than guide: It is not a "perfect" report, since each report in each area at any specific time will necessarily be unique. It is also incomplete in that it shows examples from certain sections rather than complete narratives.

This notional report is written by a brigade-level AWG and the brigade commander. It is a monthly report written at the end of August 2011, assessing progress in a notional "Sector B" that contains 15 districts and "Coastal City," a notional metropolitan area. This brigade has 16 subordinate elements reporting on each district and Coastal City; some are infantry battalions, some are PRTs that have taken control of districts, and others are special operations units filing reports from districts with no conventional military presence. This would not be an unusual circumstance in Afghanistan or in other theaters.

Note that while the summary is relatively easy to read, the detailed reports are riddled with jargon, obscure source titles, and information that is relevant only in a specific context (e.g., a single attack in a small village). Staffs and commanders must be wary of the excessive use of jargon, but the requirement to accurately source the assessment from field reports will require some technical language. As long as sources are provided, it will be possible to clear up any confusion caused by the use of military jargon. It should be easy to see where detail could be removed to create an unclassified or at least a widely releasable classified report.

This may seem like a great deal of narrative text to the nonmilitary reader. However, brigades sometimes produce ten-page intelligence summaries and equally long operational summaries on a daily basis. A daily province-level or regional intelligence summary might consist of 40 or 50 pages of text and graphics. Therefore, a 20- to 30-page monthly report assessing 15 districts and a major metropolitan area is not unusual. To mitigate the density of the text, some paragraphs might state, "no change" from the previous reporting period. However, the use of "no-change" reports is potentially dangerous because it does not force staffs to routinely reconsider their assumptions and analyses.

Third Brigade, X Division Operational Assessment for August 2010

Assessment code:
B-8-11 [in this example, Sector B, Report 8 of the current fiscal year, year 2011]

Area assessed:
Sector B (Province X) containing Districts 1 through 15
Coastal City metropolitan area

Period of assessment:
08/01/11 to 09/01/11
Information cutoff date: 09/15/11
Date of report 09/15/11

Previous assessment code with comments:
B-7-11 [sector B, month 7, year 2011]
Added District 14 and 15 since report B-7-11

Unit reporting and deployment period:
3rd Brigade, X Division
01/01/11 to 01/01/12

Assessment codes for all included subordinate assessments:
1-B-8-11; 2-B-8-11; 3-B-8-11, [and so on, with each initial number denoting a subordinate unit report]

Objectives and commander's intent:
Objectives: 3rd Brigade's objective is to bring its area of responsibility to the point that it is ready for transition to local/national security lead and local governance lead. This point should coincide with the end of the brigade's deployment period. Concurrently, the 3rd Brigade will develop the capabilities of the host-nation army and police forces to the point that they are capable of taking the lead for border security, combat operations, and local policing at the time of transition.

Commander's Intent: I intend to achieve our transition objective by leveraging a combination of local security forces and local governance to supplement host-nation capabilities. We will focus our main effort on the development of local capability in both security and governance. Our organic forces will both partner with and advise local security forces while providing security bubbles for their progressive development. Battalion commanders will weight their effort toward this mission. The Provincial Reconstruction Team (PRT) will have the lead for the development of host-nation governance, while we will take the lead for the development of local governance. We will make every effort to connect local security and governance to official host-nation programs.

Commander's comments and outlook:

Comments: This period of assessment reflects mixed results. I am satisfied that we are creating lasting security bubbles around key population centers (to include nearly all district centers) and that we have been able to develop local security forces within those bubbles. With exceptions noted below, our local security forces are operating with minimal support and are directly connected with local governance boards. This vital connection between local security and governance will help ensure that the security forces are supported in the absence of central direction and support and that local leaders can exercise some control over their own areas of responsibility.

We have been less successful in protecting lines of communication to Coastal City, as the rise in violence there indicates. I believe, however, that this rise in violence is due in great part to the increase in criminal activity as organized gangs filter into the city and not due specifically to centralized, insurgent-directed attacks. The host-nation military units are progressing quickly, but host-nation police remain a source of concern (see our separate report on security force capabilities).

The economy is developing slowly and remains largely reliant on black-market activity. Development efforts have produced limited and usually temporary effect. However, I continue to find aid and development a useful tool for stability in that it helps us support the development of local security forces. I defer to our other agency partners and the PRT to describe prospects for long-term economic development in Sector B. Some insurgent recruitment success certainly stems from the lack of economic opportunity in our sector, but we believe the majority of insurgents operating in Sector B are motivated by religion, criminal corruption, and ethnic pride (source: Intelligence Reports B-14-1, B-15-7, B-3-2). [*All source information, like the rest of the information presented in this example, is notional.*]

Based on a review of the key-leader engagement reports and sensitive intelligence reporting (source: Intelligence Reports B-KLE-12/15/19/22, B-SI-1, B-SI-29), the political will of local tribal leaders and locally elected officials seems to indicate increased willingness to support the host-nation government and government security forces. This is exemplified by the recent security agreement between Elder N and Governor M in District 6, which led to both a formal peace accord and the induction of 400 tribal members into the local security force unit there. My personal engagement with leaders across the province reinforces this assessment. There seems to have been a shift in perception among key leaders over the past three months in particular. I attribute this to our persistent operations but also to the aggressive and often ruthless behavior of the insurgents and improvements in government service delivery across the province.

Outlook: I have a fair degree of confidence that we will be able to transition nine of 15 districts to the local and host-nation security and governance lead by the end of our deployment period. In the absence of an additional battalion of host-nation army forces, the remaining six districts probably will require another four to six months of development before they are ready for transition of security lead. Coastal City can be made ready for on-time security transition with the addition of six police training teams. A written and serialized justification for these additional forces is attached to this report and to the official request for forces. With the exception of the 10th Army Brigade, Army units will be ready for transition across the board. Coastal City police will not be ready for transition in the absence of the additional police training teams.

Local government officials will be ready to assume control in 12 of 15 districts, while the three remaining districts should be ready for transition under central government control. Continued coalition security and governance support will be required in each district, to varying degrees, beyond the transition point (see below). I anticipate that all 15 districts could be under nominal central government control within two years of transition.

Coastal City, on the other hand, will probably not be ready for governance transition for an additional 12–18 months due to extreme government corruption and incompetence. This lag in progress in Coastal City cannot be addressed simply by replacing corrupt officials. It will just take time to change the culture of corruption in governance and economics. While Governor Z is performing adequately, he can neither govern nor survive without involving himself in corruption to some degree. He and his replacements will remain reliant on the system of patron-client networks that form the backbone of government in Sector B. Patron-client networks are inherently corrupt and opaque and are very difficult to transform (source: 3rd Brigade Governance Advisor Report 6-11). Long-term solutions for the governance problem in Coastal City may lie above the brigade level.

Assessment rationale including description of MtM: With the focus on developing local security and governance, as well as security bubbles around key population centers, metrics were derived from advisors, partnering teams, and security reports. Specifically, this reporting period will focus on the capabilities of local security units and their connection to local governance boards, capabilities of local governance boards (to include performance in the wake of recent local elections in eight of 15 districts), and levels of violent activity—insurgent, criminal, other—in the context of local security force development. Concerns with Coastal City governance and policing will be reflected in an increased focus on governance, corruption, and policing metrics in the Coastal City region specifically. [*The assessment could then list specific MtM or simply report them in context in the following section.*]

Contextual assessment incorporating MtM:
Overall assessment: Progress in Sector B was mixed in August, but we anticipate that nine of 15 districts (1–7, 10, and 14) will be ready for both security and governance transition by the end of the calendar year. Three additional districts (8, 9, and 15) probably will be ready for governance but not security transition by the end of the year. Districts 11, 12, and 13 continue to be heavily contested by insurgents, and all three districts suffer from endemic corruption, poor security force development, and

lack of host-nation presence. However, there are mitigating circumstances in each of these three cases that will be addressed below. Governor Z continues to support some criminal activity, but we are aware of this activity and do not believe it counterbalances his overall performance. While violence rose significantly in Sector B during the reporting period, the vast majority of these new incidents (280 of 300) were due to increased host-nation security force operations in districts 8, 9, and 15. Most of these incidents resulted in successful outcomes for host-nation forces, although Host-Nation 10th Army Brigade broke under contact and has been dismantled. Coastal City is increasingly violent. While this increased violence undermines the objective of establishing stability in Sector B, it is mostly criminal and not insurgent-related. A stronger police force should be able to reduce violence in Coastal City without significant coalition support.

Assessment by Lines of Operation—Security: Overall progress on this line of operation is mixed, with improvements in 12 of 15 districts, setbacks in three districts, and setbacks in Coastal City. Violent incident reports, which we anticipate to be 75-percent accurate at our level of analysis, rose from 220 in Assessment Period B-7-11 to 530 incidents in the current period (source: Brigade Incident Report Database 1, entry 8-11). This dramatic increase in violence stems primarily from host-nation army and police operations against insurgents and criminal gangs in districts 8, 9, and 10. Host-Nation Army Brigades 7–12 conducted sweep-and-clear operations followed by stability operations and presence patrols throughout the district centers and in the peripheries (source: subordinate reports 8 through 10-B-8-11). Although the increase in violence was expected, the level of insurgent presence and general instability in these three districts was not anticipated in intelligence reporting (source: see Brigade Intelligence Summary 15-7-10 and District Analysis Report 7-11).

Local security forces in Districts 1–7, 10, and 14 are intact and performing adequately with only a few exceptions (source: Special Operations Report B-8-11). Local forces in Districts 3 and 14 are performing exceptionally well due to strong leadership and effective use of aid programs to connect the security forces to the local governance boards and the boards to the host nation. All local security forces continue to receive threats from insurgents to a greater or lesser degree, but these threats have only proven successful in District 8. There is relatively little friction between local forces and the Army, but several incidents in District 2 between the police and local security forces have raised concerns about the future of the program there (source: District Analysis Report 2-11).

Assessment by Lines of Operation—Governance: Progress toward governance transition is excellent at district level and below but mixed to poor at the sector (provincial) level. Districts 1–7, 10, and 14 have functioning district governments. Each of these district governments is providing services to the population, including justice, health, and basic civic administration. With the exception of District 5, no district government is collecting taxes. This means that they remain reliant on host-nation government financing or PRT support. This dependence is a transition hurdle that can be overcome only with improved host-nation support. 3rd Brigade district governance assessment therefore continues to rely on the promised delivery of taxation officials and budget planners for each district from the host-nation government (source: Regional Command FRAGO [fragmentary order] 72-10). Districts 11–13 continue to suffer from the absence of district-level governance; insurgents provide justice and administrative leadership in each of these districts as of the end of the reporting period (source: District Analysis reports for 11–13 for periods 6–8, and PRT report B-8). District 15 is a unique case: It has a functioning district government, but the government is isolated in the district center due to insurgent activity on the periphery.

Governance at the village level is progressing faster than governance at the district, provincial, and metropolitan levels, particularly in Districts 1–7, 10, and 14. In each of these districts, coalition forces have facilitated the simultaneous development of local security forces with local governance boards. Local governance boards establish and control local security, a process that is fostered by the provision of development aid. Civilians in these districts report that they are generally satisfied with their local governance boards (source: District Analysis for 1–7, 10, and 14, as well as individual Atmospherics Reports from each district for periods 5–8, Infantry Patrol Reports for periods 5–8, and the Province Poll of 14 August). Local governance in districts 11–13 and 15 is either nonexistent or on the fence and hesitant to engage with the coalition or host nation.

Governance in Coastal City—both provincial and metropolitan—continues to be hindered by endemic corruption. [*The report would continue to provide depth at this point, including sourcing. Metrics might cover popular perception of justice, perception of governance, and satisfaction with official performance, among other points.*]

Assessment by Lines of Operation—Development: [*This section would mirror the other narratives on lines of operation, providing sector-level assessment of development as it relates to the brigade objectives and transition. It will be similarly sourced. Sources might include PRT reports, civil affairs reports, NGO reports, and so.*]

Assessment by District—District 1: District 1 is generally stable and is steadily progressing toward on-time transition. Violent incidents remain steady at approximately five per day, but the vast majority of these are half-hearted attacks by small insurgent or criminal groups, or they are—increasingly—failed improvised explosive device attacks that do not result in casualties (source: 1st Battalion Incident Report Roll-Up 1-B-8). The major insurgent attack in Village 23 was dramatic and caused some casualties, but the insurgent group that conducted the attack was wiped out (source: C Company, 1st Battalion Incident Report 1-B-8-82). This turned into a significant psychological victory for local security forces and for the local populations in Villages 21–27 (source: 1st Platoon, A Company, 1st Psychological Operation Battalion attachment, Incident

Report 1-B-8-7). Press reports that played up the impact of the attack were generally inaccurate (source: e.g., The Source, "Insurgent Attack in Sector B a Major Setback for Coalition," 15 August 2011).

The district continues to benefit from the presence of Host-Nation 6th Army Brigade Headquarters; the brigade helps provide local security, and its presence is a major deterrent to insurgents and criminals alike. It is possible, however, that this strong army presence has slowed the progress of police development (source: Host-Nation Security Force Report for District 1). Police units are functional on paper but will only just be ready for transition due to lingering issues of corruption (source: 1st Battalion Host-Nation Security Force Report 1-B-8). *If the Host-Nation 6th Brigade Headquarters is moved from District 1, the district will suffer a stability setback.* Local security forces in villages 1–56 are functional and can help make up for the absence of effective policing in the short term, but the long-term absence of effective police presence in the villages will probably lead to setbacks (source: Company A through D and Weapons Company Reports 1-B-5 through 1-B-8).

Governance in the district is adequate, and Governor K continues to work closely with local governance boards and with Governor Z (Intelligence Biography Report of Governor K, 1-B-8). Corruption remains endemic but, according to civil affairs reports, is manageable (source: CA reports 1-B-5 through 1-B-8). [*The report may address a few more issues on governance and development but would probably not go on beyond a page and a half of written text. Much of this text would probably be derived from existing operations and intelligence reports.*]

Assessment by District—District 2, etc.: [*Assessments of the other districts would look somewhat similar to the assessment of District 1 but would probably have some significant differences in tone, format, sourcing, and information. This should be expected and accepted as long as the reports address progress toward the 3rd Brigade's objectives. The reports for districts 11–13, where the brigade is suffering from setbacks, will probably be more detailed and will provide more information on security reporting.*]

Subordinate assessment comments and justification:

District 1: Concur, with additional information. While the attack in Village 23 was a tactical victory and a psychological victory for the population in District 1, at the brigade level, it was a psychological defeat and an operational setback. Our information operations failed to get out in front of the story, and the general consensus at the country level is that this was a major defeat for the coalition. It will not be possible to reverse this consensus without significant, visible progress in Villages 21–27. Impact: This attack may force us to shift forces into District 1 at the expense of other districts. We should be able to assess the impact of this incident in the next reporting period.

Also, it should be noted that Intelligence Biography Report of Governor K, 1-B-8, has been supplanted by Intelligence Biography Report of Governor K, 1-B-8a, which shows that he has stopped communicating with Governor Z. Impact: possible loss of funds for District 1 development projects. This should also play out over the next reporting period.

District 2, etc.: [*Remaining assessments might simply state that the brigade commander concurs with the assessment and has nothing additional to offer, or they might highlight a significant difference in analysis. In some cases, the brigade staff might have to write the district-level report due to the absence of forces in a certain district. That should be very clearly noted in both the previous section and in this section of the report.*]

References: [*Every report cited in the body text, as well as any additional reports that the consumer might find useful, should be listed here. Further, because it is often hard to find nonserialized reports (e.g., a civil affairs village report), wherever possible, these original source reports should be attached to the document or embedded in the document. At the very least, each source should contain URL to help the consumer locate the report on unclassified or classified Internet systems.*]

Phong Dinh Province Report

This appendix presents verbatim excerpts from the February 1968 CORDS report on Phong Dinh Province, Vietnam. Written by the province team, this report is a narrative that reflects input from U.S., Vietnamese, and subordinate military and civilian sources. It was written at the tail end of the VC's Tet Offensive, so it covers a period that is unusually chaotic. The report addresses all aspects of civil-military activity in the province and includes copies of subordinate district reports in their original text. The purpose of presenting this report here is to make available an example of the kind of contextual assessment that has been and can be produced by counterinsurgents. In some ways, the report parallels the CA framework outlined in Chapter Ten, and the CA framework borrows, in part, from this and other province- and district-level reports from the Vietnam, Iraq, and Afghanistan cases.

The Phong Dinh report illuminates or reinforces several aspects of the Vietnam case study, as well as key findings presented in this monograph. It shows how quantitative and qualitative data can be captured and explained through narrative context, and it reveals the complexity of the challenges faced by just a segment of the counterinsurgent force. (The Revolutionary Development program reports did not reflect other parts of the CORDS program or any tactical or operational unit activities or assessments.) Specifically, the report shows how difficult it can be to pin down quantitative data, like refugee statistics or civilian casualty statistics.

This report seems to be a status report and is not necessarily intended to address assessment requirements. Therefore, some sections might not be relevant to a holistic theater assessment. It contains recommendations listed by the province senior advisor as "required actions," so it serves both assessment and operational needs.[1] It is not clear whether a civilian or military advisor wrote this report, or whether there was collaboration between a civilian and military officer to produce it. The report describes various areas within and around Phong Dinh Province. Figure C.1 shows a contem-

---

[1] These "required actions" include a request that more ARVN forces be committed to protect a critical road, a suspension of the 1968 Revolutionary Development campaign plan to allow lagging villages to catch up, a revision of the Vietnamese command relationship, reassignment of ranks within the GVN regional force (RF) units, and a supply of seed for 6,000 hectares (MACV, 1968a, p. 18).

**Figure C.1
Map of Phong Dinh Province, Vietnam**

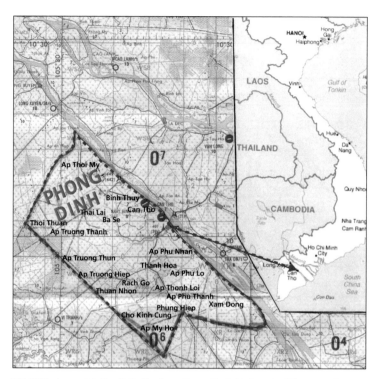

SOURCE: James M. Kraft Collection, Texas Tech University Vietnam Center
and Archives.
NOTE: Overlay added to clarify names of districts and key areas.
RAND *MG1086-C.1*

porary annotated map of Phong Dinh Province to help orient the report to the local
geography.

The report refers to specific abbreviations and terms:

RD   revolutionary development; a term used to describe villages or units
            covered by the CORDS revolutionary development program, with 1967
            RD villages having been identified as part of the program or established
            in 1967, for example. RD is also a term applied generically to projects
            associated with the Revolutionary Development program.

RF   regional forces; local civil defense units oriented at the province level

PF   popular forces; local civil defense units oriented at the district level
            New Life Hamlet: Originally "strategic hamlets," these were hamlets
            that were either cleared of VC and fortified or built for relocation of
            indigenous civilians to isolated and secure areas.

New Life Hamlet	Originally "strategic hamlets," these were hamlets that were either cleared of VC and fortified or built for relocation of indigenous civilians to isolated and secure areas.

The map and report were retrieved from the Texas Tech University Vietnam Center and Archive's online resource catalog on February 15, 2011. The declassified province report was originally archived by the University of California Indochina Archive's History of the Vietnam War on Microfilm. The report is incomplete in that it is missing several pages. Although the portions reprinted here are verbatim, some text deemed overly administrative was removed. Only three of six district reports in the document are republished here.

CORDS Report—Phong Dinh Province 02/68

Province Monthly Report
Phong Dinh Province
Period Ending February 29, 1968

1. Status of the Revolutionary Development Plan:

a. General

The RD program in Phong Dinh Province was dealt a severe setback by the Viet Cong Tet attack on Can Tho. In order to defend Can Tho City, the Province first withdrew part and then all of the 2/33 ARVN Battalion protecting the 1967 RD villages of Tan Thoi, Giai Xuan, and Cau Nhiem in the early days of the fighting. All but two RD cadre teams were pulled in to assist in the defense of Can Tho. A major portion of the RF/PF units committed to protect these 1967 RD task force areas were withdrawn leaving only the 295 RF Company in Tan Thoi Village and the 535 RF Company in Thuan Duc Village. The 826 RF Company moved from Tan Nhon New Life Hamlet to fill the position vacated by the 2/33 ARVN. There units were largely limited to protecting themselves against VC attack and so could provide little protection for the people of the area during this period. In addition, the 293 RF Company was withdrawn from Dinh Mon Village on 19 February since it could not be supported. This removed the last element shielding the 1967 RD area on its west flank.

Thus, in the face of the emergency in Can Tho, the Province denuded the RD areas of much of their external assistance in resisting the return of VC political and military units. Giai Xuan Village, a Phase II 1966 RD village, and parts of Long Tuyen, a 1966 RD village, were the scene of heavy ground fighting and subject to intense aerial bombardment since major VC units used this area as a base of operations in the period following the assault on Can Tho. The fighting in this area continued through February and numerous families were forced to flee their homes in both these villages. This situation left much of the 1966 and 1967 RD area either a battleground or a zone exposed to re-infiltration by VC terrorist groups and political cadre. Finally, the confidence of the people living in the RD areas has undoubtedly been shaken by the deep incursions made by major VC units into areas such as Can Tho. Unless early action can be taken to return RD forces to these areas in strength, many of the past accomplishments of RD in Phong Dinh may be lost.

b. Planning

The status of the Phase I 1967 RD village of Giai Xuan and the Phase II 1967 RD villages of Tan Thoi, Cau Nhiem, and Thuan Duc is unknown at this time. No member of this team has been able to visit any of these villages to assess the situation since the opening of the Tet offensive. There is no RD presence in any of these villages except Thuan Duc which has two teams in the relatively secure hamlet of An Hung, directly across the Can Tho river from Can Tho city. One battalion or more of Viet Cong occupied Nhon Loc I and Tan Nhon Hamlets (both 1967 New Life Hamlets) for at least several days before launching a surprise attack on Phong Dien District on 24 February. No one in these hamlets reported this threat to the District. A 75mm recoilless rifle was set up by the VC in the Tan Nhon RD school and the school was heavily damaged in the subsequent battle. The RD school in Tan Long was ransacked. Other damage in RD areas is unknown. Except for An Hung Hamlet in Thuan Duc Village, the VC now move through the RD areas with impunity.

The RD areas are not yet secure enough for the teams to redeploy into them at this time. Combat operations by the ARVN units are continuing within the Ba Se perimeter and it is hoped that during the month of March adequate security will be re-established permitting the RD teams to move back into the Phase I and II hamlets to resurrect the accomplishments of the 1967 RD program.

d. Village Governments

Due to the prevailing insecurity and the press of duties connected with refugee relief in Can Tho, the Public Affairs Advisor was only able to visit two village governments during February. Only one elected village government, Giai Xuan, was not functioning at the end of the month. The presence of large VC units in Giai Xuan forced the village chairman to flee to the district headquarters at Cai Rang. Chau Thanh District reported that one hamlet chief disappeared during the Tet crisis and he is presumed dead or captured. The elected hamlet chief in Thoi Trinh B Hamlet in Phuoc Thoi Village (considered secure by the GVN) was assassinated on 20 February. Except for the missing hamlet chief, all the local officials in Chau Thanh District (scene of most of the heaviest fighting) are accounted for. The large majority of hamlet officials in Long Tuyen and An Binh Village are unable to return to their hamlets and so remain near the village offices which are in secure areas. The village governments in An Binh and Long Tuyen Villages appear to be working hard to register the refugees who fled the rural sections of these villages. They have received little assistance from the province and no relief commodities. An Binh Village had to use village funds to print refugee registration forms.

3. Political/Psychological

a. General

The VC attack on Can Tho and the continuing harassment and mortaring of the city has brought home to the people, many for the first time, that there is a war going on and that, despite their desire to be left alone by both sides, they are being denied this luxury.

During the entire month the residents of Can Tho and the thousands of refugees who poured into the city have been interested primarily in survival. The fear and terror that originally drove them here have now subsided and have been replaced by more generalized feelings of insecurity. Taking advantage of these feelings are those children and old men digging up sand and dirt to put in sandbags sold for 20$ VN apiece. Many people are obviously preparing for the worst. Several are visibly upset over what they consider *the inability of the ARVN to protect them* [emphasis in original]. These same people are also *highly critical of the ARVN's engagement with the VC at the University* [emphasis in original]. When this was followed by large-scale looting by the ARVN troops, many found it difficult to consider the soldiers as their protectors. Others, however, apparently accept the destruction of the University and the surrounding area by VNAF dive bombers as a necessary corollary of war although they wish something less drastic could have been done. While the refugees are grateful for the emergency assistance provided to them, they want to know when and how the government will help them to rebuild their homes. This is a potential trouble spot which the VC might well try to exploit and the GVN should be aware that the best thing it can do to build an even more solid base of support among the refugees is to fulfill its promises of assistance to the refugees now.

As to whom the people feel was responsible for violating the Tet Truce and causing all the damage, it all depends on whom you question. Most merely reply that "you know as well as I who was responsible," which is merely a way of avoiding expressing their true feelings. It appears, however, that most of them accept the fact that it was ultimately the presence of the VC that brought down the bombs and rockets, but then these same people also consider that the ARVN used a "sledgehammer to kill a mosquito."

4. Security

a. General

The security situation has deteriorated during the past month as a direct result of the VC Tet offensive. The pacification effort has come to a complete halt. Route 4 is virtually closed to commercial traffic, although some traffic does go through on days when military road clearing operations are conducted. Inter-provincial Route 31 is closed as a result of craters and one destroyed bridge. Route 27 to the west remains open and relatively secure. The Xa No Canal remains open but constant surveillance must be maintained to forestall VC tax collection activities. The bleak security situation is improving. The ARVN, RF, and PF have apparently blunted the Tet offensive, inflicted heavy casualties on the VC, and are now regaining the initiative. Because there was never an apparent triumph, large numbers of VC infrastructure did not surface in this province as in others.

b. Enemy Situation

During the early part of the month, the VC continued their Tet offensive. In their effort, the enemy made substantial military gains. The VC forced the withdrawal of friendly forces from large portions of the Ba Se perimeter and gained the ability to launch indirect fire attacks on the IV Corps vital installations. Binh Thuy Airbase has been shelled 10 times, Can Tho Airfield 11 times, and the IV Corps ammo dump was shelled twice. Two ground attacks were directed against Can Tho airfield, and one probing action against Binh Thuy Airforce Base. . . .

There has been a degrading trend toward defensive thinking on the part of some Vietnamese leaders. The intensity of the VC attacks has caused several district chiefs to lose all offensive spirit. They seem content to wait for the VC to go away. To date RF have suffered relatively light casualties; 43 have been killed. RF and PF have accounted for 141 VC KIA and 47 Viet Cong Cadre; and they have recovered 97 weapons. Company commanders of 4 RF companies have been wounded and, because of the shallow depth of RF leadership, these companies have been hurt badly. . . .

5. New Life Development

a. Economics

The economic situation in Phong Dinh has not been seriously upset by the VC Tet offensive even though land communications routes were seriously disrupted. . . . Prices had risen considerably before Tet. Following the VC attack, prices stayed at the Tet level except for 2d grade rice which rose 2 or 3 piastres in price. Prices have now begun to decline and rice in particular is now below the pre-Tet price since the harvest is in full swing. Through the crisis adequate rice was available in Cai Rang warehouse. . . .

e. Public Health

A number of difficult public health problems were created by the VC Tet offensive. The large number of civilian war casualties created by the fighting (approximately 600 war victims were admitted during February) taxed the resources of Can Tho hospital, especially since many Vietnamese hospital staff members were unable to return from their Tet vacations for 5–6 days after the attack on Can Tho. This great influx of civilian war casualties meant that approximately 700 surgical cases were handled in February by the USAF Surgical Team and Vietnamese surgeons, a new monthly high for the hospital. . . .

f. Public Education

The use of 11 public and private schools in Can Tho as refugee centers has delayed the resumption of classes following the Tet vacation. The presence of large numbers of refugees in these schools has resulted in some damage to school facilities and general uncleanliness in and around these schools. This Team is now pressing the province to establish a realistic plan to move the refugees out of these schools and reopen them for classes as soon as possible. To date no refugees have been relocated. In the rural areas outside Can Tho such as My Khanh, An Binh, Nhon Nghia, and Nhon Ai, schools cannot be reopened at this time due to insecurity and failure of teachers to return after Tet. Many schools in rural areas will have to be repaired and several rebuilt before they can be reopened.

g. Refugees

At the height of the influx there were 12,250 refugees in 11 GVN supported relief centers in Can Tho. This figure has now decreased to 6,400. Figures are not available on refugees outside these camps but there are probably 4,000 more outside their homes. The response of the Province to the influx of refugees created by the battle for Can Tho has been erratic. At least as early as the afternoon of 1 February, GVN officials including the Deputy Province Chief for Administration, the Social Welfare Chief and several others had assembled at the Province Headquarters to decide on emergency relief measures. . . . By way of contrast, the GVN allowed military units to confiscate the major portion of several thousand loaves of bread purchased that day for distribution to the refugees. As a result, the refugees had to do without for another day. . . .

District Senior Advisor's Monthly Summary

Thuan Trung District
Period Ending February 29, 1968
District Chief: Captain Ngo Cam

The military situation in the District has improved somewhat. Viet Cong incidents have been confined to harassing fire on some of the outposts and propaganda teams entering villages and hamlets. The use of RF/PF units to provide security for the rice harvest was a great success. No Viet Cong incidents were reported in connection with the rice harvest.

Programs in the District have slowed down because of the absence of key staff officers and civilian officials who were unable to return after Tet. Many refugees in large, house-type sampans have moved into the Thoi Lai-Thoi Dong Canal. New houses have been constructed in Thoi Lai by refugees. These refugees are from the Thi Doi Canal area which has been a target of constant bombing. The only major problem at the present time is a short supply of consumer goods.

During one operation on 8 February, the 177 RF Company made contact with one VC company in Thoi Khuong (A) Hamlet only 200 meters from the center of Thoi Lai Village. As a result, 9 VC were killed, 2 VC were captured, and 11 weapons were captured. Thoi Lai Village has received several rounds of a locally manufactured rocket. Most of the rounds were duds.

Raymond G. Fischer
First Lieutenant, Artillery
Acting District Senior Advisor

District Senior Advisor's Monthly Summary

Thuan Nhon District
Period Ending February 29, 1968
District Chief: Captain Nguyen Ngoc Luu

The District town was mortared eight times and there were fifteen attacks on PF outposts during the reporting period. Over fifty roadblocks were erected on Route 31 and ten incidents of cratering and mining occurred. The 75% destruction of the Rach Goi Bridge has stopped heavy traffic from making a continuous trip through the District since 3 February. Commercial traffic on the Xa No Canal was interdicted by the VC on numerous occasions between 31 January and 18 February and merchants were searched, taxed, and turned back. Since 18 February normal traffic is slowly being restored during daylight hours. The District has been in a completely defensive status since 30 January with no participation in division or sector operations. Requests for permission to conduct sub-sector operations have been denied. In this defensive role, however, no outposts were lost or abandoned nor were any weapons lost. There were ten confirmed VC structures and approximately 25 sampans destroyed during the reporting period.

Development programs have not progressed at all and planning ceased to exist this month in this District. The one election scheduled for February has been cancelled due to the difficulties of transporting election supplies from Can Tho and travel restrictions on the voters. Rice is being harvested now with the help of RF and PF security forces. Only about 35% of potential will be realized this period. Transportation of POL supplies has been a problem this month and has curtailed the activities of the Vietnamese Navy boat platoon and generator-operated communications. ARVN's inability to provide medical evacuation of wounded soldiers and civilians has caused some morale problems and unnecessary deaths.

The understandable lack of air cover and observation aircraft in this District for the outlying areas has allowed the enemy freedom of movement and increased control over the rice harvest in these areas. This situation has eased in the last few days.

Richard C. Morris
Major, Infantry
District Senior Advisor

District Senior Advisor's Monthly Summary

Phong Dien District
Period Ending February 29, 1968
District Chief: Vo Van Dam

I. Overall Status of Pacification

1. Military situation—With the advent of the VC Tet offensive the military situation has deteriorated to its lowest point since the birth of this district in June 1966. The VC have been, and still are able to do as they desire throughout 90% of the district. This includes the placing of 2 to 3 battalions within the district, using its LOCs [lines of communication] freely in and out of the area of the Can Tho vital installations, and attacking the [outposts], and the mortaring of the district town at will.

2. The pacification program in both Tan Nhon and Tan Long hamlets has not only been halted, but has been dealt a serious setback. Prior to TET [sic], it was hoped that the RD program in these hamlets would be completed by the end of February, but now an undetermined amount of time will be necessary to repair the damages, regain the confidence of the people and complete unfinished projects.

II. Problem areas effecting [sic] the present military situation are many. For TET, approximately 25% of the PF and 70% of the sub-sector staff were allowed to take leave; the surprise offensive staged by the VC; the withdrawal of the 2/33 ARVN battalion and the 294 RF company to Can Tho; the superiority of firepower through the better weapons and a more ample ammunition supply enjoyed by the VC; the almost defensive attitude of the Vietnamese leaders within this district and the disapproval by Sector of the majority of sub-sector's operations requested have all had an influence on the current situation.

III. Efforts to revive the pacification programs shall not and can not begin in earnest until the military situation stabilizes. As to the future of the RD program (Phase II 67) as it applies to this district, all will depend upon decisions made at higher headquarters.

IV. Major problems that possibly can be remedied in part or total at this level are: 1. instilling an offensive attitude in the VN leaders; 2. the reopening of the BA SE road and Cantho [sic] River and other LOCs and 3. in a small way, revive the pacification program. Problems which must be dealt with at higher headquarters are 1. remedying the serious gap which does exist between VC and RF/PF firepower and 2. a study of the feasibility and the establishing of directives to consolidate PF [outposts] and employ an offensive means of protecting the LOCs, etc. that are now inadequately secured by the [outposts]. I have recommended such a plan to the district chief, but it will take an order from above before this will become reality.

William D. Corliss
Major, Armor
District Senior Advisor

Military Assistance Command, Vietnam, Military Assessment, September 1967

This appendix contains a memorandum bearing a telegram cable to President Lyndon Johnson from GEN William C. Westmoreland, then commanding general of MACV. The report was passed to the President by Walter W. Rostow, then Assistant for National Security Affairs (a post that is now National Security Advisor). Rostow titles his White House memorandum "General Westmoreland's Activities Report for September." This title makes the document seem like little more than an administrative report; the original cable from Westmoreland was titled, "Military Assessment for the Month of September." This is one of General Westmoreland's campaign assessment reports to the President.[1]

The excerpts reprinted here are the introduction to the cable and the section on the IV Corps area of operations. Phong Dinh Province was in IV Corps, so this appendix provides some comparison between the province-level reporting in Appendix C and theater-level reporting for the same area of operations. The time frames of the two reports are different: Westmoreland's report covers the month of September, 1967, while the Phong Dinh report in Appendix C covers the post-Tet period in 1968. This 1967 report was chosen because it is more of a routine report than the post-Tet 1968 reports; the MACV reports in the immediate post-Tet period tended to focus almost entirely on military operations and less so on other relevant campaign assessment issues.

The purpose of providing this report is to give the reader an opportunity to view a declassified military assessment. It is also intended to provide additional detail in support of the Vietnam case study in Chapter Six. The text is published verbatim in its original format, beginning with the lead for General Westmoreland's cable, skipping the I, II, and III Corps summaries, and ending with the IV Corps summary. It also does not include additional functional-area information published at the end of the report.[2] The report is marked by an optimistic tone (with some exception), and

---

[1] There was more than one type of assessment report, including the joint report written with U.S. Pacific Command referenced in Chapter Six. However, this cable was clearly titled "assessment."

[2] Included in this information is the report on 7th Air Force activities. These are primarily input indicators, such as 1,669 reconnaissance missions were flown, 65,896 tons of cargo were airlifted, 406,000 gallons of defoliant were dispersed. This information would also include what could be considered output indicators in the form

it is derived primarily from quantitative input and output indicators. While it does reflect Westmoreland's attrition strategy, there is also some focus on pacification and the development of South Vietnamese governance capacity; Westmoreland emphasizes both of these as areas for improvement in the introduction. He includes a list of military objectives in each corps reporting section, a helpful reminder that was replicated in the CA template (see Appendix A).

One of the most notable aspects of the report is that it contains a great deal of tactical detail without presenting any serious strategic analysis. Rostow's title seems more accurate than Westmoreland's. No senior policymaker should be inundated with information like, "Each airfield has established a joint command post where all internal defense, ambushes, patrols, harassment and interdiction fires and reaction operations are coordinated and controlled." The report does not tell the President what the quantitative data mean or why they are important. None of the numbers are accompanied by caveats, so the President would have to assume Westmoreland is presenting them as absolutely accurate (e.g., "Small unit operations numbered 56,864"). Assessment language is generally vague not adequately explanatory, nor is it clearly connected to objectives in a way that would help the President understand the direction of the campaign.[3] For example: "Progress was made toward improving and maintaining the security of major highways and waterways in the Delta." The report does not explain why these highways and waterways are important. It also does not explain the color-coding system used to describe each route (red, amber, or green), and it fails to explain the strategic significance of either securing or losing control of these routes to the VC. If the commanding general of MACV felt it important to tell the President that "inter-provincial Route 23 from Sa Dec to Cho Mai was changed from red to amber," then he should explain why this route was of strategic significance.[4] The report provides almost no predictive analysis, and it does not provide a justification for the allocation of resources. Instead, it is a neatly packaged list of figures that give the impression of meaning and progress without explaining much of import.

In many ways, the detail in this report is similar to the detail in the Phong Dinh Province report. However, subordinate reports are intended for an internal audience, and they are used as a platform for the development of executive reports. Detail that was appropriate in the Phong Dinh report would not be appropriate in the memorandum to the President. The Phong Dinh report also provided more analysis (albeit at the

---

of battle damage assessment: 47 artillery positions destroyed or damaged, 15 37/57-mm anti-aircraft positions destroyed, 140 sampans destroyed. It does not explain how any of these metrics are strategically relevant, nor does it explain how these numbers are relevant except that they had increased or or decreased from the previous month. The supporting information also described how the weather inhibited or did not inhibit operations.

[3]   There are exceptions to this trend, including the report on the recent elections.

[4]   It is quite possible that Johnson requested this degree of detail and that Westmoreland was simply providing what he was asked to provide. The intent of this analysis is not to blame Westmoreland for delivering an ineffective assessment; instead, it is to show what an ineffective assessment looks like.

province level) than this theater-level report. In the excerpt, text added for clarification is in brackets.

Text of Cable from General Westmoreland

Subject: Military Assessment for the Month of September [Tuesday, October 10, 1967]

The first week of last month was marked by enemy terrorist rocket and mortar attacks designed to intimidate the people of South Vietnam in order to disrupt the national elections. A determined people answered that challenge unmistakably.

Again the enemy has been defeated in his efforts to gain a major victory. Our combined forces are holding the line in some of the bitterest fighting of this conflict along the DMZ, and are seeking to protect the people and destroy the enemy throughout the country.

In order to gain and hold the people we must be able to afford them stability and protection. It is in the area of pacification and in the development of South Vietnam's civil and military leaders that we must progress.

[Skip to IV Corps]

Fourth Corps

The objectives in the Fourth Corps tactical zone for the month of September continued with emphasis on the destruction of Viet Cong main and provincial units and their principal bases; furtherance of the Revolutionary Development effort; upgrading the security and preventing interdiction of the major lines of communication; and improving the defenses of our major airfields.

Progress continues in improving the defenses of the airfields at Can Tho, Binh Thuy, Soc Trang, and Vinh Long. Day and night aerial observation of the areas and approaches to the airfields, the conduct of firefly missions nightly over these areas, and the construction of permanent and temporary revetments for aircraft protection continues. Each airfield has established a joint command post where all internal defense, ambushes, patrols, harassment and interdiction fires and reaction operations are coordinated and controlled.

The enemy situation in the Fourth Corps continues to deteriorate slowly. The Viet Cong propaganda and operational efforts against the national elections were virtually unsuccessful as evidenced by a turnout of more than 85 percent of the registered voters. Continued Government of Vietnam pressure, particularly against Viet Cong main force units and base areas, has affected their ability to significantly deter the pacification effort and interdict lines of communication. The Viet Cong have not had a major victory this year and intelligence sources confirm that they are experiencing morale problems. With the present U.S./Government of Vietnam force level in Fourth Corps, gains should continue at a slow to moderate pace. The overall morale, combat effectiveness, and fighting spirit of Fourth Corps units continue to be good.

Republic of Vietnam Armed Forces units conducted a total of 146 major unit operations of battalion size or larger, a 15 percent decrease from last month. These operations were, however, of a longer average duration. Small unit operations numbered 56,864, the highest number reported to date. Twenty-nine airmobile operations were conducted with a total of 33,970 Republic of Vietnam Armed Forces troops being airlifted. Cuu Long 63, a bilateral U.S./Republic of Vietnam Armed Forces airmobile operation conducted in Base Area 470 as part of Coronado V, resulted in 70 Viet Cong killed in action, 52 Viet Cong captured and 33 weapons captured. There were 23 additional operations penetrating seven other Viet Cong base areas resulting in 118 Viet Cong killed in action, 129 captured and 55 weapons captured. A 4.2 to 1 friendly killed versus Viet Cong killed ratio and 2.5 to 1 weapons captured versus weapons lost ratio compare favorably with previous months.

Progress was made towards improving and maintaining the security of major highways and waterways in the Delta. During the month, National Highway 4 remained open to two-way commercial traffic with only brief interruptions caused by the destruction of the An Cu bridge in Dinh Tuong Province and four cratering incidents on Route 4. The following changes in line of communication security have occurred in the past 40 days: Highway 4 from Fourth/Third Corps boundary to My Tho was changed from amber to green, inter-provincial Route 40 from Vi Thanh to Highway 4 was changed from red to amber, inter-provincial Route 23 from Sa Dec to Cho Mai was changed from red to amber, the Mang Thit-Nicholai Waterway was changed from red to amber, the My Tho River was changed from amber to green, and inter-provincial Route 30 from Kien Von to Hong Ngu (Kien Phong Province) changed from amber to green. The Cho Gao Canal in Dinh Tuong Province and the Dong Kien Canal in Kien Phong Province were reported for the first time and are carried as green lines of communications. All airfields within Fourth Corps are operational. No airfields or major lines of communication were closed by the annual flood in September.

Pacification programs remain behind schedule; however, Revolutionary Development activity in each of the provinces began to accelerate after the national election period. Red teams continued to be shifted from first semester to second semester hamlets and by September 30 approximately 60 percent of the shift was completed. There have been no reported Viet Cong attacks on first semester hamlets. ["Semesters" appear to have been levels of progress in the Revolutionary Development program for New Life hamlets.] The Viet Cong initiated five incidents against Revolutionary Development cadre resulting in five Revolutionary Development cadre, five Regional Force soldiers, and three civilians killed and 21 Revolutionary Development cadre and two Regional Force soldiers wounded.

A joint U.S./Government of Vietnam team, headed by Government of Vietnam Brigadier General Hon, completed an inspection of Revolutionary Development activities in all Fourth Corps Provinces. This team reviewed each Province's potential for completing the current 1967 program and previewed the 1968 plans to insure continuity of effort and purpose. The efforts of this team are expected to give added impetus to the pacification [effort] in Fourth Corps.

The overall effectiveness of Regional Forces and Popular Forces units remains satisfactory. The number of Regional Forces and Popular Forces desertions average approximately 990 per month, 50 percent less than last year's rate. The Popular Forces desertion rate is about double that of the Regional Forces. A study has been initiated to identify the reasons for this. In an effort to increase the effectiveness of Regional Forces and Popular Forces units, a program of instruction for in-place training of these units is being developed.

The general enemy situation deteriorated slightly as compared with July and August. Although the Viet Cong increased their activities markedly during the election period, they were unable to fulfill their plans for disruption of the elections. The rate of incidents reached a high point on September 3 and dropped appreciably following the elections. Of the 728 Viet Cong–initiated incidents occurring during the month, 320 of these happened through September 4.

Coordinated Army of the Republic of Vietnam 7th Division and U.S. 9th Division operations conducted in Western [sic] Dinh Tuong Province relieved enemy pressure directed against Highway 4. Documents captured during the period revealed a four-phase Viet Cong plan of operation in Dinh Tuong Province. One phase of this plan, interdiction of lines of communication, was accomplished only to a limited degree; phases Two, Three, and Four (attacks on rear areas, armored cavalry units and infantry units) were not accomplished.

No major Viet Cong ground attacks were initiated but harassment tactics, such as shelling of the district towns of Phong Phu, Thuan Nhon, and Ke Sach in Phong Dinh Province and Song Ong Doc in An Xuyen Province, continued. Agent reports indicate that the Viet Cong have not been able to overcome their shortage of qualified cadre and that recruitment problems continue to increase. Losses sustained by the Viet Cong totaled 944 killed, 300 captured, and 377 weapons lost compared to August losses of 1005 killed, 332 captured, and 302 weapons lost. Returnees under the Chieu Hoi program totaled 778. In addition to the Chieu Hois, there were 152 Hoa Hao [a Buddhist sect] soldiers who returned to Government of Vietnam control.

Debate over Effects-Based Operations

This section examines the debate over the use of effects-based theories in warfare and in COIN. It provides depth to the arguments referenced throughout this monograph and should help explain some of the underlying concerns behind EBA. Chapter Four described effects-based theory and EBO, so this appendix begins by presenting a critical review of effects-based theory. That discussion is followed by various defenses of EBO published in journals between 2008 and 2011. The appendix concludes with an examination of the internal contradictions in EBO doctrine and the connection between EBO and systems analysis.

This research will not settle the debate over EBO, nor does it address the applicability of effects-based theory or assessment to operations other than COIN. EBO is clearly tied to SoSA and it relies on a SoSA understanding of the operating environment. While EBO nodes and links may or may not be intended as a literal interpretation of actual people and tangible targets, doctrinal EBA is a literal interpretation of the operating environment, and it depends on precision and accuracy to be effective. There are internal contradictions in EBO theory as presented in doctrine and described in the literature.

The Debate over Effects-Based Operations in Western Literature

The debate over EBO is muddied by inconsistencies between what appear to be two loosely bounded groups of EBO proponents. The first group, represented by early advocates of RMA and EBO, including ADM Bill Owens, describes RMA (and inclusively EBO) as a means of achieving precision and accuracy through both technological advances and changes in operational methods.[1] The second group, represented by more contemporary advocates, including P. Mason Carpenter and Tomislav Ruby, contends that EBO was taken too literally by critics and is simply a means for improv-

---

[1]   See Owens, 2000.

ing and refining the time-tested theories of traditional operational art.[2] Official doctrine reflects aspects of arguments from both groups and is often internally contradictory. Some manuals clearly state that EBO accepts the immutably complex and chaotic nature of war while also calling for precision, accuracy, and mathematical rigor that seems out of reach in a complex and chaotic environment.

The most comprehensive rebuttal to EBO comes from H. R. McMaster. He builds a case against RMA, NCW, ONA, and the other concepts that EBO is associated with.[3] McMaster believes that the efforts to transform defense doctrine and policy in the 1990s were tied to a hubristic belief in the value of technology and a flawed association between capitalist business practices and military art and science.[4] He states, "The contradiction between the assumption of information superiority in future war and the 'dynamic, uncertain, and complex' security environment undermines the intellectual foundation for defense transformation."[5] He addresses strategic simulation in detail and raises concerns that, not only are military simulations based on NCW and EBO flawed, but they also reinforce what he believes are flawed theories. According to McMaster,

> The concept of effects-based operations assumed near certainty in future war; it treated the enemy as a "system" that could be fully understood through a process called "operational net assessment (ONA)." Because ONA would produce "a comprehensive system-of-systems understanding of the enemy and the environment," operations could achieve a high degree of speed as well as precision in operational effects. The enemy would be unable to keep pace with the "high rates of change" imposed on him.[6]

---

[2]   See Carpenter and Andrews, 2009, and Ruby, 2008.

[3]   For definitions of NCW and ONA see *The Implementation of Network-Centric Warfare* (DoD Office of Force Transformation, 2005). Briefly, NCW "broadly describes the combination of strategies, emerging tactics, techniques and procedures, and organizations that a fully or even partially networked force can employ to create a decisive warfighting advantage. Hannan (2005, p. 27) describes ONA as

> an analytical process designed within the Department of Defense to enhance decision-making superiority for the warfighting Commander. ONA plans to integrate people, processes, and tools using multiple information sources and collaborative analysis. The goal is a shared knowledge environment, with supporting information tools, for planners and decision-makers to focus capabilities.

[4]   According to McMaster (2003, p. 41),

> Hubris permeates the language of defense transformation and is particularly evident in the reductive fallacies of information superiority, dominant battlespace knowledge, and their various companion terms. Warnings were ignored.

[5]   McMaster, 2003, p. 9.

[6]   McMaster, 2003, p. 74.

McMaster believes that efforts to transform defense doctrine along the lines of NCW and EBO had a deleterious effect on military planning, operations, and, by extension, assessment. He lays out this argument as follows:

> Because belief in certainty or uncertainty as the dominant condition in war is rela-
> tive, so are the consequences of that belief. Unique circumstances in combat will
> shift experiences, capabilities needed, and methods along a continuum between
> extremes of certainty and uncertainty. Assuming near-certainty, however, gener-
> ates a series of derivative assumptions and predilections that are likely to lead to
> difficulties when that base assumption is proven false.[7]

In essence, McMaster claims that the proponents of EBO have set aside the commonly agreed-upon terms of warfare as expressed in MCDP 1, *Warfighting*, and FM
3-0, *Operations*; EBO is, to some extent, a rejection of the *traditionalist* understanding
of war. The arguments of McMaster and likeminded theorists gained traction as the
wars in Afghanistan and Iraq progressed, but DoD remained—and remains—rooted
in NCW and EBO.[8] Greenwood and Hammes state, "Warfare has always been interactively complex. This is not a new discovery, but admitting that DoD strayed from
this fundamental truth is."[9] The first real break with NCW and, specifically, EBO did
not come until mid-2008. On August 14, 2008, Marine Corps Gen. James N. Mattis,
then commanding general of U.S. Joint Forces Command, not only renounced EBO
theory but also rescinded official recognition of EBO in U.S. joint doctrine:

> The underlying principles associated with EBO, ONA, and SoSA are fundamen-
> tally flawed and must be removed from our lexicon, training, and operations. EBO
> thinking, as the Israelis found [in 2006 in Lebanon], is an intellectual "Magi-
> not Line" around which the enemy maneuvered. *Effective immediately, U.S. Joint
> Forces Command will no longer use, sponsor, or export the terms and concepts related
> to EBO, ONA, and SoSA in our training, doctrine development, and support of JPME*
> [joint professional military education].[10]

---

[7] McMaster, 2003, p. 91.

[8] This conclusion is based on the observation that joint doctrine published between 2001 and 2010 continued
to incorporate both NCW and EBO concepts. For an example of the persistence of the "see-all, know-all" EBO
mindset, see Nakashima and Whitlock, 2011. In this article on the new unmanned surveillance drone Gorgon
Stare, a senior U.S. general officer states, "Gorgon Stare will be looking at a whole city, so there will be no way for
the adversary to know what we're looking at, and we can see everything." This comment is in line with Owens's
stated philosophy in *Lifting the Fog of War* (Owens, 2000).

[9] Greenwood and Hammes, 2009.

[10] Mattis, 2008, p. 6 (emphasis in original). As late as mid-2010, Joint Forces Command had responsibility for
studying and developing U.S. joint doctrine.

Mattis believed that current doctrine through 2008 (as reflected in this monograph) had retained some useful elements of EBO theory. However, he was in line with Marine Corps capstone doctrine (MCDP 1) in stating that all warfare—including conventional conflict and COIN—defies the neat construct required by EBO and SoSA. He states that "[a]ll operating environments are dynamic with an infinite number of variables; therefore, it is not scientifically possible to accurately predict the outcome of action. To suggest otherwise runs contrary to historical experience and the nature of war." Mattis refers to the Israel Defense Forces's (IDF's) experience with EBO during the brief war with Hezbollah in 2006: "Although there are several factors why the IDF performed poorly during the war, various post-conflict assessments have concluded that over reliance on EBO concepts was one of the primary contributing factors for their defeat." He cites an Israeli general officer who stated that EBO doctrine was "in complete contradiction to the most important basic principles of operating an army in general . . . and is not based upon, and even ignores the universal fundamentals of war."[11]

Retired Marine Corps Lt. Gen. Paul K. Van Riper, the Bren Chair of Innovation and Transformation at Marine Corps University, reinforced Mattis's decision to unbind EBO from U.S. joint doctrine. Van Riper identifies three types of EBO theory: The first (and original) was designed to improve air-to-ground targeting, the second to improve U.S. Army fire support coordination, and the third a joint effort designed to envision and exploit a comprehensive picture of the enemy. He states that, as originally envisioned, the first type of EBO is "only effective with manmade systems that have an identifiable and tightly coupled structure, such as integrated air defenses, distribution networks, and transportation complexes."[12] The third type he labels as "egregious" and a "vacuous concept." Van Riper contends that EBO proponents do not appreciate the unpredictable nature of cascading effects:[13]

> The nearly limitless ways that an action might ricochet through an interactively complex or nonlinear system mean that for all practical purposes, the interactions within the system exceed the calculative capacities of any computer to follow, at least in any meaningful way. The numbers are so large that even the most advanced computers would take billions of years to process them. . . . In short, ONA and SoSA argue for a pseudoscientific approach to operational planning.

---

[11] Mattis, 2008, p. 4 (ellipses in original). Mattis cites Kober, 2008; Commission to Investigate the Lebanon Campaign in 2006, 2008; and Matthews, 2008, pp. 23–28, 61–65. This last comment reflects the schism that erupted over EBO in the wake of the Hezbollah campaign—some Israeli officers continue to advocate or at least defend EBO.

[12] Van Riper, 2009, p. 83.

[13] Van Riper, 2009, p. 83. He provides a chess analogy to support his statement regarding effect calculations. Even these calculations assume that the data needed to feed them are available and accurate.

Van Riper draws a distinction between a structurally complex system and an interactively complex system, with the latter defined by emergent behavior. In the structurally complex system, nodes and links are relatively static, while in an interactively complex system, the relationship between elements is "constantly in flux, and links—as conceived of by EBO advocates—are often not apparent and are frequently transitory."[14]

Like McMaster, Mattis, Van Riper, and others, Kelly and Kilcullen see COIN as a complex adaptive system rather than a closed or "equilibrium-based system."[15] They also believe that it is impossible to see the second- or third-order effects from military actions: "Such foresight would require knowledge of all possible effects, including 'effects of effects.' . . . Developing such a sequence of knowledge in military operations is almost impossible."[16] They believe that EBO will remain "at best a worthy aspiration."[17] Milan N. Vego of the U.S. Naval War College sees almost no redeeming value in EBO. In "Effects-Based Operations: A Critique," he specifically addresses some of the problems with EBO assessment that are relevant to this examination:

> [B]oth quantitative and qualitative measurements are equally subject to political manipulation, mirror-imaging, and biases. A more serious deficiency of the assessment concept is its almost total lack of sound intellectual framework. EBO proponents assume that the effects of one's actions could be precisely measured and almost instantaneously known to decisionmakers. This is highly unlikely. This heavy reliance on various quantifying measurements and fast feedback raises the issue of the utility of the effects-based approach, especially at the operational and strategic levels of war.[18]

Advocacy and defense of EBO range from efforts to refine theory and practice to outright rebuttal of McMaster, Mattis, and Van Riper's arguments. ADM (ret.) Bill Owens does not address EBO specifically but his argument in support of RMA and its associated concepts is, in essence, a claim that EBO is practicable. Owens was the official advocate for RMA in DoD in the late 1990s. In *Lifting the Fog of War*, he clearly laid out the objective of RMA and its component concepts (including EBO):

> By 2010—and earlier if we accelerate the current rate of research and procurement—the U.S. military will be able to "see" virtually everything of military significance in and above [the operating environment] all the time, in all weather conditions, and regardless of the terrain. We will be able to identify and

---

[14] Van Riper, 2009, p. 84.

[15] Kelly and Kilcullen, 2006, p. 66.

[16] Kelly and Kilcullen, 2006, p. 71.

[17] Kelly and Kilcullen, 2006, p. 72.

[18] Vego, 2006, pp. 56–57.

track—in near real time—all major items of military equipment, from trucks and other vehicles on the ground to ships and aircraft. More important, the U.S. military commander will understand what he sees.[19]

Similarly, while Alberts, Garstka, and Stein never mention EBO specifically, their detailed defense of NCW is also a defense of its associated terms and concepts. They offer a more pragmatic explanation of NCW as it relates to the traditional understanding of warfare:

> The fact that warfare will always be characterized by fog, friction, complexity, and irrationality circumscribes but does not negate the benefits that network-centric operations can provide to the forces in terms of improved battlespace awareness and access to distributed assets. While predicting human and organizational behavior will remain well beyond the state of the art, having a better near real-time picture of what is happening (in situations where this is possible from observing things that move, emit, etc.) certainly reduces uncertainty in a meaningful way.[20]

Paul K. Davis of RAND recognizes the inconsistencies between the more literal interpretations of EBO theory and the realities of the battlefield but believes that there are redeeming qualities in EBO. He recommends a family of models and mathematical games to help penetrate the fog of war.[21] Davis focuses almost exclusively on conventional warfare. In 2002, then–Air Force Lt Col Christopher W. Bowman, an effects-based approach proponent, also acknowledged uncertainty in war and believed that assessment was the "Achilles' heel" of EBO.

William J. Gregor, a social scientist at the U.S. Army Command and General Staff College, strongly defends EBO (and therefore EBA) and claims that the critics of EBO theory do not understand its premise.[22] Gregor states that EBO models are valid and can be used to see through the complexity of COIN; the military is at fault for not collecting and providing adequate data to feed effects-based operational and assessment models.[23] He also believes that senior officers have gone overboard in rejecting EBO:

---

[19] Owens, 2000, p. 119.

[20] Alberts, Garstka, and Stein, 1999, p. 11.

[21] Davis, 2001.

[22] See Gregor, 2010.

[23] The U.S. Army weaves EBO throughout the 2010 version of FM 5-0 (HQDA, 2010). Sections of that field manual provide a qualified defense of effects-based theory. While FM 5-0 clearly prescribes the use of EBA, FM 3-0 (published two years earlier and just prior to Mattis's elimination of EBO from joint doctrine) equivocates:

> Army forces conduct operations according to Army doctrine. The methods that joint force headquarters use to analyze an operational environment, develop plans, or assess operations do not change this. . . . *Army forces do not use the joint systems analysis of the operational environment, effects-based approach to planning, or effects assessment.* These planning and assessment methods are intended for use at the strategic and operational levels by

Because the effect of any action is conditioned by the environment, it is important to develop confidence in the anticipated effect by observing patterns in a large body of data collected over a long period. In both Afghanistan and Iraq, the United States should already have that data. Regrettably, it does not. . . . Currently, there are no consistent measures of effectiveness and only a small body of data with which to judge the likelihood of success. Thus, military experience cannot be generalized, and the views of generals are ideographic.[24]

Notwithstanding this last criticism, EBO and EBA are alive and well. JP 3-0 and FM 5-0 describe a hybrid of effects-based planning processes and assessment process, and effects-based terminology continues to permeate joint doctrine.[25] Mattis's edict elicited a wave of negative reaction from Air Force officers, including Col P. Mason Carpenter and Col William F. Andrews, who argued that EBO was "combat proven" and "was the basis for success of the Operation Desert Storm air campaign and Operation Allied Force."[26] They assert that "no one is suggesting certainty or absolute determinism" in EBO, so Mattis's criticisms were misdirected.[27] Air Force officer Tomislav Z. Ruby contends that Mattis is simply incorrect in his understanding of EBO and that his criticism of effects-based concepts does not, in fact, address EBO at all. He states that "the characteristics that General Mattis refers to in the memorandum [abolishing effects-based language in joint doctrine] were a 'strawman' no longer reflecting any official (read Service) position."[28] He later goes on to posit that the services were not properly applying EBO, an argument echoed by other EBO advocates. Ruby argues that some (including Mattis) confuse EBO with SoSA and that SoSA is only an "analytical tool" to help understand the enemy. The U.S. Joint Staff section responsible for joint doctrine (J7) has argued that joint doctrine does not use EBO language or refer to SoSA explicitly.[29]

---

properly resourced joint staffs. However, joint interdependence requires Army leaders and staffs to understand joint doctrine that addresses these methods when participating in joint operation planning or assessment or commanding joint forces. (JPs 3-0 and 5-0 establish this doctrine.) (HQDA, 2008, p. D-2; emphasis added)

[24] Gregor, 2010, p. 111.

[25] U.S. Joint Chiefs of Staff, 2010; HQDA, 2010.

[26] Carpenter and Andrews, 2009, p. 78. In his book *Hollow Victory: A Contrary View of the Gulf War*, Jeffrey Record states that the Gulf War "provided no genuine test of U.S. fighting power" and that it was an anomaly. He adds that the Gulf War "cannot—and should not be permitted to—serve as a model for the future" (Record, 1993, p. 135). The mixed results presented in the Gulf War Air Power Survey also call into question the claim that EBO was "proven" in the Gulf War. It would, however, be fair to say that air power had a significant impact on the success of the campaign.

[27] Carpenter and Andrews, 2009, p. 81.

[28] Ruby, 2008, p. 28.

[29] U.S. Joint Chiefs of Staff, 2009a.

But SoSA is central to EBO: It is the map of nodes and links upon which EBO planning and assessment depends. JPs 2-0 and 2-01.3, published in 2007 and 2009 respectively, contain pictures of the "interconnected operational environment" that is a near-identical copy of the SoSA map in Figure 4.2 in Chapter Four of this mono-graph.[30] The effects-based intelligence process in JPs 2-0 and 2-01.3 is literal. JP 2-0 interprets the SoSA map as a physical map, not as a conceptual prompt. It states, "A systems-oriented [analysis] effort is crucial to the identification of adversary [centers of gravity], key nodes and links." A systems-oriented understanding of the battle is derived from SoSA: "The [analyst] must also have a detailed understanding of how each aspect of the operational environment links to the others and how various per-mutations of such links and nodes may combine to form [centers of gravity]."[31] In joint intelligence doctrine, nodes and links are things or people to be identified and shaped by both kinetic and nonkinetic effects.

The language found in unofficial EBO publications and official doctrine may not be identical, but the concepts essentially are. The argument that joint doctrine does not contain EBO references is semantic and misleading. Figure E.1 shows the similari-ties between "system-of-systems analysis" as portrayed in the *Commander's Handbook*, the "interconnected operational environment" in JPs 2-0, 3-0, and 5-0, and the "sys-tems perspective of the operational environment" in JP 2-1.3. JP 3-24 incorporates the same diagram into a larger diagram depicting the joint COIN assessment process.[32] The difference in titles is rendered meaningless by the obvious similarities in the dia-grams. All three describe nodes and links, which are key elements of EBO.

Internal disagreement among the various segments of EBO literature makes an accurate assessment of effects-based theory challenging: Some literature presents EBO in a literal fashion, while some journal articles and most responses to Mattis's mem-orandum contend that critics take EBO doctrine too literally and therefore do not understand effects-based theory or practice. The original language of defense transfor-mation also tries to hedge on the nature of EBO, stating, "Rather than a new form of

---

[30] U.S. Joint Chiefs of Staff, 2007, p. IV-2; U.S. Joint Chiefs of Staff, 2009b, p. II-45.

[31] U.S. Joint Chiefs of Staff, 2007, p. IV-20.

[32] Further, JP 2-0 (U.S. Joint Chiefs of Staff, 2007, p. IV-3) clearly describes a concrete and measurable EBO process:

> Combined with a systems perspective, the identification of desired and undesired effects can help commanders and their staffs gain a common picture and shared understanding of the operational environment that promotes unified action. [Commanders] plan joint operations by developing strategic objectives supported by measur-able strategic and operational effects and assessment indicators. At the operational level, the JFC [joint force commander] develops operational-level objectives supported by measurable operational effects and assessment indicators. Joint operation planning uses measurable effects to relate higher-level objectives to component mis-sions, tasks, or actions.

Figure E.1
Comparison of Systems Analysis Diagrams in Joint Doctrinal Publications

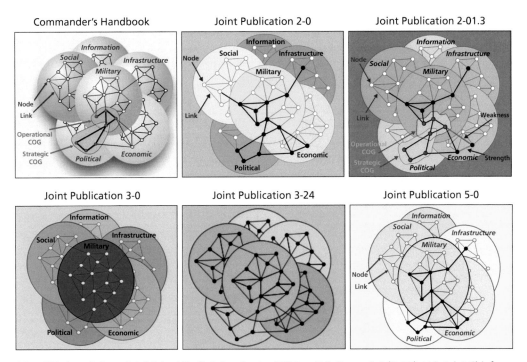

SOURCES: (top, left to right) Joint Warfighting Center, 2006, p. II-2, Figure II-1 (SoSA); U.S. Joint Chiefs of Staff, 2007, p. IV-2, Figure IV-1; and U.S. Joint Chiefs of Staff, 2009b, p. II-45, Figure II-16; (bottom, left to right) U.S. Joint Chiefs of Staff, 2010, p. II-24, Figure II-6; U.S. Joint Chiefs of Staff, 2009c, p. X-16, Figure X-4 (end-state conditions); and U.S. Joint Chiefs of Staff, 2006, p. III-17, Figure III-2.
RAND *MG1086-E.1*

warfare, EBO is a way of thinking or a methodology for planning."[33] There are two broad positions in the EBO community:

1. EBO as described in the *Commander's Handbook* (which is not official doctrine) and in some joint doctrine (e.g., JP 2.0, JP 2-01.3) is a guide for concrete action that requires large quantities of accurate and closely analyzed information, and technology can penetrate the fog of war to facilitate this process.
2. EBO is simply a way of thinking about planning and operations; it is not a strict guide for concrete action or assessment, and effects-based theory takes into account the immutably chaotic nature of war.

The latter argument is difficult to reconcile with the explicit and directive nature of most official effects-based literature, and it does not jive with concrete node-and-link analysis and targeting. There is a clear effort in joint effects-based doctrine to

---

[33] DoD Office of Force Transformation, 2003, p. 34.

shape U.S. military operations through the application and interpretation of SoSA. In most cases, the text accompanying these diagrams describes a systematic and practical process whereby staffs attempt to identify actual nodes and links in the environment, shape these nodes and links through effects, and then assess the impact of these effects on the system of systems. If the process is not intended to be taken literally—as the rebuttals to Mattis contend—then this intent is not conveyed in doctrine.

The debate over EBO in the U.S. military is ongoing and will not be settled with the conclusions and findings presented, particularly because this research effort focused specifically on COIN assessment. For the purposes of COIN assessment, however, EBO doctrine depends on the ability of military staffs to (1) see and explain all relevant events and actions with both precision and accuracy and (2) produce a clear and accurate picture of the complex adaptive COIN environment. An assessment of an effects-based operation would be able to portray and also explain past events both accurately and in detail. This places EBA on one end of a possible spectrum of assessment approaches: It assumes that sufficient accurate data are available to deliver accurate quantitative assessment from the center.

Bibliography

Historical documents in the Vietnam Center and Archive online database hosted by the Texas Tech University can be accessed via a search at http://www.virtual.vietnam.ttu.edu. Documents in the National Archives online research catalog can be accessed via a search at http://arcweb.archives.gov.

Abbaszadeh, Nima, Mark Crow, Marianne El-Khoury, Jonathan Gandomi, David Kuwayama, Christopher MacPherson, Meghan Nutting, Nealin Parker, and Taya Weiss, *Provincial Reconstruction Teams: Lessons and Recommendations*, Princeton, N.J.: Woodrow Wilson School of Public and International Affairs, January 2008.

Abella, Alex, *Soldiers of Reason: The RAND Corporation and the Rise of the American Empire*, Orlando, Fla.: Harcourt, 2008.

Abramowitz, Michael, "Congress, White House Battle over Iraq Assessment," *Washington Post*, September 10, 2007.

Ackoff, Russell L., "Towards a System of Systems Concepts [sic]," *Management Science*, Vol. 17, No. 11, July 1971, pp. 661–671. As of June 30, 2011:
http://ackoffcenter.blogs.com/ackoff_center_weblog/files/AckoffSystemOfSystems.pdf

———, "From Data to Wisdom," *Journal of Applied Systems Analysis*, Vol. 16, 1989, pp. 3–9.

———, "Systems Thinking and Thinking Systems," *System Dynamics Review*, Vol. 10, Nos. 2–3, Summer–Fall 1994, pp. 175–188.

Ahern, Thomas L. Jr., *Vietnam Declassified: The CIA and Counterinsurgency*, Lexington, Ky.: University of Kentucky Press, 2010.

Alberts, David S., John J. Garstka, and Frederick P. Stein, *Network Centric Warfare: Developing and Leveraging Information Superiority*, 2nd ed., Command Control Research Program Publications, 1999. As of June 30, 2011:
http://www.carlisle.army.mil/DIME/documents/Alberts_NCW.pdf

Allison, Paul David, *Missing Data*, Thousand Oaks, Calif.: Sage Publications, 2002.

Anderson, Edward G., Jr., "A Dynamic Model of Counterinsurgency Policy Including the Effects of Intelligence, Public Security, Popular Support, and Insurgent Experience," *System Dynamics Review*, Vol. 27, No. 2, April–June 2011, pp. 111–141.

Anderson, Joseph, "Military Operational Measures of Effectiveness for Peacekeeping Operations," *Military Review*, Vol. 81, No. 5, September–October 2001.

Armstrong, Nicholas J., and Jacqueline Chura-Beaver, "Harnessing Post-Conflict Transitions: A Conceptual Primer," Peacekeeping and Stability Operations Institute, September 2010.

Arquilla, John, and David Ronfeldt, *In Athena's Camp: Preparing for Conflict in the Information Age*, Santa Monica, Calif.: RAND Corporation, MR-880-OSD/RC, 1997. As of June 30, 2011: http://www.rand.org/pubs/monograph_reports/MR880.html

Arthur, James F., testimony before the House Armed Services Committee on hamlet pacification in Vietnam, March 1970. Available via the Vietnam Center and Archive online database.

BACM Research, *Vietnam War Document Archive (Disk 1 and Disk 2)*, set of two CDs containing official archival records, 2009.

Baranick, Mike, Center for Technology and National Security Policy, National Defense University; David Knudson, Center for Army Analysis; Dave Pendergraft, Headquarters U.S. Air Force; and Paul Evangelista, U.S. Army Training and Doctrine Command Analysis Center, "Improving Analytical Support to the Warfighter: Campaign Assessments, Operational Analysis, and Data Management, Working Group 1: Data and Knowledge Management," annotated briefing, Military Operations Research Society workshop, April 19–22, 2010. As of June 30, 2011: http://www.mors.org/UserFiles/file/2010%20IW/MORS%20IW%20WG1.pdf

Bar-Yam, Yaneer, *Complexity of Military Conflict: Multiscale Complex Systems Analysis of Littoral Warfare*, New England Complex Systems Institute, April 21, 2003. As of June 30, 2011: http://necsi.edu/projects/yaneer/SSG_NECSI_3_Litt.pdf

Bernard, H. Russell, *Research Methods in Anthropology: Qualitative and Quantitative Approaches*, 4th ed., Lanham, Md.: AltaMira Press, 2006.

Bertalanffy, Ludwig von, *General System Theory: Foundations, Development, Applications*, rev. ed., New York: G. Braziller, 1974

Beyerchen, Alan, "Clausewitz, Nonlinearity, and the Unpredictability of War," *International Security*, Vol. 17, No. 3, Winter 1992–1993, pp. 59–90.

Biddle, Stephen, *Military Power: Explaining Victory and Defeat in Modern Battle*, Princeton, N.J.: Princeton University Press, 2004.

Birtle, Andrew J., U.S. *Army Counterinsurgency and Contingency Operations Doctrine, 1860–1941*, Washington, D.C.: U.S. Army Center of Military History, 2004.

———, *U.S. Army Counterinsurgency and Contingency Operations Doctrine, 1942–1976*, Washington, D.C.: U.S. Army Center of Military History, 2006.

Blalock, Hubert M. Jr., ed., *Measurement in the Social Sciences: Theories and Strategies*, Chicago, Ill.: Aldine Publishing Company, 1974.

Boot, Max, *War Made New: Weapons, Warriors, and the Making of the Modern World*, New York: Penguin Group, 2006.

Borkowski, Piotr, and Jan Mielniczuk, "Postmodel Selection Estimators of Variance Function for Nonlinear Autoregression," *Journal of Time Series Analysis*, Vol. 31, No. 1, January 2010, pp. 50–63.

Bousquet, Antoine, *The Scientific Way of Warfare: Order and Chaos on the Battlefields of Modernity*, New York: Columbia University Press, 2009.

Bowman, Christopher W., *Operational Assessment—The Achilles Heel of Effects-Based Operations?* thesis, Newport, R.I.: U.S. Naval War College, May 13, 2002. As of June 30, 2011: http://handle.dtic.mil/100.2/ADA405868

Bracken, Jerome, Moshe Kress, and Richard E. Rosenthal, *Warfare Modeling*, Danvers, Mass.: John Wiley and Sons, 1995.

Brigham, Erwin R., "Pacification Measurement in Vietnam: The Hamlet Evaluation System," presentation at SEATO Internal Security Seminar, Manila, Philippines, June 3–10, 1968. As of June 30, 2011:
http://www.cgsc.edu/carl/docrepository/PacificationVietnam.pdf

———, "Pacification Measurement," *Military Review*, May 1970, pp. 47–55. As of June 30, 2011:
http://calldp.leavenworth.army.mil/eng_mr/txts/VOL50/00000005/art6.pdf

Brookings Institution, Saban Center for Middle East Policy, *Iraq Index: Tracking Reconstruction and Security in Post-Saddam Iraq*, Washington, D.C., last updated August 24, 2010. See archive of past reports, as of June 30, 2011:
http://www.brookings.edu/saban/iraq-index.aspx

Brown, James, Erik W. Goepner, and James M. Clark, "Detention Operations, Behavior Modification, and Counterinsurgency," *Military Review*, May–June 2009, pp. 40–47.

Bryman, Alan, Saul Becker, and Joe Sempik, "Quality Criteria for Quantitative, Qualitative and Mixed Methods Research: A View from Social Policy," *Social Research Methodology*, Vol. 11, No. 4, October 2008, pp. 261–276.

Bunker, Ellsworth, "October 2–15, 1968: The Breakthrough in Paris," official telegram from the U.S. Embassy in Saigon, Vietnam, to U.S. Department of State, Washington, D.C., October 2, 1968a, in BACM Research, Disk 2.

———, "Telegram from the Embassy in Vietnam to the Department of State," official telegram from the U.S. Embassy in Saigon, Vietnam, to the U.S. Department of State, Washington, D.C., October 19, 1968b, in BACM Research, Disk 2.

Cabayan, Hriar, *Rich Contextual Understanding of Pakistan and Afghanistan (PAKAF): A Strategic Multi-Layer Assessment*, unpublished workshop report, April 12, 2010a.

———, *Subject Matter Analysis PAKAF Rich Contextual Understanding, MG Flynn Final Progress Report*, unpublished workshop report, October 25, 2010b.

Campbell, Jason, Michael E. O'Hanlon, and Jeremy Shapiro, *Assessing Counterinsurgency and Stabilization Missions*, Washington, D.C.: Brookings Institution, policy paper 14, May 2009a. As of June 30, 2011:
http://www.brookings.edu/~/media/Files/rc/papers/2009/05_counterinsurgency_ohanlon/05_counterinsurgency_ohanlon.pdf

———, "How to Measure the War," *Policy Review*, No. 157, October–November 2009b. As of June 30, 2011:
http://www.hoover.org/publications/policy-review/article/5490

Carmines, Edward G., and Richard A. Zeller, *Reliability and Validity Assessment*, Beverly Hills, Calif.: Sage Publications, 1979.

Carpenter, P. Mason, and William F. Andrews, "Effects-Based Operations: Combat Proven," *Joint Force Quarterly*, No. 52, First Quarter 2009, pp. 78–81.

Cebrowski, Arthur K., and John J. Garstka, "Network-Centric Warfare: Its Origin and Future," *Proceedings*, January 1998, pp. 28–36.

Center for Army Analysis, "Afghanistan Consolidated Knowledge System," briefing, August 30, 2010.

CIA—*see* U.S. Central Intelligence Agency.

Cioppa, Thomas, email exchange with the author, October 8, 2010.

Cioppa, Thomas, Loren Eggen, and Paul Works, "Improving Analytical Support to the Warfighter: Campaign Assessments, Operational Analysis, and Data Management, Working Group 4: Current Ops Analysis—Tactical," briefing, Military Operations Research Society, April 19–22, 2010. As of June 30, 2011:
http://www.mors.org/UserFiles/file/2010%20IW/MORS%20IW%20WG4.pdf

Claflin, Bobby, Dave Sanders, and Greg Boylan, "Improving Analytical Support to the Warfighter: Campaign Assessments, Operational Analysis, and Data Management, Working Group 2: Campaign Assessments," working group briefing, Military Operations Research Society conference, April 19–22, 2010. As of September 15, 2010:
http://www.mors.org/UserFiles/file/2010%20IW/MORS%20IW%20WG2.pdf

Clancy, James, and Chuck Crossett, "Measuring Effectiveness in Irregular Warfare," *Parameters*, Summer 2007, pp. 88–100. As of June 30, 2011:
http://www.carlisle.army.mil/usawc/parameters/Articles/07summer/clancy.pdf

Clark, Clinton R., and Timothy J. Cook, "A Practical Approach to Effects-Based Operational Assessment," *Air and Space Power Journal*, Vol. 22, No. 2, Summer 2008. As of June 30, 2011:
http://www.airpower.maxwell.af.mil/airchronicles/apj/apj08/sum08/clark.html

Clark, Dorothy K., and Charles R. Wyman, *An Exploratory Analysis of the Reporting, Measuring, and Evaluating of Revolutionary Development in South Vietnam*, McLean, Va.: Research Analysis Corporation, 1967.

Clausewitz, Carl Von, *On War*, Colonel J. J. Graham, trans., London: N. Trübner, 1874.

Colby, William E., and Peter Forbath, *Honorable Men: My Life in the CIA*, New York: Simon and Schuster, 1978.

Commission to Investigate the Lebanon Campaign in 2006, final report, January 30, 2008. As of June 30, 2011:
http://www.mfa.gov.il/MFA/MFAArchive/2000_2009/2008/Winograd%20Committee%20submits%20final%20report%2030-Jan-2008

Connable, Ben, "The Massacre That Wasn't," in *U.S. Marines in Iraq, 2004–2008: Anthology and Annotated Bibliography*, Washington, D.C.: History Division, U.S. Marine Corps, 2010, pp. 75–81.

Connable, Ben, and Martin C. Libicki, *How Insurgencies End*, Santa Monica, Calif.: RAND Corporation, MG-965-MCIA, 2010. As of June 30, 2011:
http://www.rand.org/pubs/monographs/MG965.html

Cooper, Chester L., Judith E. Corson, Laurence J. Legere, David E. Lockwood, and Donald M. Weller, *The American Experience with Pacification in Vietnam*, Vol. 1: *An Overview of Pacification*, Institute for Defense Analysis, March 1972a.

———, *The American Experience with Pacification in Vietnam*, Vol. 2: *Elements of Pacification*, Institute for Defense Analysis, March 1972b.

Cordesman, Anthony H., *The Uncertain "Metrics" of Afghanistan (and Iraq)*, Washington, D.C.: Center for Strategic and International Studies, May 18, 2007. As of June 30, 2011:
http://www.csis.org/media/csis/pubs/070521_uncertainmetrics_afghan.pdf

———, *Analyzing the Afghan-Pakistan War*, Washington, D.C.: Center for Strategic and International Studies, draft, July 28, 2008. As of September 15, 2010:
http://csis.org/files/media/csis/pubs/080728_afghan_analysis.pdf

———, "The Afghan War: Metrics, Narratives, and Winning the War," briefing, Center for Strategic and International Studies, June 1, 2010.

Corson, William R., *The Betrayal*, New York: W. W. Norton and Company, 1968.

Cosmos, Graham A., *MACV: The Joint Command in the Years of Withdrawal, 1968–1973*, Washington, D.C.: U.S. Army Center of Military History, 2006. As of June 30, 2011: http://www.history.army.mil/html/books/macv2/CMH_91-7.pdf

Cushman, John H., *Senior Officer Debriefing Report of Major General John H. Cushman, RCS CSFOR-74*, January 14, 1972. As of June 30, 2011: http://www.dtic.mil/cgi-bin/GetTRDoc?Location=U2&doc=GetTRDoc.pdf&AD=AD0523904

Daddis, Gregory A., *No Sure Victory: Measuring U.S. Army Effectiveness and Progress in the Vietnam War*, New York: Oxford University Press, 2011.

Darilek, Richard E., Walter L. Perry, Jerome Bracken, John Gordon IV, and Brian Nichiporuk, *Measures of Effectiveness for the Information-Age Army*, Santa Monica, Calif.: RAND Corporation, MR-1155-A, 2001. As of June 30, 2011: http://www.rand.org/pubs/monograph_reports/MR1155.html

Davis, Paul K., *Effects-Based Operations (EBO): A Grand Challenge for the Analytical Community*, Santa Monica, Calif.: RAND Corporation, MR-1477-USJFCOM/AF, 2001. As of June 30, 2011: http://www.rand.org/pubs/monograph_reports/MR1477.html

DoD—*see* U.S. Department of Defense.

Donovan, David, *Once a Warrior King: Memories of an Officer in Vietnam*, New York: McGraw-Hill, 1985.

Downes-Martin, Stephen, research professor, U.S. Naval War College, interview with the author, Kabul, Afghanistan, May 5, 2010a.

———, "Assessment Process for RC(SW)," unpublished draft, Newport, R.I.: U.S. Naval War College, May 24, 2010b.

Dror, Yehezkel, *Some Normative Implications of a Systems View of Policymaking*, Santa Monica, Calif.: RAND Corporation, P-3991-1, 1969. As of June 30, 2011: http://www.rand.org/pubs/papers/P3991-1.html

Eisenstadt, Michael, and Jeffrey White, "Assessing Iraq's Sunni Arab Insurgency," *Military Review*, Vol. 86, No. 3, May–June 2006, pp. 33–51.

Eles, Philip, U.S. Central Command, interview with the author, Tampa, Fla., March 24, 2010.

Engelhardt, Tom, "Afghanistan by the Numbers: Is the War Worth It? The Cost in Dollars, Years, Public Opinion, and Lives," *Salon.com*, September 14, 2009. As of January 27, 2011: http://www.salon.com/news/feature/2009/09/14/afghanistan

Enthoven, Alain C., and K. Wayne Smith, *How Much Is Enough? Shaping the Defense Program, 1961–1969*, New York: Harper and Row/Santa Monica, Calif.: RAND Corporation, 1971/2005. As of June 30, 2011: http://www.rand.org/pubs/commercial_books/CB403.html

Fischer, Hannah, *Iraqi Civilian Death Estimates*, Washington, D.C.: Congressional Research Service, RS22537, August 27, 2008.

Flynn, Michael T., Matt Pottinger, and Paul Batchelor, *Fixing Intel: A Blueprint for Making Intelligence Relevant in Afghanistan*, Washington, D.C.: Center for a New American Security, January 2010. As of June 30, 2011: http://www.cnas.org/files/documents/publications/AfghanIntel_Flynn_Jan2010_code507_voices.pdf

Folker, Robert D. Jr., *Intelligence Analysis in Theater Joint Intelligence Centers: An Experiment in Applying Structured Analytic Methods*, Washington, D.C.: Joint Military Intelligence College, occasional paper no. 7, January 2000.

Galula, David, *Counterinsurgency Warfare: Theory and Practice*, New York: Frederick A. Praeger, (1964) 2005.

Gartner, Scott Sigmund, *Strategic Assessment in War*, New Haven, Conn.: Yale University Press, 1997.

Gates, Robert M., Secretary of Defense, statement to the U.S. Senate Appropriations Committee, Subcommittee on Defense, June 16, 2010. As of September 15, 2010:
http://appropriations.senate.gov/ht-defense.cfm?method=hearings.download&id=fac53160-5308-4ffd-9b61-bb8112599ce2

Gayvert, David, "Teaching New Dogs Old Tricks: Can the Hamlet Evaluation System Inform the Search for Metrics in Afghanistan?" *Small Wars Journal Online*, September 8, 2010. As of June 30, 2011:
http://smallwarsjournal.com/blog/journal/docs-temp/531-gayvert.pdf

Gelb, Leslie H., and Richard K. Betts, *The Irony of Vietnam: The System Worked*, Washington, D.C.: Brookings Institution Press, 1979.

Gibson, James William, *The Perfect War: Technowar in Vietnam*, New York: Atlantic Monthly Press, 2000.

Gilbert, Nigel, *Agent-Based Models*, Thousand Oaks, Calif.: Sage Publications, 2008.

Goldstein, Gordon M., *Lessons in Disaster: McGeorge Bundy and the Path to War in Vietnam*, New York: Henry Holt and Company, 2008.

Graves, Greg, Patricia Murphy, and Frederick Cameron, "Improving Analytical Support to the Warfighter: Campaign Assessments, Operations Analysis, and Data Management, Working Group 4: Current Operations Analysis—Strategic and Operational Level," briefing, Military Operations Research Society, April 19–22, 2010. As of June 30, 2011:
http://www.mors.org/UserFiles/file/2010%20IW/MORS%20IW%20WG5.pdf

Greenwood, T. C., and T. X. Hammes, "War Planning for Wicked Problems," *Armed Forces Journal*, December 2009. As of June 30, 2011:
http://armedforcesjournal.com/2009/12/4252237/

Gregersen, Hal, and Lee Sailer, "Chaos Theory and Its Implications for Social Science Research," *Human Relations*, Vol. 46, No. 7, July 1993, pp. 777–802.

Gregor, William J., "Military Planning Systems and Stability Operations," *PRISM*, Vol. 1, No. 3, 2010, pp. 99–114. As of June 30, 2011:
http://www.army.mil/-news/2010/07/22/42647-military-planning-systems-and-stability-operations/index.html

Grier, Cindy, Steve Stephens, Renee Carlucci, Stuart Starr, Cy Staniec, LTC(P) Clark Heidelbaugh, Tim Hope, Don Brock, Gene Visco, and Jim Stevens, "Improving Analytical Support to the Warfighter: Campaign Assessments, Operational Analysis, and Data Management: Synthesis Group," briefing, Military Operations Research Society, April 19–22, 2010. As of June 30, 2011:
http://www.mors.org/UserFiles/file/2010%20IW/Synthesis%20Group%20Out%20Brief%20Final.pdf

Haken, Nate, Joelle Burbank, and Pauline H. Baker, "Casting Globally: Using Content Analysis for Conflict Assessment and Forecasting," *Military Operations Research*, Vol. 15, No. 2, 2010, pp. 5–19.

Halberstam, David, *The Best and the Brightest*, New York: Random House, 1972.

Hall, CSM Michael T., and GEN Stanley A. McChrystal, commander, U.S. Forces–Afghanistan/ International Security Assistance Force, Afghanistan, "ISAF Commander's Counterinsurgency Guidance," Kabul, Afghanistan: Headquarters, International Security Assistance Force, 2009. As of June 30, 2011:
http://www.nato.int/isaf/docu/official_texts/counterinsurgency_guidance.pdf

Halmos, Paul R., *Measure Theory*, New York: Van Nostrand, 1950.

Hamilton, James D., *Time Series Analysis*, Princeton, N.J.: Princeton University Press, 1994.

Hammond, William M., *The United States Army in Vietnam: Public Affairs The Military and the Media—1962–1968*, Washington, D.C.: U.S. Government Printing Office, 1988.

Hannan, Michael J., "Operational Net Assessment: A Framework for Social Network Analysis," *IOsphere*, Fall 2005, pp. 27–32.

Harvey, Derek J., director, Afghanistan-Pakistan Center of Excellence, U.S. Central Command, discussion with the author, January 22, 2010.

Headquarters, U.S. Department of the Army, *Operations Against Guerrilla Forces*, Field Manual 31-20, Washington, D.C., 1951.

———, *Operations Against Irregular Forces*, Field Manual 31-15, Washington, D.C.: Headquarters, Department of the Army, May 1961.

———, *Counterguerrilla Operations*, Field Manual 31-16, Washington, D.C., February 1963.

———, *Civil Affairs Operation*, Field Manual, 41-1, Washington, D.C., October, 1969.

———, *Intelligence Analysis*, Field Manual 34-3, Washington, D.C., March 1990.

———, *Mission Command: Command and Control of Army Forces*, Field Manual 6-0, Washington, D.C., August 2003.

———, *Intelligence*, Field Manual 2-0, Washington, D.C., May 2004.

———, *The Army*, Field Manual 1, Washington, D.C., June 2005. As of September 17, 2010:
http://www.army.mil/fm1/

———, *The Operations Process*, Field Manual Interim 5-0.1, Washington, D.C., March 2006a.

———, *Civil Affairs Operations*, Field Manual 3-05.40, Washington, D.C., September 6, 2006b.

———, *Counterinsurgency*, Field Manual 3-24/Marine Corps Warfare Publication 3-33.5, Washington, D.C., December 2006c.

———, *2007 Army Modernization Plan*, Washington, D.C., March 5, 2007.

———, *Operations*, Field Manual 3-0, Washington, D.C., February 27, 2008.

———, *Tactics in Counterinsurgency*, Field Manual 3-24.2, Washington, D.C., April 21, 2009.

———, *The Operations Process*, Field Manual 5-0, Washington, D.C.: March 2010.

———, "Army Doctrinal Term Changes," unofficial guidelines for Army training developers, spreadsheet, April 29, 2011.

Headquarters, U.S. Marine Corps, *Small Wars Manual*, Washington, D.C., 1940.

———, *Intelligence*, Marine Corps Doctrine Publication 2, Washington, D.C., June 7, 1997a.

———, *Warfighting*, Marine Corps Doctrine Publication 1, Washington, D.C., June 20, 1997b.

———, "A Concept for Distributed Operations," white paper, Washington, D.C., April 25, 2005. As of July 21, 2011:
http://www.marines.mil/unit/tecom/mcu/grc/library/Documents/A%20Concept%20for%20Distributed%20Operations.pdf

Hitch, Charles J., *Decision-Making for Defense*, Berkeley, Calif.: University of California Press, 1965.

Hodermarsky, George T., and Brian Kalamaja, Science Applications International Corporation, "A Systems Approach to Assessments: Dealing with Complexity," briefing, 2008. As of June 30, 2011:
http://www.saic.com/sosa/downloads/Systems_Approach.pdf

Howard, Trevor, "Operational Analysis Support to OP HERRICK," briefing, Defence Science and Technology Laboratory, UK Ministry of Defence, February 13, 2007.

HQDA—*see* Headquarters, U.S. Department of the Army.

Hume, Robert, director, International Security Assistance Force Afghan Assessment Group, interview with the author, May 7, 2010.

Hunerwadel, J. P., "The Effects-Based Approach to Operations: Questions and Answers," *Air and Space Power Journal*, Spring 2006. As of June 30, 2011:
http://www.airpower.au.af.mil/airchronicles/apj/apj06/spr06/hunerwadel.html

Hylton, David, "Special Inspector General Report on Afghanistan," NATO Training Mission-Afghanistan blog, June 29, 2010. As of June 30, 2011:
http://ntm-a.com/wordpress2/?p=1732

Iklé, Fred Charles, *Every War Must End*, rev. ed., New York: Columbia University Press, 2005.

Intelligence, Surveillance, and Reconnaissance Task Force, "Open Sharing Environment: Unity Net," briefing, July 2010a.

———, "ISR TF Data Sharing Lessons," briefing, presented at Allied Information Sharing Strategy Support to International Security Assistance Force Population Metrics and Data Conference, Brunssum, Netherlands, September 1, 2010b.

Interagency Language Roundtable, homepage, undated. As of June 30, 2011:
http://www.govtilr.org/

International Security Assistance Force, "ISAF Commander Tours U.S. Detention Facility in Parwan," press release, undated. As of June 30, 2011:
http://www.isaf.nato.int/article/isaf-releases/isaf-commander-tours-u.s.-detention-facility-in-parwan.html

International Security Assistance Force Afghan Assessment Group, "Transfer of Lead Security Responsibility Effect Scoring Model," briefing, undated.

International Security Assistance Force Headquarters, *COMISAF'S Initial Assessment (Unclassified)*, August 30, 2009.

———, "Knowledge Management," briefing, August 2010.

International Security Assistance Force Headquarters Strategic Advisory Group, "Unclassified Metrics," Kabul, Afghanistan, April 2009.

International Security Assistance Force Joint Command, "District Assessment Framework Tool," spreadsheet, 2010.

International Security Assistance Force Joint Command Chief of Staff, Operational Orders Assessment Steering Group, briefing, November 19, 2009.

International Security Assistance Force Knowledge Management Office, "Afghan Mission Network," briefing, September 1, 2010.

Introduction to the Pacification Data Bank, U.S. government pamphlet, November 1969. Available via the Vietnam Center and Archive online database.

ISAF—*see* International Security Assistance Force.

ISAF AAG—*see* International Security Assistance Force Afghan Assessment Group.

Jacobson, Alvin L., and N. M. Lalu, "An Empirical and Algebraic Analysis of Alternative Techniques for Measuring Unobserved Variables," in Hubert M. Blalock, Jr., ed., *Measurement in the Social Sciences: Theories and Strategies*, Chicago, Ill.: Aldine Publishing Company, 1974, pp. 215–242.

Jaiswal, N. K., *Military Operations Research: Quantitative Decision Making*, Boston, Mass.: Kluwer Academic Publishers, 1997.

Johnson, Neil F., *Simple Complexity: A Clear Guide to Complexity Theory*, Oxford, UK: OneWorld, 2009.

Joint U.S. Public Affairs Office, *Long An Province Survey, 1966*, Saigon, South Vietnam, 1967.

Joint Warfighting Center, U.S. Joint Forces Command, J9, Standing Joint Force Headquarters, *Commander's Handbook for an Effects-Based Approach to Joint Operations*, Suffolk, Va., February 24, 2006.

Jones, Douglas D., *Understanding Measures of Effectiveness in Counterinsurgency Operations*, Fort Leavenworth, Kan.: School of Advanced Military Studies, U.S. Army Command and General Staff College, May 2006. As of June 30, 2011:
http://handle.dtic.mil/100.2/ADA450589

Kagan, Frederick, W., *Finding the Target: The Transformation of American Military Policy*, New York: Encounter Books, 2006.

Kalyvas, Stathis N., and Matthew Adam Kocher, "The Dynamics of Violence in Vietnam: An Analysis of the Hamlet Evaluation System (HES)," *Journal of Peace Research*, Vol. 46, No. 3, May 2009, pp. 335–355.

Kane, Tim, "Global U.S. Troop Deployment, 1950–2005," Heritage Foundation, Center for Data Analysis, May 24, 2006. As of June 30, 2011:
http://www.heritage.org/research/reports/2006/05/global-us-troop-deployment-1950-2005

Kaplan, Lawrence S., Ronald D. Landa, and Edward J. Drea, *The McNamara Ascendancy: 1961–1965*, Washington, D.C.: Historical Office of the Office of the Secretary of Defense, 2006.

Karnow, Stanley, *Vietnam: A History*, New York: Penguin Books, 1984.

Keaney, Thomas A., "Surveying Gulf War Airpower," *Joint Force Quarterly*, Autumn 1993, pp. 25–36.

Kelly, Justin, and David Kilcullen, "Chaos Versus Predictability: A Critique of Effects-Based Operations," *Security Challenges*, Vol. 2, No. 1, April 2006, pp. 63–73. As of June 30, 2011:
http://www.securitychallenges.org.au/ArticlePages/vol2no1KellyandKilcullen.html

Kiesling, Eugenia C., "On War Without the Fog," *Military Review*, Vol. 81, No. 5, September–October 2001, pp. 85–87.

Kilcullen, David, *The Accidental Guerrilla: Fighting Small Wars in the Midst of a Big One*, Oxford, UK: Oxford University Press, 2009a.

———, "Measuring Progress in Afghanistan," Kabul, Afghanistan, December 2009b. As of June 30, 2011:
http://hts.army.mil/Documents/Measuring%20Progress%20Afghanistan%20(2).pdf

———, *Counterinsurgency*, New York: Oxford University Press, 2010.

Kinnard, Douglas, *The War Managers*, Hanover, N.H.: University of New Hampshire Press, 1977.

Kober, Avi, "The Israeli Defense Forces in the Second Lebanon War: Why the Poor Performance?" *Journal of Strategic Studies*, Vol. 31, No. 1, February 2008, pp. 3–40.

Komer, R. W., news conference, December 1, 1967. Available via the Vietnam Center and Archive online database.

———, "Text of Ambassador Komer's News Conference," transcript, January 24, 1968a. Available via the Vietnam Center and Archive online database.

———, "Memorandum for the Record," official memorandum, November 5, 1968b, in BACM Research, Disk 2.

———, *Impact of Pacification on Insurgency in South Vietnam*, Santa Monica, Calif.: RAND Corporation, P-4443, August 1970. As of June 30, 2011:
http://www.rand.org/pubs/papers/P4443.html

Kruglanski, Arie W., and Donna M. Webster, "Motivated Closing of the Mind: 'Seizing' and 'Freezing,'" *Psychological Review*, Vol. 103, No. 2, 1996, pp. 263–283.

Kugler, Cornelius W., *Operational Assessment in a Counterinsurgency*, Newport, R.I.: U.S. Naval War College, May 10, 2006.

Langan, Patrick A., and David J. Levine, *Recidivism of Prisoners Released in 1994*, Washington, D.C.: Bureau of Justice Statistics, June 2, 2002. As of July 29, 2011:
http://bjs.ojp.usdoj.gov/index.cfm?ty=pbdetail&iid=1134

Langer, Gary, briefing on ABC News poll of Afghan perceptions, presented at the Center for Strategic and International Studies, December 2009.

Lewy, Guenter, *America in Vietnam*, New York: Oxford University Press, 1978.

Liddell Hart, B. H., *Strategy*, New York: Praeger, 1967.

Loerch, Andrew G., and Larry B. Rainey, eds., *Methods for Conducting Military Operational Analysis*, Alexandria, Va.: Military Operations Research Society, 1998.

Luck, Gary, and Mike Findlay, *Joint Operations Insights and Best Practices*, 2d ed., Joint Warfighting Analysis Center, U.S. Joint Forces Command, July 2008.

MACV—*see* U.S. Military Assistance Command, Vietnam.

Mahnken, Thomas G., *Technology and the American Way of War Since 1945*, New York: Columbia University Press, 2008.

Mann, Edward C. III, Gary Endersby, and Thomas R. Searle, *Thinking Effects: Effects-Based Methodology for Joint Operations*, Maxwell Air Force Base, Ala.: Air University Press, CADRE paper 15, October 2002.

Mao Tse-Tung, *On Guerrilla Warfare*, Samuel B. Griffith, trans., Urbana, Ill.: University of Illinois Press, 2000.

Marine Corps Order 1200.17, *Military Occupational Specialties Marine Corps Manual*, May 23, 2008. As of June 30, 2011:
http://www.marines.mil/news/publications/Documents/MCO%201200.17.pdf

Marlowe, Ann, "Defeating IEDs with Data," blog post, March 11, 2010. As of June 30, 2011: http://www.worldaffairsjournal.org/new/blogs/marlowe/Defeating_IEDs_with_Data

Marshall, S. L. A., *Vietnam Primer: Lessons Learned*, Washington, D.C.: Headquarters, U.S. Department of the Army, 1966. As of June 30, 2011: http://www.lzcenter.com/Documents/12891741-Army-Vietnam-Primer-Pamphlet.pdf

Marvin, Brett L., working group presentation at the U.S. Naval War College, Providence, R.I., October 19, 2010.

Matthews, Lloyd J., and Dale E. Brown, eds., *Assessing the Vietnam War: A Collection from the Journal of the U.S. Army War College*, McLean, Va.: Pergamon-Brassey's International Defense Publishers, 1987.

Matthews, Matt M., *We Were Caught Unprepared: The 2006 Hezbollah-Israeli War*, Fort Leavenworth, Kan.: U.S. Army Combined Arms Center, Combat Studies Institute Press, 2008. As of June 30, 2011: http://purl.access.gpo.gov/GPO/LPS104476

Mattis, Gen. James N., U.S. Marine Corps, "Assessment of Effects Based Operations," memorandum for U.S. Joint Forces Command, Norfolk, Va., August 14, 2008. As of June 30, 2011: http://smallwarsjournal.com/documents/usjfcomebomemo.pdf

McChrystal, Stanley A., *COMISAF's Initial Assessment (Unclassified)*, August 30, 2009.

McCrabb, Maris, "Effects-Based Operations: An Overview," briefing for the Air Force Research Laboratory, January 10, 2008.

McFadden, Willie J. II, and Daniel J. McCarthy, "Policy Analysis," in Andrew G. Loerch and Larry B. Rainey, eds., *Methods for Conducting Military Operational Analysis*, Alexandria, Va.: Military Operations Research Society, 1998, pp. 317–344.

McGee, Michael Calvin, "The 'Ideograph': A Link Between Rhetoric and Ideology," *Quarterly Journal of Speech*, Vol. 66, No. 1, February 1980, pp. 1–16.

McKnight, Patrick E., Katherine M. McKnight, Souraya Sidani, and Aurelio Jose Figueredo, *Missing Data: A Gentle Introduction*, New York: Guilford Press, 2007.

McLear, Michael, *The Ten Thousand Day War, Vietnam: 1945–1975*, New York: St. Martin's Press, 1981.

McMaster, H. R., *Dereliction of Duty: Lyndon Johnson, Robert McNamara, the Joint Chiefs of Staff, and the Lies That Led to Vietnam*, New York: HarperCollins, 1997.

———, "Crack in the Foundation: Defense Transformation and the Underlying Assumption of Dominant Knowledge in Future War," student issue paper, U.S. Army War College, November 2003. As of June 30, 2011: http://www.au.af.mil/au/awc/awcgate/army-usawc/mcmaster_foundation.pdf

McNamara, Robert S., transcript of a news conference, July 11, 1967. Available via the Vietnam Center and Archive online database.

———, "Southeast Asia Operations: Statement by Secretary of Defense McNamara Before Senate Armed Services Committee (excerpts)," testimony on the fiscal year 1969–1973 defense program and the 1969 defense budget, February 1, 1968, pp. 263–289. Available via the Vietnam Center and Archive online database.

———, "Affidavit of Robert S. McNamara," *General William C. Westmoreland v. CBS Inc., et al.*, December 1, 1983. Available via the Vietnam Center and Archive online database.

McNamara, Robert S., with Brian VanDeMark, *In Retrospect: The Tragedy and Lessons of Vietnam*, New York: Times Books, 1995.

McNamara, Robert S., James G. Blight, and Robert K. Brigham, with Thomas J. Biersteker and Herbert Y. Schandler, *Argument Without End: In Search of Answers to the Vietnam Tragedy*, New York: Public Affairs, 1999.

Meharg, Sarah Jane, *Measuring Effectiveness in Complex Operations: What Is Good Enough*, Calgary, Alberta: Canadian Defence and Foreign Affairs Institute, October 2009. As of June 30, 2011: http://www.cdfai.org/PDF/Measuring%20Effectiveness%20in%20Complex%20Operations.pdf

Menkhaus, Kenneth J., "State Fragility as a Wicked Problem," *Prism*, Vol. 1, No. 2, March 2010, pp. 85–100.

Miller, Delbert C., and Neil J. Salkind, *Handbook of Research Design and Social Measurement*, 6th ed., Thousand Oaks, Calif.: Sage Publications, 2002.

Miller, John H., and Scott E. Page, *Complex Adaptive Systems: An Introduction to Computational Models of Social Life*, Princeton, N.J.: Princeton University Press, 2007.

Ministry of Supply and War Office: Military Operational Research Unit, successors and related bodies, WO 291, undated.

Mission Coordinator, U.S. Embassy, Saigon, South Vietnam, *MACCORDS Field Reporting System*, July 1, 1969. Available via the Vietnam Center and Archive online database.

Mitchell, Robbyn, "Mattis Takes Over as CENTCOM Chief," *St. Petersburg Times*, August 12, 2010, p. 1. As of June 30, 2011: http://www.tampabay.com/news/mattis-takes-over-as-centcom-chief/1114800

Mowery, Samuel, Warfighting Analysis Division, J8, U.S. Department of Defense, "A System Dynamics Model of the FM 3-24 COIN Manual," briefing, 2009. Slide shown in Figure 3.1, Chapter Three, as of June 30, 2011: http://www.guardian.co.uk/news/datablog/2010/apr/29/mcchrystal-afghanistan-powerpoint-slide

Moyar, Mark, *A Question of Command: Counterinsurgency from the Civil War to Iraq*, New Haven, Conn.: Yale University Press, 2009.

Multi-National Force–Iraq, "Charts to Accompany the Testimony of General David H. Petraeus," briefing slides, September 10–11, 2007. As of April 1, 2011: http://smallwarsjournal.com/documents/petraeusslides.pdf

———, "Charts to Accompany the Testimony of General David H. Petraeus," briefing slides, April 8–9, 2008. As of April 1, 2011: http://www.defense.gov/pdf/Testimony_Handout_Packet.pdf

Murray, William S., "A Will to Measure," *Parameters*, Autumn 2001, pp. 134–147. As of June 30, 2011: http://www.carlisle.army.mil/USAWC/Parameters/Articles/01autumn/Murray.htm

Nakashima, Ellen, and Craig Whitlock, "With Air Force's Gorgon Drone 'We Can See Everything,'" *Washington Post*, January 2, 2011. As of January 15, 2011: http://www.washingtonpost.com/wp-dyn/content/article/2011/01/01/AR2011010102690.html

National Commission for the Protection of Human Subjects of Biomedical and Behavioral Research, *The Belmont Report: Ethical Principles and Guidelines for the Protection of Human Subjects Research*, Washington, D.C.: U.S. Government Printing Office, 1978. As of March 10, 2011: http://www.videocast.nih.gov/pdf/ohrp_belmont_report.pdf

National Intelligence Council, *Estimative Products on Vietnam: 1948–1975*, Pittsburgh, Pa.: U.S. Government Printing Office, April 2005.

Nitschke, Stephen G., *Vietnam: A Complex Adaptive Perspective*, U.S. Marine Corps Command and Staff College, thesis, 1997.

Oberdorfer, Don, *Tet! The Turning Point in the Vietnam War*, Baltimore, Md.: Johns Hopkins University Press, 2001.

Office of the Director of National Intelligence, *Analytic Standards*, Intelligence Community Directive No. 203, June 21, 2007. As of June 30, 2011:
http://www.dni.gov/electronic_reading_room/ICD_203.pdf

Office of the Secretary of Defense, "Metrics Conference Outbrief to IJC/IDC," briefing, March 18, 2010.

Office of the Special Inspector General for Afghanistan Reconstruction, *Actions Needed to Improve the Reliability of Afghan Security Force Assessments*, Arlington, Va., June 29, 2010. As of June 30, 2011:
http://www.washingtonpost.com/wp-srv/hp/ssi/wpc/sigar.pdf?sid=ST2010062805531

Olson, James S., and Randy Roberts, *Where the Domino Fell: America and Vietnam, 1945–1990*, New York: St. Martin's Press, 1991.

Owens, Bill, with Ed Offley, *Lifting the Fog of War*, Baltimore, Md.: Johns Hopkins University Press, 2000.

Palmer, Bruce, Jr., *The 25-Year War: America's Military Role in Vietnam*, Lexington, Ky.: University Press of Kentucky, 1984.

Paul, Christopher, Colin Clarke, and Beth Grill, *Victory Has a Thousand Fathers: Sources of Success in Counterinsurgency*, Santa Monica, Calif.: RAND Corporation, MG-964-OSD, 2010. As of June 30, 2011:
http://www.rand.org/pubs/monographs/MG964.html

Paul, Christopher, William M. Mason, Daniel McCaffrey, and Sarah A. Fox, "A Cautionary Case Study of Approaches to the Treatment of Missing Data," *Statistical Methods and Applications*, Vol. 17, No. 3, July 2008, pp. 351–372.

Perry, Walter L., "Linking Systems Performance and Operational Effectiveness," in Andrew G. Loerch and Larry B. Rainey, eds., *Methods for Conducting Military Operational Analysis*, Alexandria, Va.: Military Operations Research Society, 2007.

———, email exchange with the author, February 3, 2011.

Perry, Walter L., Robert W. Button, Jerome Bracken, Thomas Sullivan, and Jonathan Mitchell, *Measures of Effectiveness for the Information-Age Navy: The Effects of Network-Centric Operations on Combat Outcomes*, Santa Monica, Calif.: RAND Corporation, MR-1449-NAVY, 2002. As of June 30, 2011:
http://www.rand.org/pubs/monograph_reports/MR1449.html

Perry, Walter L., and John Gordon IV, *Analytic Support to Intelligence in Counterinsurgencies*, Santa Monica, Calif.: RAND Corporation, MG-682-OSD, 2008. As of June 30, 2011:
http://www.rand.org/pubs/monographs/MG682.html

Phillips, Rufus, *Why Vietnam Matters: An Eyewitness Account of Lessons Not Learned*, Annapolis, Md.: Naval Institute Press, 2008.

Pool, Ithiel de Sola, Gordon Fowler, Peter McGrath, and Richard Peterson, *Hamlet Evaluation System Study*, Cambridge, Mass.: Simulmatics Corporation, May 1, 1968. As of June 30, 2011:
http://handle.dtic.mil/100.2/AD839821

Prince, W. B., and J. H. Adkins, *Analysis of Vietnamization: Hamlet Evaluation System Revisions*, Ann Arbor, Mich.: Bendix Aerospace Systems Division, February 1973. As of June 30, 2011: http://www.dtic.mil/cgi-bin/GetTRDoc?AD=AD908686&Location=U2&doc=GetTRDoc.pdf

Public Law 110-28, U.S. Troop Readiness, Veterans' Care, Katrina Recovery, and Iraq Accountability Appropriations Act, May 25, 2007.

Quade, E. S., ed., *Analysis for Military Decisions*, Santa Monica, Calif.: RAND Corporation, R-387-PR, 1964. As of June 30, 2011: http://www.rand.org/pubs/reports/R0387.html

Race, Jeffrey, *War Comes to Long An: Revolutionary Conflict in a Vietnam Province*, Berkeley and Los Angeles, Calif.: University of California Press, 1972.

Ramjeet, T. J., *Operational Analysis in Support of HQ RC(S), Kandahar, Afghanistan, September 2007 to January 2008, in Cornwallis XIII*, Farnborough, UK: Defence Science and Technology Laboratory, 2008.

Reconstruction and Stabilization Policy Coordinating Committee, *Interagency Conflict Assessment Framework*, Washington, D.C., 2009. As of July 29, 2011: http://www.state.gov/documents/organization/161781.pdf

Record, Jeffrey, *Hollow Victory: A Contrary View of the Gulf War*, New York: Brassey's, Inc., 1993.

Rehm, Allan, ed., *Analyzing Guerrilla Warfare*, conference proceedings, September 24, 1985. Available via the Vietnam Center and Archive online database.

Rennie, Ruth, Sudhindra Sharma, and Pawan Kumar Sen, *Afghanistan in 2009: A Survey of the Afghan People*, Kabul, Afghanistan: Asia Foundation, December 2009. As of June 30, 2011: http://www.asiafoundation.org/resources/pdfs/Afghanistanin2009.pdf

"Rescinding ASCOPEs for PMESII-PT at the Tactical Level; Possibly Good in Theory, but What About in Application?" *U.S. Army Combined Arms Center Blog*, April 20, 2009. No longer available.

Ricchiardi, Sherry, "Whatever Happened to Iraq?" *American Journalism Review*, June–July 2008. As of November 9, 2010: http://www.ajr.org/article.asp?id=4515

Rittel, Horst W. J., and Melvin M. Webber, "Dilemmas in a General Theory of Planning," *Policy Sciences*, Vol. 4, 1973, pp. 155–169.

Rogers, Simon, "The McChrystal Afghanistan PowerPoint Slide: Can You Do Any Better?" *Guardian Data Blog*, April 29, 2010. As of June 30, 2011: http://www.guardian.co.uk/news/datablog/2010/apr/29/mcchrystal-afghanistan-powerpoint-slide

Rosenau, William, and Austin Long, *The Phoenix Program and Contemporary Counterinsurgency*, Santa Monica, Calif.: RAND Corporation, OP-258-OSD, 2009. As of June 30, 2011: http://www.rand.org/pubs/occasional_papers/OP258.html

Roush, Maurice D., "The Hamlet Evaluation System," *Military Review*, September 1969.

Roy-Bhattacharya, Joydeep, *The Storyteller of Marrakesh*, New York: W. W. Norton and Company, 2011.

Rubin, Alissa J., "U.S. Report on Afghan War Finds Few Gains in 6 Months," *New York Times*, April 29, 2010a. As of June 30, 2011: http://www.nytimes.com/2010/04/30/world/asia/30afghan.html

———, "Petraeus Says Taliban Have Reached Out to Karzai," *New York Times*, September 27, 2010b. As of June 30, 2011: http://www.nytimes.com/2010/09/28/world/asia/28afghan.html

Ruby, Tomislav Z., "Effects-Based Operations: More Important Than Ever," *Parameters*, Autumn 2008, pp. 26–35.

Rumsfeld, Donald H., interview with Steve Inskeep, *Morning Edition*, National Public Radio, March 29, 2005. As of June 30, 2011:
http://www.defense.gov/transcripts/transcript.aspx?transcriptid=2551

Schaffer, Marvin B., *On Rescoring the Hamlet Evaluation System*, Santa Monica, Calif.: RAND Corporation, 1968, not available to the general public.

Schrecker, Mark, "U.S. Strategy in Afghanistan: Flawed Assumptions Will Lead to Ultimate Failure," *Joint Force Quarterly*, No. 59, 4th Quarter, 2010, pp. 75–82.

Schroden, Jonathan J., "Measures for Security in a Counterinsurgency," *Journal of Strategic Studies*, Vol. 32, No. 5, October 2009, pp. 715–744.

———, email exchange with the author, February 1, 2010a.

———, email exchange with the author, September 29, 2010b.

———, briefing at the U.S. Naval War College, Newport, R.I., October 19, 2010c.

———, email exchange with the author, October 27, 2010d.

Shadid, Anthony, "In Iraq, Western Clocks, but Middle Eastern Time," *New York Times*, August 14, 2010. As of June 30, 2011:
http://www.nytimes.com/2010/08/15/weekinreview/15shadid.html

Sharp, U. S. G., and William C. Westmoreland, *Report on the War in Vietnam as of 30 June 1968*, Washington, D.C.: U.S. Government Printing Office, 1968.

Sheehan, Neil, *A Bright Shining Lie: John Paul Vann and America in Vietnam*, New York: Vintage Books, 1988.

Shimko, Keith L., *The Iraq Wars and America's Military Revolution*, New York: Cambridge University Press, 2010.

Shrader, Charles R., *History of Operations Research in the United States Army*, Vol. 1: 1942–62, Washington, D.C.: Office of the Deputy Under Secretary of the Army for Operations Research, U.S. Army, Center for Military History publication 70-102-1, August 11, 2006. As of June 30, 2011:
http://purl.access.gpo.gov/GPO/FDLP521

Smith, Douglas S., commander, 2nd Battalion, 47th Infantry, 9th Infantry Division, interview with Robert L. Keeley, commander, 19th Military History Detachment, Bien Phuoc, Republic of Vietnam, July 1, 1969. As of June 30, 2011:
http://www.history.army.mil/documents/vietnam/vnit/vnit457f.htm

Smith, Edward Allen, *Effects-Based Operations: Applying Network Centric Warfare to Peace, Crisis, and War*, Washington, D.C.: U.S. Department of Defense Command and Control Research Program, November 2002.

Soeters, Joseph, "Measuring the Immeasurable? The Effects-Based Approach in Comprehensive Peace Operations," draft revision of a paper presented at the tenth European Research Group on Military and Society conference, Stockholm, Sweden, June 23, 2009.

Sok, Sang Am, Center for Army Analysis, "Assessment Doctrine," briefing presented at the Allied Information Sharing Strategy Support to ISAF Population Metrics and Data Conference, Brunssum, Netherlands, September 1, 2010.

Sorley, Lewis, ed., *The Abrams Tapes: 1968–1972*, Lubbock, Tex.: Texas Tech University Press, 2004.

State of Washington, "Recidivism of Adult Felons 2004," web page, December 2005. As of May 10, 2011:
http://www.sgc.wa.gov/PUBS/Recidivism/Adult_Recidivism_Cy04.pdf

Stolzenberg, Ross M., ed., *Sociological Methodology*, Vol. 26, 2006.

Sullivan, Thomas J., Iraq Reconstruction Management Office, U.S. Embassy Baghdad, "Long Term Plan for IRMO Metrics," memorandum, August 2004.

Sweetland, Anders, *Item Analysis of the HES (Hamlet Evaluation System)*, Santa Monica, Calif.: RAND Corporation, D-17634-ARPA/AGILE, 1968. As of June 30, 2011:
http://www.rand.org/pubs/documents/D17634.html

Sykes, Charles S., *Interim Report of Operations, First Cavalry Division: July 1965 to December 1966*, Albuquerque, N.M.: First Cavalry Division Association, undated. Available via the Vietnam Center and Archive online database.

Terriff, Terry, Frans Osinga, and Theo Farrell, *A Transformation Gap? American Innovations and European Military Change*, Stanford, Calif.: Stanford University Press, 2010.

Thayer, Thomas C., ed., *A Systems Analysis View of the Vietnam War, 1965–1972*, Vol. 1: *The Situation in Southeast Asia*, Washington, D.C.: Office of the Assistant Secretary of Defense for Systems Analysis, Southeast Asia Intelligence Division, 1975a.

———, *A Systems Analysis View of the Vietnam War, 1965–1972*, Vol. 9: *Population Security*, Washington, D.C.: Office of the Assistant Secretary of Defense for Systems Analysis, Southeast Asia Intelligence Division, 1975b.

———, *A Systems Analysis View of the Vietnam War, 1965–1972*, Vol. 10: *Pacification and Civil Affairs*, Washington, D.C.: Office of the Assistant Secretary of Defense for Systems Analysis, Southeast Asia Intelligence Division, 1975c.

———, *War Without Fronts: The American Experience in Vietnam*, Boulder, Colo.: Westview Press, 1985.

Tho, Tran Dinh, *Pacification*, Indochina Monographs, Washington, D.C.: U.S. Army Center of Military History, 1980. As of June 30, 2011:
http://cgsc.cdmhost.com/cdm/fullbrowser/collection/p4013coll11/id/1409/rv/singleitem/rec/2

Thompson, Robert, *No Exit from Vietnam*, New York: David McKay Company, Inc., 1970.

Toevank, Freek-Jan, "Methods for Collecting and Sharing Data," plenary presentation, Allied Information Sharing Strategy Support to ISAF Population Metrics and Data Conference, Brunssum, Netherlands, September 1, 2010.

Tufte, Edward R., *The Visual Display of Quantitative Information*, 2nd ed., Cheshire, Conn.: Graphics Press, 2006.

Tunney, John V., *Measuring Hamlet Security in Vietnam: Report of a Special Study Mission*, Washington, D.C.: U.S. Government Printing Office, December 1968.

Ulrich, Mark, U.S. Army and Marine Corps Joint Counterinsurgency Center, "Center of Influence Analysis: Linking Theory to Application," briefing presented at the National Defense University, 2009.

U.S. Agency for International Development, *Measuring Fragility: Indicators and Methods for Rating State Performance*, Washington, D.C., June 2005. As of June 30, 2011:
http://pdf.usaid.gov/pdf_docs/PNADD462.pdf

U.S. Army, "FM 5-0 Overview," Combined Arms Directorate, briefing, April 1, 2010.

U.S. Army Combined Arms Center, "Understanding the Operational Environment in COIN," briefing, February 13, 2009.

U.S. Army Training and Doctrine Command, *Commander's Appreciation and Campaign Design*, version 1.0, pamphlet 525-5-500, January 28, 2008. As of June 29, 2011:
http://www.tradoc.army.mil/tpubs/pams/p525-5-500.pdf

U.S. Bureau of Labor Statistics, "How the Government Measures Unemployment," web page, last updated October 16, 2009. As of June 30, 2011:
http://www.bls.gov/cps/cps_htgm.htm

U.S. Central Command, "CJOA—Afghanistan TF 236 Overview," unpublished briefing, September 1, 2010.

U.S. Central Intelligence Agency, *Analysis of the Vietnamese Communists' Strengths, Capabilities, and Will to Persist in Their Present Strategy in Vietnam*, Annex VI: *The Morale of the Communist Forces*, August 26, 1966. As of March 15, 2011:
http://www.vietnam.ttu.edu/star/images/041/04114192001e.pdf

———, *The Communist's Ability to Recoup Their Tet Military Losses*, memorandum, March 1, 1968. As of March 15, 2011:
http://www.vietnam.ttu.edu/star/images/041/0410586008.pdf

———, *A Compendium of Analytic Tradecraft Notes*, Vol. 1, Washington, D.C., February 1997. As of June 30, 2011:
http://www.au.af.mil/au/awc/awcgate/cia/tradecraft_notes/contents.htm

U.S. Defense Logistics Agency, *A Study of Strategic Lessons Learned in Vietnam,* Vol. VI: *Conduct of the War,* Book 2: *Functional Analysis*, May 2, 1980. As of June 30, 2011:
http://handle.dtic.mil/100.2/ADA096430

U.S. Department of Defense, *Measuring Security and Stability in Iraq*, Washington, D.C., November 2006. As of January 13, 2011:
http://www.defense.gov/home/features/iraq_reports/index.html

———, *Measuring Security and Stability in Iraq*, Washington, D.C., December 2007. As of June 30, 2011:
http://www.defense.gov/home/features/iraq_reports/index.html

———, *Report on Progress Toward Security and Stability in Afghanistan: Report to Congress in Accordance with the 2008 National Defense Authorization Act (Section 1230, Public Law 110-181)*, Washington, D.C., semiannual report 2, January 2009a.

———, *Measuring Stability and Security in Iraq*, Washington, D.C., September 2009b. As of June 30, 2011:
http://www.defense.gov/home/features/iraq_reports

———, *Report on Progress Toward Security and Stability in Afghanistan: Report to Congress in Accordance with the 2008 National Defense Authorization Act (Section 1230, Public Law 110-181)*, Washington, D.C., April 2010a.

———, *Report on Progress Toward Security and Stability in Afghanistan: Report to Congress in Accordance with the 2008 National Defense Authorization Act (Section 1230, Public Law 110-181)*, Washington, D.C., November 2010b.

———, *Measuring Stability and Security in Iraq*, Washington, D.C., June 2010c. As of June 30, 2011:
http://www.defense.gov/home/features/iraq_reports

————, "DoD News Briefing with Vice Adm. Harward from Afghanistan," November 30, 2010d. As of May 10, 2011:
http://www.globalsecurity.org/military/library/news/2010/11/mil-101130-dod02.htm

U.S. Department of Defense Office of Force Transformation, *Military Transformation: A Strategic Approach*, Washington, D.C., 2003.

————, *The Implementation of Network-Centric Warfare*, Washington, D.C., January 5, 2005. As of September 16, 2010:
http://purl.access.gpo.gov/GPO/LPS57633

U.S. Department of Defense Office of the Inspector General's Inspection and Evaluation Directorate, "Combined Forces Command—Afghanistan Management Decision Model," Arlington, Va., 2005.

U.S. Government, *A Tradecraft Primer: Structured Analytic Techniques for Improving Intelligence Analysis*, March 2009. As of June 30, 2011:
https://www.cia.gov/library/publications/publications-rss-updates/tradecraft-primer-may-4-2009.html

U.S. Government Interagency Counterinsurgency Initiative, *U.S. Government Counterinsurgency Guide*, Washington, D.C.: U.S. Department of State, Bureau of Political-Military Affairs, January 2009. As of June 30, 2011:
http://www.state.gov/documents/organization/119629.pdf

U.S. Joint Chiefs of Staff, *Joint Operation Planning*, Joint Publication 5-0, Washington, D.C., December 26, 2006. As of June 27, 2011:
http://www.dtic.mil/doctrine/new_pubs/jp5_0.pdf

————, *Joint Intelligence*, Joint Publication 2-0, Washington, D.C., June 22, 2007. As of June 27, 2011:
http://www.dtic.mil/doctrine/new_pubs/jp2_0.pdf

————, Joint Doctrine and Education Division Staff, "Effects-Based Thinking in Joint Doctrine," *Joint Force Quarterly*, No. 53, 2nd Quarter 2009a, p. 60.

————, *Joint Intelligence Preparation of the Operational Environment*, Joint Publication 2-01.3, Washington, D.C., June 16, 2009b.

————, *Counterinsurgency Operations*, Joint Publication 3-24, Washington, D.C., October 5, 2009c. As of June 27, 2011:
http://www.dtic.mil/doctrine/new_pubs/jp3_24.pdf

————, *Joint Operations*, Joint Publication 3-0, incorporating change 2, Washington, D.C., March 22, 2010. As of June 27, 2011:
http://www.dtic.mil/doctrine/new_pubs/jp3_0.pdf

————, *Department of Defense Dictionary of Military and Associated Terms*, Joint Publication 1-02, Washington, D.C., November 8, 2010, as amended through May 15, 2011. As of June 27, 2011:
http://purl.access.gpo.gov/GPO/LPS14106

U.S. Marine Corps Intelligence Activity, *Cultural Generic Information Requirements Handbook (C-GIRH)*, August 2008.

————, *Cultural Intelligence Indicators Guide*, October 2009.

U.S. Military Assistance Command, Vietnam, *CORDS Report—Phong Dinh Province 2/68*, province monthly report, February 1968a. Available via the Vietnam Center and Archive online database.

————, *MACCORDS-OAD Fact Sheet: RD Cadre Evaluation System*, November 15, 1968b. Available via the Vietnam Center and Archive online database.

————, *Commander's Summary of the MACV Objectives Plan*, 1969a. Available via the Vietnam Center and Archive online database.

————, *Directive on MAC-CORDS Field Reporting System*, July 1, 1969b. Available via the Vietnam Center and Archive online database.

————, *System for Evaluating the Effectiveness of Republic of Vietnam Armed Forces (SEER)*, November 27, 1969c. Available via the National Archives online catalog.

————, *Form Used for Hamlet Evaluation Survey (HES), Report by CORDS Employees*, 1969d. Available via the Vietnam Center and Archive online database.

————, *Hamlet Evaluation System District Advisors' Handbook*, 1971a. Available via the Vietnam Center and Archive online database.

————, *The Hamlet Evaluation System (HES) Reports*, official memorandum, January 22, 1971b. Available via the Vietnam Center and Archive online database.

————, *Hamlet Evaluation System Advisors Handbook*, June 1971c. Available via the Vietnam Center and Archive online database.

————, *Order of Battle Summary*, Vols. 1 and 2, March 1972. Available via the Vietnam Center and Archive online database.

U.S. National Institute of Justice, "Measuring Recidivism," web page, February 20, 2008. As of May 10, 2011:
http://www.nij.gov/topics/corrections/recidivism/measuring.htm

U.S. Naval Postgraduate School, *NPS Academic Catalog*, September 10, 2010. As of June 30, 2011:
http://www.nps.edu/admissions/Doc/Academic_Catalog_10_SEPT_2010.pdf

U.S. Senate Foreign Relations Committee, *Vietnam Hearings: Answers to Questions for the Hearing Record on the CORDS Program*, transcript, March 1970. Available via the Vietnam Center and Archive online database.

USAID—*see* U.S. Agency for International Development.

Vanden Brook, Tom, "IEDs Kill 21,000 Iraqi Civilians 2005–2010," *USA Today*, January 11, 2011. As of January 13, 2011:
http://www.usatoday.com/news/world/iraq/2011-01-12-1Aied12_ST_N.htm

Van Riper, Paul K., "EBO: There Was No Baby in the Bathwater," *Joint Force Quarterly*, No. 52, 1st Quarter 2009, pp. 82–85.

Vego, Milan N., "Effects-Based Operations: A Critique," *Joint Force Quarterly*, No. 41, 2nd Quarter 2006, pp. 51–57.

Vincent, Etienne, Philip Eles, and Boris Vasiliev, "Opinion Polling in Support of Counterinsurgency," in *The Cornwallis Group XIV: Analysis of Societal Conflict and Counter-Insurgency*, Cornwallis Group, 2010, pp. 104–125. As of April 1, 2011:
http://www.thecornwallisgroup.org/cornwallis_2009/7-Vincent_etal-CXIV.pdf

Wagenhals, Lee W., Alex Levis, and Sajjad Haider, *Planning, Execution, and Assessment of Effects-Based Operations (EBO)*, Rome, N.Y.: Air Force Research Laboratory, AFRL-IF-RS-TR-2006-176, May 2006. As of June 30, 2011:
http://handle.dtic.mil/100.2/ADA451493

Walker, David M., *DoD Should Provide Congress and the American Public with Monthly Data on Enemy-Initiated Attacks in Iraq in a Timely Manner*, Washington, D.C.: U.S. Government Accountability Office, GAO-07-1048R, September 28, 2007. As of June 30, 2011:
http://purl.access.gpo.gov/GPO/LPS87929

Wass de Czege, Huba, "Systemic Operational Design: Learning and Adapting in Complex Missions," *Military Review*, January–February 2009, pp. 2–12. As of June 30, 2011:
http://usacac.army.mil/CAC2/MilitaryReview/Archives/English/MilitaryReview_20090228_art004.pdf

Westmoreland v. CBS, 752 F.2d 16, 2d Cir., November 2, 1984.

Westmoreland, William C., "General Westmoreland's Activities Report for September," memorandum for the President of the United States, October 10, 1967a.

———, "Military Briefing by General William Westmoreland, USA, Commander, Military Assistance Command, Vietnam," November 22, 1967b.

Wheeler, Earle G., Chairman, U.S. Joint Chiefs of Staff, , "Khe Sanh," memorandum to the President of the United States, February 3, 1968. As of June 27, 2011:
http://www.vietnam.ttu.edu/star/images/001/0010113001A.pdf

White House, *Iraq Benchmark Assessment Report*, Washington, D.C.: U.S. Government Printing Office, September 17, 2007.

Wilder, Andrew, "Losing Hearts and Minds in Afghanistan," *Viewpoints—Afghanistan, 1979–2009: in the Grip of Conflict*, Washington, D.C.: Middle East Institute, December 2, 2009, pp. 143–146. As of June 30, 2011:
http://www.mei.edu/Portals/0/Publications/Afghanistan%20VP.pdf

Williams-Bridgers, Jacquelyn, *Iraq and Afghanistan: Security, Economic, and Governance Challenges to Rebuilding Efforts Should Be Addressed in U.S. Strategies*, testimony before the U.S. House of Representatives Committee on Armed Services, March 25, 2009, Washington, D.C.: U.S. Government Accountability Office, GAO-09-476T, 2009. As of June 30, 2011:
http://purl.access.gpo.gov/GPO/LPS113358

Wirtz, James J., *The Tet Offensive: Intelligence Failure in War*, Ithaca, N.Y.: Cornell University Press, 1994.

Zehna, Peter W., ed., *Selected Methods and Models in Military Operations Research*, rev. ed., Honolulu, Hawaii: University Press of the Pacific, 2005.

Zhai, Qiang, *China and the Vietnam Wars, 1950–1975*, Chapel Hill, N.C.: University of North Carolina Press, 2000.

Zhu, Hongwei, and Richard Y. Yang, "An Information Quality Framework for Verifiable Intelligence Products," Massachusetts Institute of Technology, undated. As of June 30, 2011:
http://mitiq.mit.edu/Documents/Publications/Papers/2007/An%20IQ%20Framework%20for%20Verifiable%20Intelligence%20Products.pdf

Ziliak, Stephen Thomas, and Dierdre N. McCloskey, *The Cult of Statistical Significance: How the Standard Error Costs Us Jobs, Justice, and Lives*, Ann Arbor, Mich.: University of Michigan Press, 2008.